PERVASIVE COMPUTING AND NETWORKING

PERVASIVE COMPUTING AND NETWORKING

Editors

Mohammad S. Obaidat
Monmouth University, USA

Mieso Denko
University of Guelph, Canada

Isaac Woungang
Ryerson University, Canada

WILEY
A John Wiley and Sons, Ltd., Publication

Library of Congress Cataloging-in-Publication Data

Pervasive computing and networking / Mohammad S. Obaidat, Mieso Denko, Isaac Woungang (eds.).
    p. cm.
  Includes bibliographical references and index.
  ISBN 978-0-470-74772-8 (cloth)
 1. Ubiquitous computing. 2. Computer networks. 3. Mobile computing. I. Obaidat, Mohammad S. (Mohammad Salameh),
1952– II. Denko, Mieso K. III. Woungang, Isaac.
  QA76.5915.P4553 2011
  004.165–dc22

    2011002202

A catalogue record for this book is available from the British Library.

Print ISBN: 9780470747728 (H/B)
ePDF ISBN: 9781119970439
oBook ISBN: 9781119970422
ePub ISBN: 9781119971429
eMobi ISBN: 9781119971436

Set in 9/11pt Times by Aptara Inc., New Delhi, India

This book is dedicated to our Dear friend and colleague, late Dr. Mieso Denko, with whom we started this book project. He was an extremely active researcher in the area of wireless pervasive networking and communications. We will never forget him.

Contents

List of Contributors   xv

About the Editors   xix

PART ONE   PERVASIVE COMPUTING AND SYSTEMS   1

1 Introduction   3

Mohammad S. Obaidat and Isaac Woungang

1.1   Pervasive Computing and Its Significance   3

1.2   Research Trends in Pervasive Computing and Networking   3

1.3   Scanning the Book   4

1.4   Target Audience   6

1.5   Supplementary Resources   6

1.6   Acknowledgments   6

     References   6

2 Tools and Techniques for Dynamic Reconfiguration and Interoperability of Pervasive Systems   9

Evens Jean, Sahra Sedigh, Ali R. Hurson, and Behrooz A. Shirazi

2.1   Introduction   9

2.2   Mobile Agent Technology   10

    2.2.1   Introduction   10

    2.2.2   Mobile Agent Security   11

    2.2.3   Mobile Agent Platforms   12

2.3   Sensor Networks   13

    2.3.1   Introduction   13

    2.3.2   Sensor Network Applications   14

    2.3.3   Dynamic Reconfiguration of Sensor Networks   14

2.4   Collaboration and Interoperability Among Sensor Networks   17

2.5   Applications   18

    2.5.1   A Pervasive System for Volcano Monitoring   19

    2.5.2   A Pervasive Computing Platform for Individualized Higher Education   22

2.6   Conclusion   24

     References   24

3 Models for Service and Resource Discovery in Pervasive Computing   27

Mehdi Khouja, Carlos Juiz, Ramon Puigjaner, and Farouk Kamoun

3.1   Introduction   27

3.2   Service Oriented Architecture   28

3.3   Industry and Consortia Supported Models for Service Discovery   28

    3.3.1   Existing Models for Service Discovery   28

    3.3.2   Suitability of Existing Models for Pervasive Systems   32

3.4   Research Initiatives in Service Discovery for Pervasive Systems   33
    3.4.1   Related Work   33
    3.4.2   Our Proposition: Using Semantic in Context Modeling to Enhance Service Discovery   34
3.5   Conclusions   35
    References   36

4 Pervasive Learning Tools and Technologies   **37**
Neil Y. Yen, Qun Jin, Hiroaki Ogata, Timothy K. Shih, and Y. Yano
4.1   Introduction   37
4.2   Pervasive Learning: A Promising Innovative Paradigm   38
    4.2.1   Historical Development of Computing and IT in Education   38
    4.2.2   Past Experience and Issues   40
    4.2.3   Practice and Challenge at Waseda E-School   41
4.3   Emerging Technologies and Systems for Pervasive Learning   42
    4.3.1   Emerging Computing Paradigms for Education   42
    4.3.2   Pervasive Learning Support Systems and Technologies   43
4.4   Integration of Real-World Practice and Experience with Pervasive Learning   45
    4.4.1   Ubiquitous Learning   45
    4.4.2   UPS (Ubiquitous Personal Study)   46
4.5   Nature of Pervasive Learning and Provision of Well-Being in Education   46
    4.5.1   Ubiquitous and Pervasive   46
    4.5.2   The Possible Trend of Pervasive Technology in Education   46
4.6   Conclusion   48
    References   48

5 Service Management in Pervasive Computing Environments   **51**
Jiannong Cao, Joanna Siebert, and Vaskar Raychoudhury
5.1   Introduction   51
5.2   Service Management in Pervasive Computing Environments   52
    5.2.1   Introduction   52
    5.2.2   Pervasive Computing Environments   53
    5.2.3   Service Management Framework   53
    5.2.4   General Components of a Service Management System   54
    5.2.5   System Support Components   55
    5.2.6   Service Management Challenges   56
5.3   Techniques for Service Management in PvCE   57
    5.3.1   Introduction   57
    5.3.2   Classification of Service Discovery Protocols   57
    5.3.3   Service Discovery in Infrastructure-Based Networks   58
    5.3.4   Service Discovery in Infrastructure-Less Networks   59
    5.3.5   Multiprotocol Service Discovery   62
    5.3.6   Service Discovery Approaches   63
5.4   Service Composition   63
    5.4.1   Service Composition Functions   63
    5.4.2   Survey of Methods in Service Composition Process   65
    5.4.3   Service Composition Approaches   66
5.5   Conclusions   67
    References   68

6 Wireless Sensor Cooperation for a Sustainable Quality of Information   **71**
Abdelmajid Khelil, Christian Reinl, Brahim Ayari, Faisal Karim Shaikh, Piotr Szczytowski,
Azad Ali, and Neeraj Suri
6.1   Introduction   71
6.2   Sensing the Real World   72
6.3   Inter-Sensor Cooperation   74

	6.3.1	Sensor Cooperation for Self-Organization	74
	6.3.2	Cooperation for Quality of Information Provisioning	75
	6.3.3	Sensor Node Cooperation for Information Transport	76
6.4	Mobile Sensor Cooperation		79
	6.4.1	Mobility to Enhance Functionality (gMAP)	79
	6.4.2	Mobility to Enhance Dependability	83
6.5	Cooperation Across Mobile Entities		85
	6.5.1	Cooperative Path Planning	85
	6.5.2	Data-Based Agreement for Coordination	88
6.6	Inter-WSN Cooperation		92
	6.6.1	The Sensor-Map Integration Approaches	92
	6.6.2	Generalized Map-Based Cooperation	93
6.7	Conclusions and Future Research Directions		94
	References		95

7 An Opportunistic Pervasive Networking Paradigm: Multi-Hop Cognitive Radio Networks **101**
Didem Gözüpek and Fatih Alagöz
7.1	Introduction		101
7.2	Overview of Multi-Hop Cognitive Radio Networks MAC Layer		102
	7.2.1	MAC Design Challenges in MHCRNs	102
	7.2.2	Comparison of Multi-Channel Networks and MHCRN	105
7.3	Proposed Mac Layer Protocols		105
	7.3.1	POMDP Framework for Decentralized Cognitive MAC (DC-MAC)	106
	7.3.2	DCR-MAC	106
	7.3.3	Cross-Layer Based Opportunistic MAC (O-MAC)	107
	7.3.4	HC-MAC	107
	7.3.5	C-MAC	108
	7.3.6	DOSS-MAC	109
	7.3.7	SCA-MAC	109
7.4	Open Issues		110
7.5	Conclusions		111
	References		111

8 Wearable Computing and Sensor Systems for Healthcare **113**
Franca Delmastro and Marco Conti
8.1	Introduction		113
8.2	The Health Body Area Network		114
8.3	Medical and Technological Requirements of Health Sensors		115
8.4	Wearable Sensors for Vital Signals Monitoring		117
8.5	Wearable Sensors for Activity Recognition		120
8.6	Sensors and Signals for Emotion Recognition		122
8.7	Intra-BAN Communications in Pervasive Healthcare Systems: Standards and Protocols		124
	8.7.1	IEEE 802.15.4 and ZigBee	125
	8.7.2	Bluetooth	126
	8.7.3	Bluetooth Low Energy	127
	8.7.4	Integrated and Additional Solutions for Health BAN Communications	128
8.8	Conclusions		128
	References		131

9 Standards and Implementation of Pervasive Computing Applications **135**
Daniel Cascado, Jose Luis Sevillano, Luis Fernández-Luque, Karl Johan Grøttum, L. Kristian Vognild, and T. M. Burkow
9.1	Introduction		135
	9.1.1	Pervasiveness and Mobility in Computing and Communications	135
	9.1.2	Context Awareness	136
	9.1.3	Heterogeneity	136

9.2   Wireless Technologies and Standards   137
      9.2.1 A Simple Classification of Wireless Networks   138
      9.2.2 Concluding Remarks   145
9.3   Middleware   145
      9.3.1 Future Trends: Beyond the Middleware   151
9.4   Case Studies   151
      9.4.1 Pervasive Computing in Extreme Areas; The Hiker's Personal Digital Assistant   152
      9.4.2 Pervasive Computing in Personal Health Systems; The MyHealthService Approach   154
      References   156

PART TWO   PERVASIVE NETWORKING SECURITY   **159**

10 Security and Privacy in Pervasive Networks   **161**
     Tarik Guelzim and Mohammad S. Obaidat
     10.1   Introduction   161
     10.2   Security Classics   162
            10.2.1 Perimeter Security   162
            10.2.2 Access Control   162
     10.3   Hardening Pervasive Networks   162
            10.3.1 Pervasive Computational Paradigms   162
            10.3.2 Pervasive Hardware   163
            10.3.3 Pervasive Networking and Middleware   163
            10.3.4 Pervasive Applications   163
            10.3.5 Pervasive Distributed Application   164
            10.3.6 Logic Based Level Security   164
            10.3.7 Deterministic Access Models   165
            10.3.8 Predictive Statistical Schemes   166
     10.4   Privacy in Pervasive Networks   169
            10.4.1 Problem Definition   169
            10.4.2 Challenges to Privacy Protection   170
            10.4.3 Location Dependency   170
            10.4.4 Data Collection   170
            10.4.5 Internet Service Provider (ISP) Role   170
            10.4.6 Data Ownership   170
            10.4.7 Private Systems   170
            10.4.8 Quality of Privacy (QoP)   171
            10.4.9 Open Issues in Privacy of Systems   172
            10.4.10 'Sharing' in Personal Networks   172
     10.5   Conclusions   172
            References   172

11 Understanding Wormhole Attacks in Pervasive Networks   **175**
     Isaac Woungang, Sanjay Kumar Dhurandher, and Abhishek Gupta
     11.1   Introduction   175
     11.2   A Wormhole Attack   175
     11.3   Severity of a Wormhole Attack   176
     11.4   Background   177
     11.5   Classification of Wormholes   178
            11.5.1 In Band and Out Band Wormholes   179
            11.5.2 Hidden and Participation Mode Wormholes   179
            11.5.3 Physical Layer Wormhole   179
     11.6   Wormhole Attack Modes   179
            11.6.1 Using the Existing Wireless Medium   179
            11.6.2 Using the Out-of- Band Channel   180

		11.6.3	Using High Transmission Power	181
		11.6.4	Using Packet Relays	181
	11.7	Mitigating Wormhole Attacks		181
		11.7.1	Pro-Active Counter Measures	181
		11.7.2	Reactive Countermeasures	182
	11.8	Discussion of Some Mitigating Solutions to Avoid Wormhole Attacks		182
		11.8.1	Immuning Routing Protocols Against Wormhole Attacks	182
		11.8.2	De Worm: A Simple Protocol to Detect Wormhole Attacks in Wireless Ad Hoc Networks	183
		11.8.3	Wormhole Attacks in Wireless Ad Hoc Networks – Using Packet Leashes to Prevent Wormhole Attacks	185
	11.9	Conclusion and Future Work		186
		References		187

12 An Experimental Comparison of Collaborative Defense Strategies for Network Security   **189**

Hao Chen and Yu Chen

	12.1	Introduction		189
	12.2	Background		190
	12.3	Small-World Network Based Modeling Platform		191
		12.3.1	Small World Network	191
		12.3.2	Structure of Three-Layered Modeling Platform	192
	12.4	Internet Worm Attack and Defense		194
		12.4.1	Modeling a Worm Attack	194
		12.4.2	Modeling Defense Schemes	195
	12.5	Experiments and Performance Evaluation		195
		12.5.1	Experimental Setup	195
		12.5.2	Experimental Results and Discussion	196
		12.5.3	Simple SI Model	197
		12.5.4	The SIR Model	197
	12.6	Conclusions		201
		References		201

13 Smart Devices, Systems and Intelligent Environments   **203**

Joaquin Entrialgo and Mohammad S. Obaidat

	13.1	Introduction		203
	13.2	Smart Devices and Systems		204
		13.2.1	Definition and Components	204
		13.2.2	Taxonomy	205
	13.3	Intelligent Environments		207
		13.3.1	Definition and Components	207
		13.3.2	Taxonomy	208
	13.4	Trends		210
	13.5	Limitations and Challenges		211
	13.6	Applications and Case Studies		211
		13.6.1	Smart Everyday Objects	212
		13.6.2	Smart Home	212
		13.6.3	Smart Office	213
		13.6.4	Smart Room	213
		13.6.5	Supply Chain Management	214
		13.6.6	Smart Car	214
		13.6.7	Smart Laboratory	215
		13.6.8	Smart Library	215
	13.7	Conclusion		216
		References		216

PART THREE   PERVASIVE NETWORKING AND COMMUNICATIONS   **219**

14 Autonomic and Pervasive Networking   **221**
  Thabo K. R. Nkwe, Mieso K. Denko, and Jason B. Ernst
  14.1 Introduction   221
  14.2 Ubiquitous/Pervasive Networks   223
     14.2.1 Introduction   223
     14.2.2 Ubiquitous/Pervasive Networks Architecture   223
     14.2.3 Ubiquitous/Pervasive Networks Applications   223
  14.3 Applying Autonomic Techniques to Ubiquitous/Pervasive Networks   224
     14.3.1 Introduction   224
     14.3.2 Autonomic Networking and Computing Paradigms   224
     14.3.3 Cross-Layer Interactions   225
     14.3.4 Limitations of Existing Networks   227
  14.4 Self-* (star) In Autonomic and Pervasive Networks   228
     14.4.1 Self-Protection   228
     14.4.2 Self-Organization   230
     14.4.3 Self-Management   231
     14.4.4 Autonomic Wireless Mesh Networks   233
  14.5 Autonomic and Pervasive Networking Challenges   234
  14.6 Conclusions and Future Directions   234
        References   234

15 An Adaptive Architecture of Service Component for Pervasive Computing   **237**
  Fei Li, Y. He, Athanasios V. Vasilakos, and Naixue Xiong
  15.1 Introduction   237
  15.2 Motivation   238
     15.2.1 Scenario   238
     15.2.2 Requirements For Component Adaptation   238
  15.3 An Overview of the Delaying Adaptation Tool   239
     15.3.1 Framework   239
     15.3.2 Adaptation Table   240
     15.3.3 Working Process   241
     15.3.4 Supplementary Mechanism   242
  15.4 Case Study   243
     15.4.1 Background   243
     15.4.2 Case Process   243
     15.4.3 Discussion   245
     15.4.4 Summary   247
  15.5 Related Work   247
  15.6 Conclusions   247
        References   248

16 On Probabilistic *k*-Coverage in Pervasive Wireless Sensor Networks   **249**
  Habib M. Ammari
  16.1 Introduction   249
  16.2 The Coverage Problem   250
     16.2.1 Computing Density for Coverage   250
     16.2.2 Overview of Coverage Protocols   251
  16.3 Coverage Configuration Problem   252
     16.3.1 Key Assumptions   252
     16.3.2 Definitions   252
     16.3.3 Fundamental Results   252
     16.3.4 Protocol Description   252

16.4 Stochastic *k*-Coverage Protocol   254
    16.4.1 Stochastic Sensing Model   255
    16.4.2 k-Coverage Characterization   255
    16.4.3 Protocol Description   257
16.5 Conclusion   261
    References   261

**17 On the Usage of Overlays to Provide QoS Over IEEE 802.11b/g/e Pervasive
and Mobile Networks**   **263**
Luca Caviglione, Franco Davoli, and Piergiulio Maryni
17.1 Introduction   263
17.2 A Glance at P2P Overlay Networks and QoS Mechanisms   264
17.3 Design of Overlays to Support QoS   265
    17.3.1 Overlays for Guaranteeing Loose-QoS   265
    17.3.2 Overlays for Guaranteeing Strict QoS   267
    17.3.3 Bootstrapping and Overlay Maintenance   267
17.4 Performance Evaluation   269
    17.4.1 Performance Evaluation in the Absence of QoS Support From the MAC Layer   269
    17.4.2 Performance Evaluation When the MAC Supports QoS   272
17.5 Conclusions and Future Developments   274
    Appendix I. The Distributed Algorithm for Bandwidth Management   275
    References   276

18 Performance Evaluation of Pervasive Networks Based on WiMAX Networks   **279**
Elmabruk Laias and Irfan Awan
18.1 Introduction   279
18.2 IEEE 802.16 Architecture and QoS Requirements   281
18.3 Related Work   281
18.4 Proposed QoS Framework   283
    18.4.1 Customized Deficit Round Robin Uplink Scheduler (CDRR)   283
    18.4.2 Proposed Call Admission Control   286
    18.4.3 Proposed Frame Allocation Scheme   287
18.5 Simulation Experiments and Numerical Results   288
18.6 Summary   298
    References   298

19 Implementation Frameworks for Mobile and Pervasive Networks   **301**
Bilhanan Silverajan and Jarmo Harju
19.1 Introduction   301
19.2 Correlating Design to Implementations   302
    19.2.1 Evolving to meet Mobile and Pervasive Networks   303
    19.2.2 Moving from Models to Implementations   304
19.3 Challenges for Implementation Frameworks   305
    19.3.1 Obtaining and Providing Device Characteristics   305
    19.3.2 Supporting Context Awareness   306
    19.3.3 Managing User and Quality of Service Policies   306
    19.3.4 Discovering Resources and Services   306
    19.3.5 Interworking and Interoperability   307
19.4 State of the Art in Implementation Frameworks   308
    19.4.1 Adopting High-Level Specifications   308
    19.4.2 Adopting a Micro-Protocol Approach   308
    19.4.3 Adopting a Design Pattern Approach   308
    19.4.4 Adopting a Protocol Repository Approach   309
19.5 Current Frameworks Research for Network Protocols and Applications   309
    19.5.1 DOORS Framework Architecture   310
    19.5.2 Description of Selected Framework Components   311

     19.5.3 Supporting Service Discovery for Mobile Devices   311
     19.5.4 Support for Basic Movement Detection   312
     19.5.5 Addressing the Need to Furnish Platform Characteristics   312
     19.5.6 Analysis of the DOORS Framework   313
  19.6 Evaluating Frameworks and Implementations   314
     19.6.1 Simulation-Based Evaluation   314
     19.6.2 Testbed-Based Evaluation   314
     19.6.3 Metrics-Based Evaluation   315
  19.7 Conclusion   316
     References   316

Index   **319**

List of Contributors

Fatih Alagoz
Dept. of Computer Engineering
Boaziçi University
P.K. 2 TR-34342 Bebek, Istanbul, Turkey
fatih.alagoz@boun.edu.tr

Azad Ali
Dept. of Computer Science and Engineering
Technische Universität Darmstadt
Hochschulstr 10, 64289 Darmstadt, Germany
azad@cs.tu-darmstadt.de

Habib M. Ammari
Dept. of Computer Science
Hofstra University
211 Adams Hall, Hofstra University, Hempstead NY
11549, USA
Habib.M.Ammari@Hofstra.edu

Irfan Awan
Informatics Research Institute
University of Bradford
Richmond Road, Bradford, West Yorkshire, BD7 1DP,
Bradford, U.K.; Computer Science, KSU
Soudi Arabia
i.u.awan@Bradford.ac.uk

Brahim Ayari
Dept. of Computer Science and Engineering
Technische Universität Darmstadt
Hochschulstr 10, 64289 Darmstadt, Germany
brahim@cs.tu-darmstadt.de

T. M. Burkow
Norwegian Centre for Integrated Care and
Telemedicine
P.O. Box 6060 N-9038 Tromsø, Norway
tatjana.m.burkow@telemed.no

Jiannong Cao
Dept. of Computing
Hong Kong Polytechnic University
PQ806, Mong Man Wai Building, Hong Kong
csjcao@comp.polyu.edu.hk

Daniel Cascado
Robotics and Computer Technology Laboratory
University of Sevilla
Av. Reina Mercedes, s/n. 41012, Sevilla, Spain
danicas@us.es

Luca Caviglione
Institute of Intelligent Systems for Automation
Via Giovanni Amendola, 122/D-I - 70126 Bari BA
Puglia, Genova, Italy
luca.caviglione@ge.issia.cnr.it

Hao Chen
Dept. of Electrical and Computer Engineering
State University New York (SUNY)
State University Plaza, 353 Broadway, Albany,
New York 12246
hchen8@binghamton.edu

Yu Chen
Dept. of Electrical and Computer Engineering
State University New York (SUNY)
State University Plaza, 353 Broadway, Albany,
New York 12246
ychen@binghamton.edu

Marco Conti
Institute of Informatics and Telematics
Aula 40 – "Franco Denoth" Via G. Moruzzi 1 Pisa,
Pisa 56124, Italy
marco.conti@iit.cnr.it

Franco Davoli
Dept. of Communications Computer and Systems
Science (DIST)
University of Genova
Via all'Opera Pia 13 16145, Genova Italy
franco@dist.unige.it)

Franca Delmastro
Institute of Informatics and Telematics (CNR)
Aula 40 - "Franco Denoth" Via G. Moruzzi 1 Pisa,

Pisa 56124, Italy
franca.delmastro@iit.cnr.it

Mieso Denko
Dept. of Computing and Information Science
University of Guelph
50 Stone Rd. East, Guelph, ON., N1G 2W1 , Canada
mdenko@uoguelph.ca

Sanjay Kumar Dhurandher
University of Delhi, Netaji Subas Institute of
Technology
Azad Hing Fauj Marg Sector–3, Dwarka
(Pappankalan)
New Delhi - 110078, India
dhurandher@rediffmail.com

Joaquin Entrialgo
Dept. of Computer Science and Engineering
University of Oviedo
Edificio departamental 1, Campus de Viesques s/n
33204, Gijón, Spain
joaquin@uniovi.es

Jason B. Ernst
Dept. of Computing and Information Science
University of Guelph
50 Stone Rd. East, Guelph, ON., N1G 2W1 , Canada
jernst@uoguelph.ca

Luis Fernández-Luque
Northern Research Institute
Postboks 6434 Forskningsparken, 9294 Tromsø,
Norway
luis.luque@norut.no

Didem Gözüpek
Dept. of Computer Engineering
Boaziçi University
P.K. 2 TR-34342 Bebek, Istanbul, Turkey
didem.gozupek@boun.edu.tr

Tarik Guelzim
Dept. of Computer Science and Software Engineering
Monmouth University
West Long Branch, NJ 07764, USA
guelzimtr@gmail.com

Abhishek Gupta
University of Delhi, Netaji Subas Institute of
Technology
Azad Hing Fauj Marg Sector–3, Dwarka
(Pappankalan)
New Delhi - 110078, India
abhishekg.nsit@gmail.com

Jarmo Harju
Institute of Communications Engineering
Tampere University of Technology
P.O. Box 553, FI-33101, Tampere, Finland
jarmo.harju@tut.fi

Yanxiang He
School of Computer Science
Wuhan University
Wuhan, Hubei 430072, China
yxhe@wh.edu.cn)

Ali R. Hurson
Dept. of Computer Science
Missouri University of Science and Technology
1870 Miner Circle, Rolla, MO 65409, USA
hurson@mst.edu

Evens Jean
Dept. of Computer Science and Engineering
Pennsylvania State University
University Park, PA 16802, USA
jean@cse.psu.edu

Qun Jin
Dept. of Human Informatics and Cognitive Sciences
Waseda University
2-579-15 Mikajima, Tokorozawa-shi, Saitama
359-1192, Japan
jin@waseda.jp

Karl Johan-Grøttum
Northern Research Institute
Postboks 6434 Forskningsparken, 9294 Tromsø,
Norway
karl.johan.grottum@norut.no

Carlos Juiz
Dept. of Mathematics and Computer Science
University of the Balearic Islands
07122 Palma de Mallorca, Illes Balears, Spain
cjuiz@uib.es

Farouk Kamoun
National School of Computer Science
La Manouba University
Postal Code 2010, Tunisia
frk.kamoun@planet.tn

Abdelmajid Khelil
Dept. of Computer Science and Engineering
Technische Universität Darmstadt
Hochschulstr 10, 64289 Darmstadt, Germany.
khelil@cs.tu-darmstadt.de

Medhi Khouja
Dept. of Mathematics and Computer Science
University of the Balearic Islands
07122 Palma de Mallorca, Illes Balears, Spain
mehdi.khouja@uib.es

Elmabruk Laias
Informatics Research Institute
University of Bradford
Richmond Road, Bradford, West Yorkshire, BD7 1DP,
Bradford, U.K.
emlaias@yahoo.com

Fei Li
School of Computer Science
Wuhan University
Wuhan, Hubei 430072, China
kevin.lifei@gmail.com

Piergiulio Maryni
Datasiel S.p.A.
Via Angelo Scarsellini 16149, Genova, Italy
p.maryni@datasiel.net

Thabo K. R. Nkwe
Dept. of Computing and Information Science
University of Guelph
50 Stone Rd. East, Guelph, ON., N1G 2W1 ,
Canada
tnkwe@uoguelph.ca

Mohammad S. Obaidat
Dept. of Computer Science and Software
Engineering Monmouth University
West Long Branch, NJ 07764, USA
Obaidat@monmouth.edu

Hiroaki Ogata
Dept. of Information Science and Intelligent
Systems Tokushima Universityy
Minamijosanjima, Tokushima 770-8506,
Japan
ogata@is.tokushima-u.ac.jp

Ramon Puigjaner
Dept. of Mathematics and Computer Science
University of the Balearic Islands
07122 Palma de Mallorca, Illes Balears, Spain
putxi@uib.cat

Vaskar Raychoudhury
Dept. of Computing
Hong Kong Polytechnic University
PQ806, Mong Man Wai Building, Hong Kong
csvray@comp.polyu.edu.hk

Christian Reinl
Dept. of Computer Science and Engineering
Technische Universität Darmstadt
Hochschulstr 10, 64289 Darmstadt, Germany
reinl@sim.tu-darmstadt.de

Sahra Sedigh
Dept. of Electrical and Computer Engineering
Missouri University of Science and Technology
1870 Miner Circle, Rolla, MO 65409, USA
sedighs@mst.edu

Jose Luis Sevillano
Robotics and Computer Technology Laboratory
University of Sevilla
Av. Reina Mercedes, s/n. 41012, Sevilla, Spain
sevi@atc.us.es

Faisal Karim Shaikh
Dept. of Computer Science and Engineering
Technische Universität Darmstadt
Hochschulstr 10,?64289 Darmstadt, Germany
faisal@cs.tu-darmstadt.de

Timothy K. Shih
Dept. of Computer Science and Information
Engineering
National Central University
No.300, Jung-da Rd,, Chung-li, Taoyuan, Taiwan,
R.O.C.
timothykshih@gmail.com

Behrooz A. Shirazi
School of Electrical Engineering and Computer
Science
Washington State University
Pullman, WA 99164-2752, USA
shirazi@wsu.edu

Joaquin Siebert
Dept. of Computing
Hong Kong Polytechnic University
PQ806, Mong Man Wai Building, Hong Kong
csjsiebert@comp.polyu.edu.hk

Bilhanan Silverajan
Institute of Communications Engineering
Tampere University of Technology
P.O. Box 553, FI-33101, Tampere, Finland
Bilhanan.Silverajan@tut.fi

Neeraj Suri
Dept. of Computer Science and Engineering
Technische Universität Darmstadt
Hochschulstr 10, 64289 Darmstadt, Germany
suri@cs.tu-darmstadt.de

Piotr Szczytowski
Dept. of Computer Science and Engineering
Technische Universität Darmstadt
Hochschulstr 10, 64289 Darmstadt, Germany
piotr@cs.tu-darmstadt.de

Athanasios V. Vasilakos
Dept. of Computer and Telecommunications
Engineering
University of Western Macedonia
2 Eleftherioy Venizeloy street, Kozani, GR 50100,
Greece
vasilako@ath.forthnet.gr

L. Kristian Vognild
Northern Research Institute
Postboks 6434 Forskningsparken, 9294 Tromsø,
Norway
lars.kristian.vognild@norut.no

Isaac Woungang
Dept. of Computer Science
Ryerson University
350 Victoria Street, Toronto, ON., M5B 2K3, Canada
iwoungan@scs.ryerson.ca

Naixue Xiong
Dept. of Computer Science
Georgia State University
P.O. Box 3994, Atlanta, GA 30302-3994
nxiong@cs.gsu.edu

Y. Yano
Dept. of Information Science and Intelligent Systems
Tokushima University
Minamijosanjima, Tokushima 770-8506, Japan
yano@is.tokushima-u.ac.jp

Neil Y. Yen
Dept. of Human Informatics and Cognitive Sciences
Waseda University
2-579-15 Mikajima, Tokorozawa-shi, Saitama
359-1192, Japan
neil219@gmail.com

About the Editors

Professor Mohammad S. Obaidat is an internationally known academic/researcher/scientist. He received his Ph.D and M.S. degrees in Computer Engineering with a minor in Computer Science from The Ohio State University, Columbus, Ohio, USA. He is currently a full Professor of Computer Science at Monmouth University, NJ, USA. Among his previous positions are Chair of the Department of Computer Science and Director of the Graduate Program at Monmouth University and a faculty member at the City University of New York.

He has received extensive research funding and has published numerous books and numerous refereed technical articles in scholarly international journals and proceedings of international conferences, and is currently working on three more books.

Professor Obaidat has served as a consultant for several corporations and organizations worldwide. He is the Editor-in-Chief of the International Journal of Communication Systems published by John Wiley & Sons Ltd, and Editor-in-Chief of the FTRA Journal of Convergence. He served as Editor of IEEE Wireless Communications from 2007–2010. Between 1991–2006, he served as a Technical Editor and an Area Editor of Simulation: Transactions of the Society for Modeling and Simulations (SCS) International, TSCS. He also served on the Editorial Advisory Board of Simulation. He is now an editor of the Wiley Security and Communication Networks Journal, Journal of Networks, International Journal of Information Technology, Communications and Convergence, IJITCC, Inderscience. He served on the International Advisory Board of the International Journal of Wireless Networks and Broadband Technologies, IGI-global. Professor Obaidat is an associate editor/editorial board member of seven other refereed scholarly journals including two IEEE Transactions, Elsevier Computer Communications Journal, Kluwer Journal of Supercomputing, SCS Journal of Defense Modeling and Simulation, Elsevier Journal of Computers and EE, International Journal of Communication Networks and Distributed Systems, The Academy Journal of Communications, International Journal of BioSciences and Technology and International Journal of Information Technology. He has guest edited numerous special issues of scholarly journals such as IEEE Transactions on Systems, Man and Cybernetics, SMC, IEEE Wireless Communications, IEEE Systems Journal, SIMULATION: Transactions of SCS, Elsevier Computer Communications Journal, Journal of C & EE, Wiley Security and Communication Networks, Journal of Networks, and International Journal of Communication Systems, among others. Professor Obaidat has served as the steering committee chair, advisory Committee Chair and program chair of numerous international conferences including the IEEE Int'l Conference on Electronics, Circuits and Systems, IEEE International Phoenix Conference on Computers and Communications, IEEE Int'l Performance, Computing and Communications Conference, IEEE International Conference on Computer Communications and Networks, SCS Summer Computer Simulation Conference, SCSC'97, SCSC98-SCSC2005, SCSC2006, the International Symposium on Performance Evaluation of Computer and Telecommunication Systems since its inception in 1998, International Conference on Parallel Processing, Honorary General Chair of the 2006 IEEE Intl. Joint Conference on E-Business and Telecommunications, ICETE2006. Professor Obaidat served as General Co-Chair of ICETE 2007-ICETE 2010. He has served as the Program Chair of the International Conference on Wireless Information Networks and Systems from 2008-present.

Professor Obaidat is the co-founder and Program Co-Chair of the International Conference on Data Communi-cation Networking, DCNET since its inception in 2009. Professor Obaidat has served as the General Chair of the 2007 IEEE International Conference on Computer Systems and Applications, AICCSA2007, the IEEE AICCSA 2009 Conference and the 2006 International Symposium on Adhoc and Ubiquitous Computing (ISAHUC'06). He is the founder of the International Symposium on Performance Evaluation of Computer and Telecommunication Systems, SPECTS and has served as the General Chair of SPECTS since its inception. Obaidat has received a recognition certificate from IEEE. From 1994–1997, Professor Obaidat served as a distinguished speaker/visitor of IEEE Computer Society. Since 1995 he has been serving as an ACM Distinguished Lecturer. He is also an SCS distinguished Lecturer. From 1996–1999, Professor Obaidat served as an IEEE/ACM program evaluator of the Computing Sciences Accreditation Board/Commission, CSAB/CSAC. Professor Obaidat is the founder and first Chairman of SCS Technical Chapter (Committee) on PECTS (Performance Evaluation of Computer and Telecom-munication Systems). He has served as the Scientific Advisor for the World Bank/UN Digital Inclusion Workshop – The Role of Information and Communication Technology in Development. From 1995–2002, he served as a member of the board of directors of the Society for Computer Simulation International. From 2002–2004, he served as Vice President of Conferences of the Society for Modeling and Simulation International SCS. From 2004–2006, Professor Obaidat served as Vice President of Membership of the Society for Modeling and Simulation Interna-tional SCS. From 2006–2009, he served as the Senior Vice President of SCS. Currently, he is the President of SCS. One of his recent co-authored papers has received the best paper award in the IEEE AICCSA 2009 international conference. He also received the best paper award for one of his papers accepted in IEEE GLOBCOM 2009 con-ference. Professor Obaidat received very recently the prestigious Society for Modeling and Simulation Intentional (SCS) McLeod Founder's Award in recognition of his outstanding technical and professional contributions to modeling and simulation.

Professor Obaidat has been invited to lecture and give keynote speeches worldwide. His research interests are: wireless communications and networks, telecommunications and Networking systems, security of network, information and computer systems, security of e-based systems, performance evaluation of computer systems, algorithms and networks, high performance and parallel computing/computers, applied neural networks and pattern recognition, adaptive learning and speech processing.

Recently, Professor Obaidat has been awarded a Nokia Research Fellowship and the distinguished Fulbright Scholar Award. During 2004/2005, he was on sabbatical leave as Fulbright Distinguished Professor and Advisor to the President of Philadelphia University in Jordan, Dr. Adnan Badran. The latter became the Prime Minister of Jordan in April 2005 and served earlier as Vice President of UNESCO. Prof. Obaidat is a Fellow of the Society for Modeling and Simulation International SCS, and a Fellow of the Institute of Electrical and Electronics Engineers (IEEE).

 **Professor Mieso Denko** came originally from Ethiopia. He received his MSc degree from the University of Wales, UK, and his Ph.D from the University of Natal, South Africa, both in Computer Science. Dr. Denko was an Associate Professor in the Department of Computing and Information Science, University of Guelph, Ontario, Canada, from November 2002 until his sudden death late April 2010. He was the founding director of the Pervasive and Wireless Networking Laboratory (PerWiN) at University of Guelph. Professor Denko was an active member of IEEE ComSoc and energetic volunteer. Dr. Denko received the best paper award for one of his papers authored with Professor Mohammad S. Obaidat in IEEE GLOBECOM 2009. He published numerous journal and conference papers and authored or co-authored several books. He was also on the editorial board of several international journals.

His research interests included performance evaluation of computer and telecommunication systems, mobile and wireless networks, pervasive and mobile computing and autonomic networks. Dr. Denko was a senior member of IEEE and a member of ACM and SCS. Dr. Mieso Denko passed away in the middle of finalizing this book.

Dr. Isaac Woungang received his M.S. and Ph.D degrees in Mathematics from the Université de la Méditerranée-Aix Marseille II, France, and the Université du Sud, Toulon et Var, France, in 1990 and 1994 respectively. In 1999, he received a M.S from the INRS-Materials and Telecommunications, University of Quebec, Montreal, Canada. From 1999–2002, he worked as a Software engineer at Nortel Networks. Since 2002, he has been with Ryerson University, where he is now an Associate Professor of Computer Science. In 2004, he founded the Distributed Applications and Broadband NEtworks Laboratory (DABNEL) R&D group. His research interests include network security, computer communication networks, mobile communication systems, computational intelligence applications in telecommunications and Coding theory. Dr. Woungang serves as Editor-in-Chief of the International Journal of Communication Networks and Distributed Systems (IJCNDS), Inderscience, UK, and the International Journal of Information and Coding Theory (IJICoT), Inderscience, UK, as Associate Editor of the International Journal of Communication Systems (IJCS), John Wiley & Sons, Ltd. Dr. Woungang has edited several books in the areas of wireless ad hoc networks, wireless sensor networks, wireless mesh networks, communication networks and distributed systems and Information and coding theory, published by reputable publishers such as Springer, Elsevier and World Scientific.

Part One

Pervasive Computing and Systems

1

Introduction

Mohammad S. Obaidat[1] and Isaac Woungang[2]
[1]Department of Computer Science and Software Engineering, Monmouth University West Long Branch, NJ 07764, USA.
[2]Department of Computer Science, Ryerson University, 350 Victoria Street, Toronto, Ontario, M5B 2K3, Canada.

1.1   Pervasive Computing and Its Significance

Ubiquitous computing (nowadays also referred to as pervasive computing) was a revolutionary paradigm and technology introduced nearly a decade ago in a seminal 1991 paper by Mark Weiser [1] in these terms: 'the method of enhancing computer use by making many computers available throughout the physical environment, but making them invisible to the user', based upon the following vision [1]: 'The most profound technologies are those that disappear. They weave themselves into the fabric of everyday life until they are indistinguishable from it'. The essence of this vision was the dream of having an environment where traditional networking technologies will complement new advanced computing and wireless communication capabilities, while being integrated gracefully with human users' needs.

Thanks to the Internet and the ubiquitous presence of wearable computers, sensor networks, radio frequency identification (RFIDs) tags, and embedded devices, this vision is now heading towards the reality of a world where using information and communication technologies in our daily lives will not be limited only to high speed distributed computers, but will also extend to intelligent and smart devices [2]. Examples of such devices are scientific instruments, home appliances and entertainment systems, personal digital assistants, mobile phones, coffee mugs, key chains, digital libraries, human body, to name a few, interconnected anytime, seamlessly, and available transparently anywhere, constituting our novel computing network infrastructure. Pervasive computing is aiming at improving significantly the human experience and quality of life [3] without explicit awareness of the underlying computing technologies and communications.

1.2   Research Trends in Pervasive Computing and Networking

In recent years, there have been a number of research developments and technologies that have emerged in areas such as Internet technologies, mobile and distributed computing, handheld devices, computer hardware, wireless communication networks, embedded systems and computing, wireless sensor networks, software agents, human-computer interfaces, and the like. These advances have led to the emergence of several pervasive computing and networking applications. A typical example of such applications is the introduction of pervasive healthcare systems [4], where RFIDs and sensor network technologies have enabled the introduction of computing and communicating capabilities into devices that were considered traditionally as passive physical objects [5], allowing their ubiquitous presence in an environment not originally designed to handle them. Of course, this type of integration and advantage also poses several research challenges that are yet to be addressed [6]. Indeed, the research path towards making pervasive computing a complete reality is still long and winding.

Current research in pervasive computing [7] includes, but is not limited to: (1) heterogeneity and interoperability of computing devices, communication technologies, and software services – today's computing systems are made of various types of entities, mandating the need for designing incentive schemes for ensuring cooperation and collaboration among them [8]; (2) autonomic concepts of pervasive computing and networks [9] – in today's networking environment, enabling a network with self management and self-healing capabilities, and allowing it to cope with the rapid growth of the Internet and their complexities, is a key concern; (3) transparency and pro-activeness [8], [10], in existing computing devices – the development of computing tools has led to the introduction of situation-awareness requirements [11] in the computing world, where it is now envisaged that users of a system can negotiate for a quality of service that accommodates their profiles and applications; (4) location-awareness, scalability, and mobility [11] – in today's computing world, having explicit operator control when dealing with the interaction of entities is no longer a necessary requirement, and context-awareness has been proposed as an innovative novel paradigm for this type of intelligent computing model; (5) security, privacy and trust [12–17] – in today's computing environments, information exchange among the various entities involved brings a means of collaboration, context-based and other types of services, that can lead to a high risk of privacy breach when collaborators use their private information or objects. Protecting each entity as well the environment and information exchange are but a few of the challenges.

1.3   Scanning the Book

The book is organized into 19 chapters, each chapter written by experts on the topic concerned. These chapters are grouped into *three parts*.

PART 1 is devoted to topics related to the design, implementation, and/or management of pervasive computing applications and systems. It is composed of nine chapters: Chapters 1–9.

Chapter 1 introduces the book's content, organization and features, and its target audience.

Chapter 2 promotes the idea that interoperability among independently designed and deployed systems is a critical precursor to the development of pervasive systems. An overview of the tools and techniques that can be utilized to this end is presented, with emphasis on mobile agent technologies and platforms for dynamic reconfiguration and interoperability of sensor networks.

Chapter 3 focuses on the need for discovery mechanisms as a prerogative for accessing resources and services in a pervasive system. The existing approaches and models for discovery of services are discussed, as well as their suitability for pervasive systems.

Chapter 4 focuses on the potential offered by pervasive computing and networking technologies in the area of education, by proposing a thorough review of existing and emerging pervasive learning tools, technologies and applications for mobile and pervasive education.

Chapter 5 deals with service management in pervasive computing environments. The approaches and techniques for managing services in such environments are reviewed thoroughly and a novel framework for analysing the functionalities of service management is proposed.

Chapter 6 promotes the idea of using wireless sensor cooperation as a key enabling technology for objects to cooperate in pervasive computing environments. The techniques for sensor and mobile sensor cooperation in an intra-wireless sensor network are presented, as well as methods for enabling coordination across mobile entities and wireless sensor networks.

Chapter 7 presents multi-hop cognitive radio networks as a vital paradigm in opportunistic pervasive communications. Several MAC layer protocols for multi-hop cognitive radio networks are surveyed, along with related design challenges and open research issues.

Chapter 8 focuses on the design and development of wearable sensor networks for pervasive healthcare systems. A thorough review of available solutions is presented, as well as an analysis of the technological aspects of such designs. This topic is presented at a level of detail that is not found elsewhere in the literature.

Chapter 9 describes the main standards and technologies that are currently available for pervasive computing applications, focusing on wireless connections for the lower layers and middleware for the higher layers. Two examples of pervasive applications are illustrated. The first concerns access to computing services in a remote area and the second deals with home-based telemedicine systems.

Part 2 focuses on topics related to pervasive networking security. It is composed of four chapters: Chapters 10–13.

Chapter 10 discusses in depth the aspects and issues of security and privacy of pervasive networks. Prototype systems that attempt to solve these issues are also presented.

Chapter 11 focuses on wormhole attacks in pervasive wireless ad hoc and sensor networks. An analysis of this type of attack is presented, and current mitigating solutions designed to avoid them are discussed.

Chapter 12 discusses the concept of collaborative defense against Internet worm attacks. A comparative study of two major collaboration schemes for distributed defense is reported, leading to the design of a novel three-layered network model suitable for the evaluation of collaborative schemes. The impact of these schemes on network infrastructure security at the system level is also discussed.

Chapter 13 discusses the role of smart devices and intelligent systems in fulfilling the vision of pervasive computing from the perspective of a user's context. The components of these systems are analysed, and a taxonomy is proposed based on predefined criteria.

Part 3 focuses on pervasive networking and communications issues. It is composed of six chapters: Chapters 14–19.

Chapter 14 focuses on the current state of research addressing autonomic concepts in pervasive networks. An overview of the architectures and applications of ubiquitous and pervasive networks is presented, along with the application of autonomic computing principles. The benefits of cross-layer design approaches with autonomic capabilities are also discussed.

Chapter 15 promotes the idea of using component adaptation as a key solution to eliminate mismatches between existing components and their particular reuse contexts in a pervasive computing system. A framework in the form of an adaptive architecture that can be used to resolve functional dependency among components while enabling delay adaptation is introduced.

Chapter 16 focuses on the problem of sensor scheduling in order to guarantee sensing coverage in pervasive wireless sensor networks. A survey of the existing protocols for computing sensor spatial density to achieve coverage or k-coverage in such networks is proposed.

Chapter 17 deals with the problem of quality of service (QoS) provisioning – in terms of bandwidth, access and transfer delay – in pervasive computing environments. A discussion of the architectural blueprints and mechanisms to support QoS in a self-organizing framework – both automatically and configuration-free – is provided.

Chapter 18 addresses the issues of QoS for fixed Point-to-Multi-Point 802.16 systems, by proposing a novel framework consisting of an uplink scheduler, a call admission control module and a frame allocation scheme in order to resolve these issues.

Chapter 19 reports on some of the major challenges for implementation frameworks that can be anticipated when used for pervasive networking. A survey of a few representative approaches to using frameworks in implementing protocols and services is presented.

Below are some of the important features of this book, which, we believe, make it a valuable resource for our readers:

- This book is designed, in structure and content, with the intention of making it useful at all levels of learning.
- The chapters are authored by prominent academicians/researchers and practitioners, with solid experience in wireless networking and pervasive computing, who have been working in these areas for many years and have a thorough understanding of the concepts and practical applications of these fields.

- The authors are distributed worldwide in a large number of countries and most of them are affiliated with institutions with a global reputation. This gives this book an international flavour.
- The authors have attempted to provide a comprehensive bibliography, which should greatly assist readers interested in delving deeper into the topics.
- Throughout the chapters, most of the core research topics of pervasive computing and networking are covered from both theoretical and practical viewpoints. This makes the book particularly useful for industry practitioners working directly with the practical aspects that enable the technologies in the field.
- To make the book useful for pedagogical purposes, all of the chapters are accompanied by a corresponding set of presentation viewgraphs. The viewgraphs can be obtained as a supplementary resource by contacting the publisher, John Wiley & Sons Ltd., UK.

We have tried to make the chapters of the book look as coherent and consistent as possible. However, it cannot be denied that owing to the fact that the chapters were written by different authors, it was not possible to achieve this task 100%. We believe that this applies to all edited books.

1.4   Target Audience

The book is aimed primarily at the student community. This includes students at both undergraduate and graduate level – as well as students having an intermediate level of knowledge of the topics, and those having extensive knowledge about many of the topics. To achieve this goal, we have attempted to design the overall structure and content of the book in such a manner that makes it useful at all learning levels. The secondary audience for this book is the research community, in academia or in the industry. Finally, we have also taken into consideration the needs of those readers, typically from the industries, who desire insight into the practical significance of the topics, expecting to learn how the spectrum of knowledge and the ideas is relevant to the real-life applications of pervasive computing and networking.

1.5   Supplementary Resources

As mentioned earlier, this book comes with *presentation viewgraphs* for each chapter, which can be used for classroom instruction by instructors who adopt the book as a text. Instructors are requested to contact the publisher, John Wiley & Sons Ltd., UK, for access to these supplementary resources.

1.6   Acknowledgments

We are extremely grateful to the 61 authors of the 19 chapters of this book, who have worked very hard to create this unique resource for the aid of students, researchers and community practitioners. As the individual chapters of this book are written by different authors, the responsibility for the contents of each of the chapters lies with the authors concerned.

   We are also very grateful to the publishing and marketing staff of John Wiley & Sons, for taking a special interest in the publication of this book, and for recognizing the current global market need for such a book. In particular, we would like to thank Ms Sarah Tilley, Ms Anna Smart, and Ms Susan Barclay, who worked so efficiently with us in the publication process. Special thanks go to our institutions, students and research colleagues who in one way or another contributed to this book. Finally, we would also like to thank our families, for their patience and for the continuous support and encouragement they have offered during the course of this project.

References

[1] M. Weiser (1991) 'The Computer for the Twenty-First Century' *Scientific American* **265**(3): 94–104.
[2] M. Kumar, B. Shirazi, S. K. Das, M. Singhal, B. Sung and D. Levine (2003) 'Pervasive Information Communities Organization PICO: A Middleware Framework for Pervasive Computing' *IEEE Pervasive Computing*, pp. 72–9.
[3] U. Hansmann, L. Merk, M. S. Nicklous and T. Stober (2003) *Pervasive Computing: The Mobile World*, 2nd edn, Springer-Verlag, Berlin.
[4] V. Upkar (2009) 'Pervasive Healthcare Computing', *EMR/EHR, Wireless and Health Monitoring*, Springer.

[5]   P. Bellavista, A.Corradi and C. Stefanelli (2000) 'A Mobile Agent Infrastructure for the Mobility Support', *Proc. of the ACM Symposium on Applied Computing*, pp. 239–45.

[6]   http://oxygen.lcs.mit.edu/ (accessed 29 November, 2010).

[7]   F.M.M. Neto and P.F.R. Neto (2010) *Designing Solutions-Based Ubiquitous and Pervasive Computing: New Issues and Trends*, IGI Publishing Ltd.

[8]   A.Hopper (1999) Sentient computing, The Royal Society Clifford Patterson, Lecture, http://www.uk.research.att.com/ ~hopper/publications.html.

[9]   M.K. Denko, L.T Yang and Y. Zhang (2009) *Autonomic Computing and Networking*, 1st edn, Springer Publishing.

[10] A-E. Hassanien, J. H. Abawajy. A. Abraham and H. Hagras (eds) (2009) *Pervasive Computing: Innovations in Intelligent Multimedia and Applications*, Springer.

[11] S.K. Das, A. Bhattacharya, A. Roy and A. Misra (2003) 'Managing Location in 'Universal' Location-Aware Computing', *Handbook of Wireless Internet*, B. Furht and M. Ilyas eds, CRC Press, Chap. 17, pp. 407–25.

[12] A. Soppera and T. Burbridge (2004) 'Maintaining Privacy in Pervasive Computing -Enabling Acceptance of Sensor-Based Services' *BT Technology Journal* **22**(3): 106–7.

[13] R. Campbell, J Al-Muhtadi, G. Sampemane and M. D. Mickunas (2002) 'Towards Security and Privacy for Pervasive Computing', *Proc. of the 2002 Mext-NSF-JSPS Intl. Conference on Software Security: Theories and Systems (ISSS'02)*, Tokyo, Nov. 8–10.

[14] Z. Li, X. Fu, H. Su, M. Jiang and S. T. Xiao (2006) 'Research of Protecting Private Information in Pervasive Computing Environment', *Proc. of 1st Intl. Symposium on Pervasive Computing and Applications*, Urumqi, Aug. 3–5, pp. 561–6.

[15] N. Iltaf, M. Hussain and F. Kamran (2009) 'A Mathematical Approach Towards Trust Based Security in Pervasive Computing Environment', *Advances in Information Security and Assurance, LNCS*, Vol. **5576**, pp. 702–11.

[16] A. Boukerche and Y. Ren (2008) 'A Trust-Based Security System for Ubiquitous and Pervasive Computing Environments' *Computer Communications* **31**(18): 4343–51, 2008.

[17] P. D. Giang, L. X. Hung, R. Ahmed Shaikh, Y. Zhung, S. Lee, Y-K. Lee and H. Lee (2007) 'A Trust-Based Approach to Control Privacy Exposure in Ubiquitous Computing Environments', *IEEE Intl. Conference on Pervasive Services*, July 15–20, Istanbul, pp. 149–52, Aug.

2

Tools and Techniques for Dynamic Reconfiguration and Interoperability of Pervasive Systems[1]

Evens Jean,[1] Sahra Sedigh,[2] Ali R. Hurson,[3] and Behrooz A. Shirazi[4]

[1]*Department of Computer Science and Engineering, The Pennsylvania State University, University Park, PA, USA.*
[2]*Department of Electrical and Computer Engineering, Missouri University of Science and Technology, Rolla, MO, USA.*
[3]*Department of Computer Science, Missouri University of Science and Technology, Rolla, MO, USA.*
[4]*School of Electrical Engineering and Computer Science, Washington State University, Pullman, WA, USA.*

2.1 Introduction

Pervasive systems embody *all the time, everywhere, transparent* services, such as those provided by modern critical infrastructure systems, computer-supported health care networks, and smart living environments [1]. As the science and technology of such systems advances, we approach realization of the vision of an interconnected infrastructure that creates ambient intelligence, allowing anytime, anywhere, unobtrusive services that are gracefully and non-invasively integrated into humans' daily activities. The infrastructure envisioned is composed of heterogeneous computing devices, ranging from supercomputers and powerful workstations, to small devices such as sensors, PDAs, and cell phones, augmented by software and middleware. Central to this visionary computing environment is a ubiquitous, secure, reliable, and often wireless infrastructure that cooperatively, autonomously, and intelligently collects, processes, integrates, and transports information, with adaptability to the spatial and temporal context, while satisfying constraints such as just-in-time operation and sustained performance.

Several shortcomings of existing technology impede the development of the cohesive infrastructure required for large-scale deployment of pervasive systems capable of providing a diverse and dynamic range of services. Foremost among these shortcomings is the lack of *interoperability* among independently designed and deployed pervasive systems. In the design of such systems, the primary focus has not been on interoperability, but rather on ease of network deployment and configuration, energy efficiency, data processing, reliable data transport, security, and other concerns that pertain to the system as a rather isolated entity. This approach may shorten the design cycle and achieve localized efficacy, but could result in needless redundancy that would be avoided by prudent

---

[1] This work was supported in part by the U.S. National Science Foundation under Grant No. IIS-0324835.

interoperation of pervasive systems with overlapping coverage. A broader perspective facilitates interoperability by careful design of interfaces, development of and adherence to standards, and a modular design approach that enables dynamic assembly of services and simplifies reconfiguration.

A pervasive system subsists in a dynamic environment; however, traditionally, the tasks of its hardware and software components are generally static, and these components cannot adapt to changes in application or user requirements. Pervasive systems should be able to support a more diverse array of tasks, varying with the needs of users or with environmental stimuli. Lack of flexibility in the design and implementation of the underlying hardware, software, and middleware is another significant roadblock to realizing a large-scale pervasive computing environment. Currently, once the components are deployed, they coordinate with each other to accomplish the targeted single/multiple task(s), and react to events only within the confines of what was predicted in the original design. As the tasks of nodes are static, multiple systems may need to be deployed in one area to support heterogeneous tasks, even when these tasks have similar, if not identical requirements.

Tools and techniques for dynamic reconfiguration and interoperability of pervasive systems are the main themes of this chapter. Section 2.2 provides an introduction to mobile agent technology, which can be employed in achieving interoperability among multiple interacting pervasive systems. Sensor networks, which support a considerable fraction of the pervasive systems currently deployed, are discussed in Section 2.3, along with software and hardware approaches to their dynamic reconfiguration. The focus of Section 2.4 is collaboration and interoperability among independently deployed sensor networks. Section 2.5 presents two examples of successful utilization of the methods described in earlier sections. Section 2.6 provides a summary and concludes the chapter.

2.2 Mobile Agent Technology

Rapid advances in computing technologies have led to the availability of a plethora of new services to users in recent years. A new outlook on computing applications has emerged, where the growth of pervasive and ubiquitous systems has necessitated the provision of access to data and services at any time, from any location. The dominant approach to adapting to these novel requirements is *mobile agent technology* – a programming paradigm centered on the ability of a program to halt its execution in a particular environment, and then move to a new environment where execution can then be resumed. The success of the approach is due to the inherent aptness of the mobile agent paradigm in providing transparency, adaptability, and robustness of operation, all of which are defining attributes of pervasive systems.

This section provides an overview of mobile agent technology, with emphasis on the role of agents as intermediaries that facilitate interaction among the components of a system. Of special note is the discussion of security, which is intended to alleviate concerns that arise from the movement of agents within the computing infrastructure.

2.2.1 Introduction

An *agent* is a computer program that acts autonomously on behalf of a person or organization [2]. A *mobile agent* is an agent that can autonomously migrate from host to host through a potentially heterogeneous computing infrastructure, and interact with other agents [3]. The use of mobile agents covers a wide spectrum of applications, ranging from the retrieval of information from multiple sources to the administration of complex distributed systems. The advantages of mobile agent technology in supporting disconnected operation, load balancing, and reducing network traffic in global information-sharing have been extensively studied in the literature [4]. These advantages make the mobile agent paradigm especially well-suited to the development of pervasive systems, where transparency is a defining feature.

In general, mobile agents are software entities that roam a network to carry out a task. These agents are perceivably intelligent and autonomous entities that can cooperate with each other to achieve their respective goals, which may align as a common goal. Any mobile agent system is composed of two primary components, namely, the execution environment provided by the hosts, and the mobile agents that travel to various environments on a network [5].

Mobile agents find their applications in environments where there is a need to collect data from multiple sources over a network. The use of mobile agents provides programmers with a new computational model that deviates from the traditional client-server approach, yet yields significant improvements in performance, as agents take the computation to the data, thereby reducing network traffic [1, 6]. The ability of a user to dispatch an agent to roam a network in search of travel tickets has been cited in the research community as one possible application of this programming paradigm [7]. After deployment, such an agent would then be able to make a decision as to which ticket to recommend to the user for purchasing, and may even be able to purchase the ticket. It has been reported that mobile agents generally lend themselves well to searching and computational tasks that require parallel processing [1, 6].

2.2.2   Mobile Agent Security

The mobility of agents may depend on a predetermined itinerary or intermediate results of computation. Along with flexibility in system design, agent mobility also introduces security concerns, which can impede the feasibility or prudence of interaction among pervasive systems. The security requirements of agent systems are identical to those of traditional computing environments [8]; and are classified as confidentiality, integrity, availability, authentication and non-repudiation. *Confidentiality* refers to the protection of information against the possibility of being disclosed to unauthorized parties. *Integrity* ensures that third parties cannot modify relayed information, if any such modification would be undetectable. *Availability* requires that attacks do not prevent information and system resources from performing their intended purposes. *Authentication* is concerned with ensuring that the identity of any entity in the system has been verified. Lastly, *non-repudiation* is intended to prevent any party from being able to deny accountability for an action, by providing mechanisms to prove that such actions have indeed originated from the specified party.

The violation of any security requirement of an agent system constitutes a threat to the security of the system as a whole. It has been noted that security threats to the mobile agent paradigm stem from insecure networks, malicious agents, malicious hosts, or any other malicious entities with access to the network [7–11]. Using the term *agent platform* to refer to the agent's host or execution environment; the security threats to an agent platform have been categorized into four main categories [7]:

1. Agent-to-platform.
2. Agent-to-agent.
3. Platform-to-agent.
4. Other-to-platform.

Agent-to-platform threats encompass issues arising from an agent violating the security requirements of the executing environment through masquerading, denial of service or unauthorized access to system resources. *Agent-to-agent* threats stem from violations of an agent's security requirements by another agent in order to exploit any security weaknesses. *Agent-to-agent* threats can occur through denial of service, masquerading, repudiation or unauthorized access. *Platform-to-agent* threats arise in instances where the platform attacks the agents through masquerading, denial of service, eavesdropping, or alteration of code or data, to cite a few. Lastly, *Other-to-platform* threats occur when the platform's security is compromised by entities external to the agent system. Such threats can occur through masquerading, denial of service, and unauthorized access.

Proposals to secure agent systems have focused on protecting either the hosts or the agents. The security requirements of the two entities are not complementary; as the mobile agent may require anonymity, which may conflict with the requirements of hosts [10]. The execution environments of hosts provide the basic mechanisms for transmission and reception of mobile agents; this is generally achieved through interpreters. The use of interpreters serves the two-fold purpose of providing support for mobile code portability and that of executing mobile agents in a sandbox for security purposes [10]. The use of an interpreted script or programming language can allow the host to deny execution of potentially harmful commands [8]. Protection of hosts can also be achieved through path histories and code signing to verify the authenticity and source of the mobile code. The latter is instrumental in satisfying the host's security needs for authentication and access control.

Protecting agents from malicious hosts involves protecting their data, while ensuring the privacy and integrity of the agent's execution, which encompasses the agent's code and its state [7]. Bierman *et al.* [9] have classified proposals put forth to address the issues of securing agent entities into four categories, namely, trust-based computing, recording and tracking, cryptographic techniques, and time techniques.

Within *trust-based computing*, a host is considered trustworthy if it adheres to its published security policy; protection of the agent is achieved through provision of tamper-resistant hardware or trusted execution environments, which restrict the hosts to which an agent can travel.

Table 2.1 Countermeasures to deter security threats to agent systems

Origin of Threats	Proposed Countermeasures
Agents	Code signing
	Interpreters
	Authentication
	Access control
Platforms	Time techniques
	Partial result encapsulation
	Cryptographic tracing
	Anonymous itinerary
	Server replication
	Path histories

Recording and tracking an agent's itinerary represents the second category of approaches to agent security, and relies on mechanisms such as anonymous itinerary, server replication, or path histories to protect the agent. *Path histories* refer to the maintenance of a record of all platforms visited by the agents. Within the implementation of path histories, each host adds a signed entry to the record, containing its identification along with that of the next host to be visited by the agent [7, 9]. *Server replication* is a mechanism that allows detection of tampering by executing multiple copies of an agent on various execution environments [7, 9].

Cryptographic techniques rely on encryption/decryption algorithms to address various threats. Cryptographic tracing and partial result encapsulation represent two of the mechanisms that fall under this category. *Cryptographic tracing* occurs through the generation of a signed execution log of the agent on a host [9]. The current host passes the signed log on to the next host in the agent's itinerary, and maintains a copy locally for future verification by the agent's owner. *Partial result encapsulation* encrypts the result of the agent's execution on each host using the owner's public key [7, 8]. The incrementally encrypted data can later be retrieved using the owner's private key.

Time techniques protect agents by restricting the time an agent spends on any particular host to prevent evaluation or reverse engineering of the agent by a malicious host. It is worth noting that restricting the execution time of an agent may place unrealistic constraints on some agent applications. Table 2.1 provides a summary of a subset of countermeasures that have been proposed to address agent security. Despite the threats plaguing the paradigm, numerous platforms have been released to support agent-based applications; a discussion of such platforms follows in Section 2.2.3.

2.2.3 Mobile Agent Platforms

As mentioned earlier (Section 2.2.2), the execution environment of agents is generally provided through the use of interpreted programming languages or scripts to provide code portability. Available agent platforms have been implemented through the use of Scheme and Tcl, as well as Java; the latter representing the dominant approach [12]. Altmann *et al.* ranked the Java-based mobile agent platforms based on security, availability, environment, development and characteristic properties [13]. The *security* criterion evaluated platforms based on support for encryption and provision of a secure execution environment; the *availability* parameter refers to the ease of acquiring and using the platform. The *environment* criterion evaluates platforms based on supported operating systems and available documentation; while the *development* criterion focuses on rating efficiency in designing, implementing and deploying agent applications on the platform. Lastly, the *characteristic properties* of the platforms are measured based on support for mobility of agents and adherence to standards of the Foundation for Intelligent Physical Agents (FIPA) [14] and the Object Management Group's MASIF [15]. Altman's study concluded that Grasshopper, Jumping Beans and Aglets represent the top three Java-based agent platforms, respectively.

Grasshopper [16] integrates the traditional client/server and mobile agent paradigms, and conforms to both the FIPA standards and MASIF; furthermore, it provides support for Secure Sockets Layer (SSL) [17] and X.509 Certificates. Jumping Beans [18], while not a mobile agent system per se, provides the framework to build an agent system by allowing applications to 'jump' between hosts on a network. The framework automatically encapsulates the code and data of jumping applications in order to bypass issues relating to software/tools requirements on the receiving host. Lastly, Aglet, initially released by IBM to support the development of mobile code [2, 19], is currently available as an open-source project. Aglets run on the Tahiti Server within the Aglets' context, which is responsible for enforcing the security restrictions of the mobile code. The term *Aglet* is used interchangeably in the literature to refer to each individual mobile agent as well as the platform. Within this chapter, the term Aglet will be followed by the term platform when referencing the actual mobile agent platform; the term will otherwise be a reference to individual agents.

The use of the Java programming language provides platforms with the ability to secure hosts through sandboxing; however, the security of the host is only as effective as the security policies put in place. Furthermore, hosts are still susceptible to denial of service attacks from agents unless limitations are imposed on the processor, memory, and external resources allocated to any migrating agent [8].

As another mobile agent platform, the Pervasive Information Community Organization (PICO) is a middleware framework specifically designed to meet the requirements of time-critical applications of pervasive systems, including autonomy, availability, robustness, and transparent operation in dynamic, heterogeneous environments [20]. The mobile agents in PICO, denoted as *delegents* (intelligent delegates), are members of mission-oriented dynamic computing communities that perform tasks on behalf of the users or computing devices. The delegents are autonomous software entities capable of migrating among and executing on hosts (hardware devices) denoted as *camileuns* (connected, adaptive, mobile, intelligent, learned, efficient, ubiquitous nodes). Camileuns can vary in complexity, as well as communication and computing capabilities. Examples include simple sensing devices (such as a heat sensor), embedded systems that serve as nodes on a wired or wireless network, or a state-of-the-art workstation.

As compared to other mobile agent platforms, the uniqueness of PICO is in the community aspect, i.e., the proactive collaboration of delegents in dynamic information retrieval, content delivery, and facilitation of interfacing, the latter of which is instrumental to interoperability.

2.3   Sensor Networks

A considerable fraction of pervasive systems rely on an underlying network of sensing devices for information about their operating environment. This information is key to proactive and transparent operation of the pervasive system, and facilitates adaptation to dynamic conditions. This section provides an overview of sensor networks and their prevalent applications, with emphasis on dynamic reconfiguration of their functionality to adapt to changing requirements and operating conditions.

2.3.1   Introduction

Sensor networks result from the possibly random deployment of multiple devices equipped with sensing apparatus, in a particular area, to perform a task through coordination and communication. Sensor networks are most often utilized in monitoring designated physical parameters, e.g. temperature or water level, in a particular environment, with the objective of facilitating the appropriate reaction to the occurrence of events of interest, e.g. fire or flooding. In this context, the devices and their sensing apparatus are typically referred to as *sensor nodes*. A sensor network is generally composed of four basic components, namely:

- *sensor nodes*, which are equipped with sensors for one or more physical phenomena, such as seismic, heat, motion, infrared sensors, to cite a few;
- a *networking infrastructure*, which is typically wireless;
- a *sink* or *base station*, to which collected information is relayed; and
- *computing resources* at the sink, or beyond, which perform data mining and correlation.

The nodes in the network are generally comprised of a transceiver, a memory unit and an embedded processor for local processing. Nodes in a sensor network are typically low-power devices with memory capacity on the order of kilobytes, and highly constrained computational power. They are typically inexpensive, possibly to the extent of being considered disposable in the event of failure of destruction. Furthermore, the nodes may be mobile, if mounted on a robot. On the other hand, the base station is usually assumed to be equipped with greater computational resources and data storage capacity, and is not necessarily equipped with any sensing apparatus.

Using the sensing apparatus of the sensor nodes, the network can monitor its coverage area and react to events of interest. The task is accomplished through relaying of the sensed information from the nodes to the sink for processing. Note that the transfer of information can be initiated from the sink or from the sensor nodes, depending on the implementation of the network and the task at hand. The network may be composed of thousands of nodes that have been programmed before being deployed in the area of interest. The deployment of the nodes may be random, or the nodes may be placed in specific points of interest, depending upon the application at hand and the ease of access to the terrain.

2.3.2   Sensor Network Applications

Sensor networks are well-suited to applications that require data collection from a particular environment, often to facilitate reaction to the occurrence of specific events that can be deduced from the data. These applications can be classified into several main categories, including military, environmental monitoring, home and office, habitat, and medical applications [21].

In military applications, sensor networks are used to monitor friendly or enemy forces, assess damages on a battlefield, or detect biological or chemical attacks, amongst other uses.

Within environmental monitoring, the aim is to detect environmental incidents such as flood, fire, seismic activities, or biological events in the area of interest [22]. Sensor networks can also be used in support of agriculture, to facilitate more efficient irrigation of farmland, track animals, or monitor the temperature in a barn. A related category is structural health monitoring, where sensor networks are used to monitor indicators of the safety of civil infrastructure, e.g. a bridge. Sensing devices are embedded at the time of construction or retrofitted on existing structures to measure phenomena such as strain, acceleration, and tilt [23].

The ubiquitous coverage offered by inexpensive sensor nodes is useful in home and office applications, where the sensing nodes can be integrated into household appliances and configured to respond to environmental stimuli or user commands issued locally or remotely, possibly over the Internet.

In habitat monitoring applications, nodes in a sensor network can be used to observe the breeding pattern of wild animals or the life cycle of plants, without disturbing the environment they are deployed to monitor.

Medical applications depend on sensor nodes to carry a patient's vital information in order to reduce errors; they can also be used to monitor a patient and react to physical events or to the patient's vital signs.

The compendium of sensor network applications is not solely represented in the aforementioned categories. Numerous applications of sensor network do not fall under any single category. Notable examples include the use of sensor nodes to detect suspicious individuals or survivors in a disaster, to interact with humans in a classroom setting, or to track a moving object in a designated environment.

2.3.3   Dynamic Reconfiguration of Sensor Networks

Reconfigurable sensor networks have been introduced to support dynamic tasking of sensor nodes and to allow for the network to concurrently support multiple applications. One approach to achieving a reconfigurable sensor network is to consider the sensor nodes as a set of data-stores into which queries can be injected, to collect information that can be used by the sink for a given purpose. Collecting information from the sensor nodes is inadequate in applications where the nodes need to interact with each other in order to reach a conclusion in real time, as would be the case in distributed target tracking applications or any applications that require the use of distributed algorithms. Two salient approaches to reconfiguration of sensor networks are described in this section. The first approach, which is based on the concept of active sensors, reconfigures sensor nodes through software, via abstractions of the runtime environment. In contrast, the second approach focuses on dynamic reconfiguration of the hardware and utilizes Field Programmable Gate Arrays (FPGAs) to this end.

2.3.3.1   Software Approaches to Reconfiguration of Sensor Nodes

The *active sensors* approach typically makes use of virtual machines, script interpreters and mobile agents to render sensor nodes reprogrammable. Related research has led to the development of a number of platforms for dynamic sensor networks, including Maté [25, 26], SensorWare [24], Deluge [27, 28], Agilla [29], a mobile agent framework developed at UC Davis [30], ActorNet [31], and SOS [32]. In general, these platforms have targeted their developments to suit applications requiring low-cost reprogrammable nodes with no restrictions on the maximum physical size of the node. With the exception of SensorWare, the storage requirements of these systems are such that they can inhabit the Berkeley Mica motes, which have a 4 MHz microprocessor and 136 KB total memory (Flash, SRAM, and EEPROM). The application domains of these systems range from military to environmental and habitat monitoring; they do not, however, span to applications requiring particularly small sensors or those that do not benefit from reprogrammable nodes.

Maté aims at providing sensor networks with a flexible architecture upon which application specific scripting environments can be built [25, 26]. Maté consists of three major components: *contexts* (units of concurrent execution), *operations* (units of execution functionality), and *capsules* (units of code propagation). A Maté *virtual machine* (VM) component can be either part of the basic template, which is general, or part of the specific VM tailored to an application domain. Maté makes use of Trickle [33], a protocol designed to address the issue of code maintainability in sensor networks, to update the network.

Maté suffers from the assumption that any reprogramming occurs over all nodes in the network. It also assumes, restrictively, that at any given time, all nodes are coordinated for the execution of a specific application. Furthermore, Maté views the network as an isolated entity and does not address issues of interoperability with other networks.

SensorWare, introduced as an attempt to address the issue of reconfigurable sensor networks, runs on top of an *Operating System (OS) Layer,* which handles the standard functions and services of a multi-threaded environment [24]. The *SensorWare Layer* is comprised of the language as well as the run-time environment for the mobile scripts in the network. The SensorWare scripting language is based on the widely popular scripting language Tcl, augmented with functionalities suitable for sensor network environments. The SensorWare language is event-based and can be considered a state machine influenced by external events. Each event is tied to a specific handler that executes when the event occurs. An event may trigger one or more subsequent events or change the state of the system as it executes. A SensorWare script waits on events and invokes the appropriate handler when an event occurs; the script can then wait on a new set of events or loop around and wait on the same set of events after the execution of the handler.

SensorWare enables a sensor network to run multiple scripts simultaneously; as such, unlike Maté, it does not assume that the whole network is focused on only one task at any point in time. On the other hand, just as Mate, it ignores issues of interoperability. The latest implementation of the system required 179 KB of space with the core accounting for 30 KB, which makes it very unsuitable for environments populated with nodes having very few storage capabilities.

Deluge has been designed to handle the dissemination of large data objects over wireless sensor networks [27, 28]. Deluge is aware of the network density, and is built to handle the unpredictable availability of nodes by representing data objects as a set of fixed-sized pages, which allows for multiplexing and incremental upgrades. Deluge, just as Maté, is based on Trickle. Trickle focuses on single packet dissemination, while Deluge addresses the multiple-packet aspect. However, Deluge suffers from the same restrictive assumption that all nodes in the network need to be programmed, and as such is unable to select a subset for reconfiguration. Furthermore, it does not address interoperability or the need to support multiple tasks.

Agilla [29] allows each node to support multiple agents, which may or may not be cooperating to accomplish a task. A unique assumption made by Agilla is that each node knows its geographical location, which is used as the address of the node. Agents in Agilla can clone themselves, or move to another location, carrying with them either their code and state, or just their code. Agilla agents die at the completion of their task to allow for efficient memory usage. Similar to the other reconfigurable sensor network platforms discussed above, Agilla does not provide support for collaboration among different sensor networks, nor does it provide services to migrating agents.

Researchers at UC Davis have introduced a mobile agent framework for sensor networks built on top of the Maté virtual machine, to allow use of the agent-programming paradigm within a sensor network environment [30]. The framework allows agents to execute within an interpreter that implements the basic functionalities of agents, such as forwarding, so as to minimize the size of agent code that needs to be transferred from node to node. The advantage of this framework over Maté is that selective reprogramming can be carried out on a subset of the nodes in a network. Interoperability among heterogeneous networks has not been addressed, and interaction among networks is contingent upon their use of a common agent platform and protocols.

ActorNet [31] is a mobile agent system for wireless sensor networks that supports an asynchronous communication model, context-switching, multi-tasking, agent coordination as well as virtual memory. The agent system can be thought of as two entities: the agent language and the platform design. The ActorNet platform is a virtual machine that can support multiple actors (agents) per node. Similar to Agilla and the UC Davis platform, ActorNet employs mobile agents to selectively reprogram nodes, rather than the network as whole. The platform, however, does not allow for interoperability of heterogeneous networks.

SOS is a sensor network operating system that supports run-time software reconfiguration [32]. The introduction of SOS is meant to allow the update of modules without interrupting sensor operation, while providing the flexibility of virtual machines without the associated cost of interpreted languages. SOS is composed of a statically-compiled kernel that provides system services to dynamically loadable binary modules for the implementation of drivers, user programs, and the like. Although the services provided by SOS allow for sensor nodes to be dynamically reconfigured, the platform still views sensor networks as isolated entities. Furthermore, the nodes in the network do not perform processing of acquired data at the point of collection, instead opting for the relay of such data to a base station.

2.3.3.2   Hardware Approaches to Reconfiguration of Sensor Nodes

Recent efforts in utilizing FPGAs for reconfiguration of sensor nodes have been motivated by the need to increase the computational power of nodes, in order to allow some local processing of data. The *Virtual Architecture for Partially Reconfigurable Embedded Systems (VAPRES)* has been put forth to that end, based on the observation that FPGAs can outperform the microprocessors typically found in sensor nodes [34]. The introduction of VAPRES is also motivated by the inability of Agilla (see Section 2.3.3.1) to handle video feeds and other advanced sensor data. Using VAPRES, advanced sensor data can be processed without halting execution of the device. The proposed architecture relies on the ability of some FPGAs to be partially reconfigured by modules in order to react to environmental observations. VAPRES handles inter-module communication and consists of a central controlling agent, a flash controller core to read and store partial bit streams, and peripherals for communication.

The VAPRES approach, while efficient, suffers from the same shortcoming as the software approaches described in Section 2.3.3.1, all of which consider a sensor node an isolated entity and thus do not address issues of interaction among existing networks.

Motivated by the need to provide in-network data aggregation, Commuri *et al.* also adopted the notion of FPGA-based sensor nodes [35]. In their approach, *Reconfigurable Cluster Heads (RCHs)* are used to aggregate data from other nodes in the network and relay it to the base station for processing. The election of RCHs is done based on the energy available at the participating nodes, with the RCH being the node with the most energy.

The reconfiguration of RCHs is query-based, in that RCHs are reconfigured based on specific aggregation algorithm of incoming queries. This represents a considerable drawback to the proposed work, as the rate of arriving queries and their heterogeneity may require a drastic number of reconfigurations to be performed. While the proposed approach is limited, as it does not take advantage of the power of FPGAs to process the data close to the point of collection; it does however allow for the possible bridging of sensor networks with established infrastructure through the RCHs, hence enabling the foundation of an interoperable system.

Table 2.2 provides a comparative summary of the systems discussed in this section.

Table 2.2   Comparison of approaches to reconfiguration of sensor networks

System	Platform	Heterogeneous Tasks	Interoperability of Networks	Service-Oriented Infrastructure
Maté	Virtual machine	No	No	No
SensorWare	Run-time environment	Yes	No	No
Deluge	Network programming	Yes	No	No
Agilla	Agent system	Yes	No	No
UC Davis Framework	Agent system	Yes	No	No
ActorNet	Agent system	Yes	No	No
SOS	Binary modules	Yes	No	Yes
VAPRES	FPGA-based	Yes	No	No
RCH	FPGA-based	No	Yes	No

2.4   Collaboration and Interoperability Among Sensor Networks

To date, research on sensor networks has been focused on issues related to deployments intended for a single application. The prevalence of relatively inexpensive, commercially available sensor nodes such as Mica, Intel, and TMote Sky motes has facilitated such deployments. Such motes typically have a limited set of onboard sensors, but support the interfacing of a diverse array of external sensing devices. However, in the bulk of research studies on sensor networks, a fixed set of sensing modalities is selected prior to deployment per the requirements of the application, resulting in a static system configuration that is incapable of adapting to changing environmental conditions or application requirements.

The lack of widely accepted standards is the main impediment to interoperability. Standardization efforts related to sensor networks have been focused on communication protocols, e.g. the Zigbee specification [36] based on IEEE 802.15.4 [37], or interfaces between sensors and the network, e.g. IEEE 1451 [38]. The need for data interoperability has been recognized, as evidenced by groups such as the Open Geospatial Consortium (OGC), which aims to develop open standards for geographic information systems [39]. Leveraging this effort is SensorNet [40,41], which connects the sensor networks of strategic testbeds to each other and to operations centers for emergency dispatch and mass notification services. The individual networks are heterogeneous; however, they have been designed per SensorNet specifications and standards, which considerably alleviates challenges associated with interoperability. Secure and redundant links are available for connecting the networks. This is rarely the case with sensor networks deployed independently by different owners.

Another framework based on OGC is the Semantic Sensor Web (SSW) [42], which aims to increase the situational awareness of sensor networks by annotating sensor data with spatial, temporal, and thematic semantic metadata. SSW enables interoperability through the use of this metadata and contextual information from networks, by building an ontology-based hierarchical system that allows access to sensor data through web applications. Such initiatives are yet to be adopted on a wide scale, are typically domain-specific, and cannot be retrofitted' to legacy systems already deployed.

The interoperability of sensor networks with enterprise networks motivated the introduction of *Edge Servers*, which filter raw sensor data in an effort to alleviate the computing burden placed on application servers [43]. They have been proposed as a means of interconnecting sensor networks, but the application-specific nature of the code restricts their use to the limited set of enterprise networks for which they were originally conceived.

This limitation is overcome by a related platform, the *Global Sensor Network (GSN)*, which has been proposed as middleware for connecting heterogeneous sensor networks [44]. No assumptions are made in GSN regarding the underlying network infrastructure, except for the existence of a sink connected to a base computer through a *GSN Wrapper*. The main abstraction defined is a *virtual sensor*, which can take as input several data streams from physical or virtual sensors and deliver a single output data stream. A virtual sensor can be anything from a physical sensor to a sink or set of physical sensor nodes. This abstraction achieves separation of concerns, as it hides the details of the physical sensors and the fashion in which they are accessed. The emphasis of the GSN platform is on efficient distributed query processing. Membership of the virtual sensors is determined a priori and cannot be changed to support dynamic service composition. In GSN internetwork communication is through the base computer, as a result, interoperability is at the high level, as each base computer becomes the communication portal to the underlying network.

IrisNet is similarly focused on query processing [45], as it aims to provide an interface for users to query a vast amount of data collected by a collection of possibly heterogeneous sensor networks. The approach taken by IrisNet is to view a sensor network as an entity capable of providing services to consumers. The platform utilizes *sensing agents (SAs)* for collection and pre-processing of data from the sensors, and *organizing agents (OAs)* for storing the data in a distributed database. The sensor networks are assumed to be under the same ownership, eliminating the considerable challenges associated with interoperability and control of access privileges. Heterogeneity of the underlying communication infrastructure has not been addressed in IrisNet.

The first step towards dynamic service composition from multiple sensor networks is discovery of the services and resources offered by each network. The data-centric nature of sensor networks and energy concerns differentiate the problem from the general case of distributed resource discovery, which is well-studied. Optimizations have been proposed for sensor networks, e.g. [46], but they assume a homogeneous sensor network. Challenges associated with heterogeneity of the sensor networks have been articulated in [47], with Dynamic Resource Discovery (DRD) proposed as a solution. In this approach, resource discovery in sensor networks is divided into the tasks of a) identifying the resources that need to be tracked and b) querying the network in an energy-efficient manner. The former is accomplished with the collection of metadata that provides information regarding communication

Table 2.3   Comparative analysis of interoperable sensor networks

System	Service-Oriented Infrastructure	Internetwork Communication	Shared Communication Protocol
SensorNet	No	Direct	Yes
Semantic Sensor Web	No	Indirect	Not required
Edge Servers	No	Indirect	Not required
Global Sensor Network	No	Indirect	Not required
IrisNet	Yes	Indirect	Not required
Dynamic Resource Discovery	No	Not specified	Not specified
Virtual Sensor Network	No	Direct	Yes
Tiny Web	Yes	Direct	Yes

protocols, message formats, and other information pertinent to interoperability. Cluster heads are used to hold resource attributes and respond to queries, allowing other sensors to conserve energy. The study does not venture into how various networks communicate in exchanging metadata.

Other solutions proposed to the problem of resource discovery in sensor networks include UPnP bridges [48] and service-oriented platforms [49, 50], which are better suited for static sensor networks of limited scale. Clustering techniques, e.g. [51], are better suited to the large-scale networks that typically serve as the infrastructure for pervasive systems, and lend themselves well to an agent-based implementation.

Enabling the sharing of physical resources by a dynamic subset of sensor nodes led to the introduction of *virtual sensor networks* (VSN) [52]. VSNs are formed using subsets of one or more physical sensor network nodes in order to accomplish a common task. As a VSN may not be fully interconnected nor connected to a base station, the physical nodes not comprising the VSN are used to provide communication support to the VSN. A VSN performs two major functions – VSN maintenance and membership maintenance. The former includes the addition and deletion of nodes, merging or splitting VSNs, and determination of boundaries. The latter manages the role of each sensor in the VSN, which can be either communication or processing. VSNs are most beneficial in areas where geographically overlapping sensor networks are deployed. However, data from the VSN is still relayed to a base station for processing and there exists the underlying assumption in the proposal that cooperating networks use the same communication medium and thus can easily exchange messages. The emphasis of the approach is on enabling remote control of the sensing infrastructure, rather than dynamic composition of services from disparate and heterogeneous sensor networks intended for dedicated applications.

Tiny Web aims at providing interoperability through adoption of the notion of web services to the sensor network environment [53]. Tiny Web addresses interoperability at both the network and the application layers of the communication stack. Using the *Web Service Description Language (WSDL)*, nodes in Tiny Web advertise their interfaces to applications. Tiny Web relies on a standardized mean of communication between networks and focuses on dealing with data representation in order to provide interoperability.

Table 2.3 provides an evaluation summary of the interoperable platforms described in this section.

2.5   Applications

The previous three sections of this chapter presented tools and techniques for dynamic configuration of and interoperability among pervasive systems, with emphasis on sensor networks, which play a significant role in enabling proactive, adaptive, and transparent operation for the majority of pervasive systems. This section illustrates the application of these tools and techniques in two different pervasive computing contexts.

The first application presented is a volcano monitoring system that serves as a successful example of interoperability among a terrestrial sensor network and a satellite component. Due to the hostile operating environment, strict timing constraints, and exceedingly high computational and communication requirements, this application also serves as an example of the extremes of pervasive systems. In the second application, pervasive computing is employed in support of higher education, to facilitate customization of courses and degree programs to the needs, interests, and backgrounds of individual students. This application highlights both the interoperability of the diverse components involved and the dynamic reconfiguration of the system, based on the collective intelligence gained through the interaction of these components.

2.5.1 A Pervasive System for Volcano Monitoring

The availability of increasingly powerful computing resources at decreasing costs has enabled the integration of more robust and reliable techniques into sophisticated pervasive computing applications. The case study presented in this section is a real-time pervasive computing system for volcano monitoring, denoted as *OASIS* (for Optimized Autonomous Space In-situ Sensorweb). The primary objectives of OASIS are to integrate complementary space and in-situ ground sensors into an autonomous pervasive system, to optimize power and communication resource management of this pervasive system, and to provide mechanisms for seamless and scalable fusion of future space and in-situ components [54]. Interoperability among the components comprising the monitoring system is a fundamental requirement of the application.

The OASIS application is designed to monitor Mount St. Helens, an active volcano in the state of Washington, and provide feedback and decision support to Earth scientists engaged in critical activities such as evacuation planning and air traffic routing. The development of this Earth-hazard-monitoring system demonstrates the ability to mitigate volcano hazards through an intelligent in-situ network. Key to this approach is the establishment of a continuous feedback loop between two primary components: a ground in-situ component and a space component.

The *in-situ ground sensor network* is composed of a suite of tiny Imote2 wireless sensor motes, each of which is connected to an array of deformation monitoring sensors. These deformation monitoring sensor arrays consist of GPS (Global Positioning System), seismic, infrasonic, RSAM (Real-time Seismic-Amplitude Measurement), and lightning sensors. In order to protect the sensors from the harsh environment, each mote and its sensor array are housed in a box. The boxes will be dropped by helicopter onto Mount St. Helens. In order to allow for proper placement of the boxes on the crater, each is equipped with a metal tripod-like structure, called a *spider* designed and created by Earth Scientists at Cascades Volcanic Observatory. Figure 2.1 depicts one such spider immediately before its deployment.

Although the wireless sensor nodes are tiny compact devices, each is equipped with complex, yet lightweight, software architecture. Of particular interest is the application layer of this architecture, which is the top-most

Figure 2.1   Deployment of Sensing Device by Helicopter [54].

layer of the communication stack and composed of a data sensing module, a network management module, and a situation awareness module. The sensing module is vital to the overall reliability of the network, as it controls the clock synchronization and time stamping of packets, making sure that the network is in sync. The Earth scientists interact with the network management module, which gives them the ability to monitor the current status of the network, as well as make necessary adjustments to network parameters.

During active periods when bandwidth demands are highest, the network prioritizes information flow and reserves bandwidth for high-priority data. For instance, if during volcanic activity, gas measurements are deemed the highest priority; other data may be buffered to make more bandwidth available for gas data. Cluster coordinators are able to automatically identify and select the minimum set of sensors that will provide mission critical data. Bayesian network techniques were applied for sensor selection [55].

To further optimize bandwidth utilization, in-situ data reduction, compression, and aggregation are driven by science requirements. For example, when necessary, seismic data, typically recorded at 100 Hz, are reduced by two orders of magnitude at the node level by reporting an average RSAM parameter, which is an established measurement of both earthquakes and volcanic tremors [56]. In addition, continuous seismic data is streamed into a buffer at each node, and when seismic events are detected, the buffered waveform with precise time markers will be compressed and delivered to the control center for higher level processing.

The functionality provided by the situation awareness module is to ensure that the network is conscious of its environment and capable of dynamically responding to changes. If a physical phenomenon, such as an eruption, occurs within a particular geographic region of the network, that region is categorized as an area of interest. The categorization of a physical phenomenon can take place either automatically or manually. Automatic categorization is done through a set of predefined situations described by the Earth scientists and identified by physical changes detected either by the ground sensors or another external source, such as the satellite. Alternatively, the Earth scientists can manually categorize an area as a region of interest.

Sensor nodes within a region of interest are given a higher priority, guaranteeing that data and communication from those sensors is processed promptly and without loss, even in the presence of network saturation and/or congestion. For example, if lava is emitted from a portion of the volcano, that area of the network is classified as high priority.

The space component of the volcano monitoring system consists of the Jet Propulsion Laboratory (JPL) sensor-web ground software and the EO-1 satellite. The EO-1 satellite is the first mission in NASA's New Millennium Program Earth Observing series, managed from Goddard Space Flight Center. Designed as a testbed for the next-generation of advanced land imaging instruments, EO-1 carries three instruments: the Advanced Land Imager (ALI), the hyper-spectral Hyperion Imager, and the Atmospheric Corrector (AC). The ALI combines novel wide-angle optics with a highly-integrated multispectral and panchromatic spectrometer. The Hyperion is a high-resolution imager capable of resolving 220 spectral bands (from 0.4 to 2.5 μm) with a 30-meter spatial resolution. Finally, the EO-1 AC provides the first space-based test of an atmospheric corrector that is designed to compensate for atmospheric absorption and scattering, which allows for increased accuracy of surface reflectance estimates.

Figure 2.2 depicts the OASIS volcano monitoring system throughout the data collection and communication stages that lead to prioritization of one or more nodes.

The requirements of the space component are twofold. First, it is responsible for responding to inquiries regarding generic sensor capabilities, such as providing information on the data pedigree or the signal-to-noise ratio of the data. Second, it is responsible for tasking all requests made by Earth scientists for data acquisition, processing the data, and generating any alerts derived from the data. Specifically, the space component receives alerts from the in-situ sensors.

Upon receipt of an alert, the space component attempts to issue a request to EO-1 to acquire data from the region of interest. Once the data has been acquired onboard EO-1, a search is carried out for features of interest, e.g. thermal activity. The results of the analysis are later downlinked in an engineering telemetry stream, while onboard they may have generated a follow-up observation activity.

In the interest of interoperability with other environmental monitoring systems, all service-oriented interfaces were developed in conformance with the OGC Sensor Web Enablement initiative, described in Section 2.4, which includes the following services:

1. *Sensor Planning Service (SPS)*: used to determine if the sensor is available to acquire requested data.
2. *Sensor Observation Service (SOS)*: used to retrieve engineering or science data. This includes access to historical data, as well as data requested and acquired from the SPS.

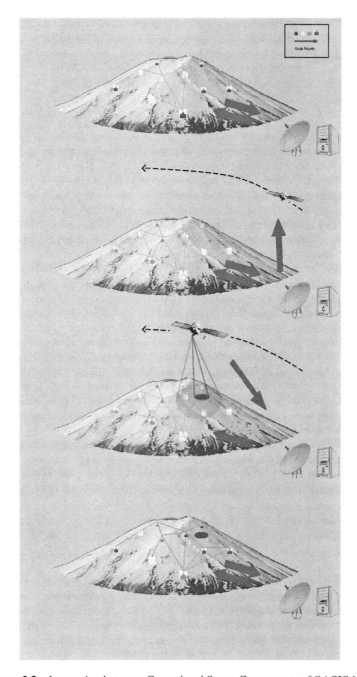

Figure 2.2   Interaction between Ground and Space Components of OASIS [54].

3. *Web Processing Service (WPS)*: used to perform calculations on the acquired remote sensing data. This includes processing the raw data into derivative products such as vegetation indices, soil moisture, burn areas, lava flows and effusions rates.
4. *Sensor Alert Service (SAS)*: used to publish and subscribe to alerts from sensors. Users register with this service and provide conditions for alerts. When these conditions are met by the acquired data, alerts containing the data along with the time and location of the events are automatically issued to the user.

5. *A description of the sensor and their associated products and services using the Sensor Markup Language (SensorML).* SensorML provides a high-level description of sensors and observation processes using an XML schema methodology. It also provides the functionality for users to discover instruments on the web, along with services to task and acquire sensor data (such as the SPS, SOS, SAS, and WPS).

2.5.2   A Pervasive Computing Platform for Individualized Higher Education

Advances in database management, distributed computing; computational intelligence; and especially pervasive systems, which allow anytime, anywhere access to information; provide fertile ground for radical changes in pedagogy. This section describes a pervasive platform that builds on these technologies to facilitate customization of course content to the needs, interests, and backgrounds of individual students. Teaching tools, animation techniques, and remote access to the course content are utilized to present the same information in different ways, to accommodate differences in learning styles; encourage active, rather than passive learning; allow self-pacing, privacy, and flexibility; and ensure efficient utilization of resources [57]. The success of the platform is contingent upon interoperability of the numerous components involved, as well as the ability to dynamically reconfigure the system to best serve the needs of each student.

As described in Sections 2.1 and 2.2, pervasive computing has the potential to significantly improve the handling of dynamic situations and exceptions in disparate environments, through the use of software agents that monitor and record events and take proactive and collaborative actions to ensure the availability of accurate information in an adaptive fashion. This adaptability and flexibility of pervasive computing environments can facilitate a radical change in the way students interact with professors and peers.

In the proposed educational platform, pervasive computing and communications are utilized at various levels, through the use of the PICO mobile agent platform, which was described in Section 2.2.3. As mentioned, PICO consists of software agents, denoted as delegents, which are created by the user, application, or another delegent. These delegents self-organize into dynamic communities with the purpose of sharing data, processing different sources of information, and making context-aware decisions. In the context of the educational platform, PICO creates mission-oriented dynamic computing entities that autonomously perform tasks on behalf of students and faculty, and provides a framework for adaptively composing course and curriculum content for each student. The pervasive computing environment provided by PICO enables continuous curriculum methodology, and in the process, enriches course delivery, improves the quality of course content, and encourages interaction.

The objective of the educational platform is the customization of courses and curricula to the needs of each student. To this end, a degree program is viewed as being comprised of three sets of entities:

- the set of instructors/advisors, I;
- the set of students, S; and
- the set of courses, C.

An instructor/advisor, $i \in I$, has expertise in one or more subjects. A student, $s \in S$, is studying towards a degree and is required to take courses from C, in an orderly fashion, to satisfy degree requirements and objectives. Each $c \in C$ represents a course in the curriculum. The courses in C are interrelated and the structure of the curriculum determines relationships among the courses. In addition, courses are tagged according to degree requirements, as required or elective. Each set, I, S, and C, respectively is represented by a community of software agents that communicate and negotiate with each other according to the defined tasks, viz., advising a student, scheduling courses, or individualizing the content of a course.

Fundamental to the proposed educational platform is the modular approach to course development. Course modules are self-contained, self-paced, and designed to promote active, rather than passive, learning. Available multimedia tools are used to design a module to be interactive and flexible for the students, who can navigate freely through the program. Self-assessment questions at the end of each module take a variety of forms, including objects to be dragged into place, and multiple-choice or short-answer questions.

The most interactive components of a course module are the design challenges, in which the student must make decisions and respond to hints and other feedback, in order to find a solution for the problem at hand. Graphics and animation are used to demonstrate complex concepts. Text, narration, charts, and diagrams are used to reinforce concepts during animation. All courses are 'mix-and-match,' in the sense that each course consists of several modules on interrelated topics that are, in turn, interrelated across the whole curriculum.

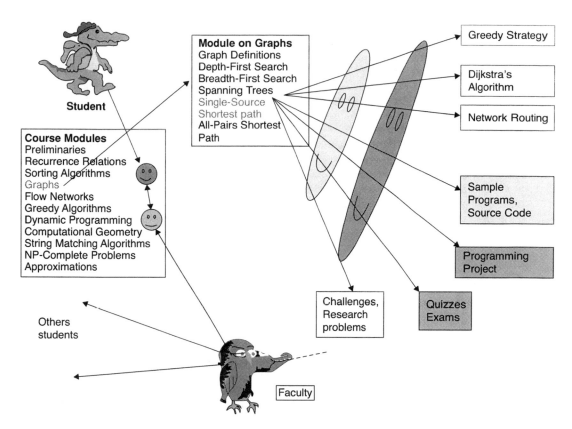

Figure 2.3   Sample environment for a course on design and analysis of algorithms.

Each course is divided into several content modules, a number of which are mandatory and dictated by the curriculum. The remaining content modules, typically smaller in number, include elective content that can be chosen to supplement the student's knowledge of prerequisites, or to engage an interested student in more advanced topics. Each content module has prerequisites, lecture notes, questions, exercise problems, sample solutions, programming or laboratory assignments, exams, quizzes, and evaluations. The content modules for each course can be supplemented by experiential modules, which are intended to reemphasize key issues covered in the content modules through hands-on individual and group projects.

A delegent is associated with each module of a course. This delegent is also linked to other modules in the same course, and can be linked to related modules in other courses using the platform. Figure 2.3 illustrates a sample environment for a course on design and analysis of algorithms. For example, a module on 'graphs' in a course on algorithms will be closely linked to other modules of the course – sorting, greedy algorithms, dynamic programming, and complexity analysis. Furthermore, the module representing 'routing' in a course on computer networks will be linked to the module on 'graphs,' from the algorithms course. The delegent associated with a module directly reports to the instructor. In other words, this course module delegent serves as a virtual guide to the student taking the course. Delegents representing various modules within a course form a community among themselves, representing that course.

When a student, s, is required to take a course, x, a delegent, D_{sx} is created to represent the student in that course. D_{sx} acquires information about the student's academic background and determines the student's major degree, interests, and accumulated prerequisite knowledge, among other needed information. D_{sx} interacts with the delegents of course modules to customize the course material. Furthermore, D_{sx} ensures that the student peruses required material – notes, sample programs, exercises, and the like. D_{sx} also alerts the student to timelines, class schedules, learning/discussion schedules, project deadlines, appointments with the instructor, and corresponding preparations. The instructor delegent for the course, D_{ix}, ensures that the student meets all requirements for each mandatory module, and collaborates with D_{sx} to ensure that the student is supplied with all required course material.

D_{ix} also informs the instructor about the progress of the student and provides alerts to the instructor whenever a student is progressing too slowly or too rapidly.

The proposed teaching practice aims to shift the focus of classes from lectures to interactive problem-solving based on the course modules, which the students are required to study beforehand. Each class begins with a quiz on the major concepts. The remainder of the class period is devoted to addressing students' questions and discussing and analyzing the more complex issues covered in the mandatory modules. Traditional periodic homework assignments, projects, and examinations will be utilized to enforce learning and determine students' overall knowledge of the course material.

One of the main objectives of the proposed educational platform is to create a virtual environment that supports a continuous, subject-based, one-to-one student-faculty ratio for the purposes of mentoring, advising, guiding, and educating students within the program. The division of courses into modules allows the system to view the curriculum at a finer granularity, and can hence minimize the redundancy of topics and maximize the adaptability of the course contents as a student progresses towards his or her degree. A student agent, D_s, in communication with his or her academic advisor agent, D_y, utilizes the student's profile to determine an appropriate course schedule that fulfills degree requirements while catering to the student's individual needs and interests to the extent possible.

2.6   Conclusion

The two themes explored in this chapter were interoperability and dynamic reconfiguration in pervasive systems, both of which are pivotal to the efficient provision of diverse services. Mobile agent platforms were discussed as enabling technologies, with emphasis on their role as intermediaries among the heterogeneous entities that comprise a pervasive system. Mechanisms were discussed for alleviating security concerns related to the migration of agents among hosts in the system. Sensor networks, whose data collection capabilities are critical to the proactive and transparent operation of pervasive systems, were introduced, along with software and hardware approaches to dynamically reconfiguration of sensing nodes in response to changes in their operating environment or application requirements. Several platforms intended to achieve interoperability and cooperation among independent sensor networks were introduced and discussed. Finally, two significantly different application of pervasive computing, namely, volcano monitoring and individualized higher education, were presented to illustrate the successful utilization of the tools and techniques described earlier in the chapter.

Emerging applications of pervasive systems, such as disaster recovery or elder care, can improve the safety and quality of life of a significant fraction of the population. The success of these applications is contingent on reliable access to rich data sets, which necessitates flexibility and interoperability of the sensor networks and other pervasive computing resources supplying this data. Careful design of interfaces, development of and adherence to standards, and modular designs approach that enable dynamic service composition will expedite the successful deployment of large-scale pervasive systems, while increasing their efficiency and reliability.

References

[1] Y. Jiao and A.R. Hurson (2004) 'Performance Analysis of mobile agents in mobile database access systems – A quantitative case study, '*Journal of Interconnection Networks* **5**(3): 351–72.

[2] D. Lange and M. Oshima (1998) *Programming and Developing Java Mobile Agents with Aglets*. Reading, MA: Addison-Wesley.

[3] R. Gray, D. Kotz, G. Cybenko, and D. Rus (2002) 'Mobile agents: motivations and state-of-the-art systems,' in *Handbook of Agent Technology*, J. Bradshaw, Ed. AAAI/MIT Press, pp. 1–30.

[4] A. Hurson, E. Jean, M. Ongtang, X. Gao, Y. Jiao, and T. Potok (2007) 'Recent advances in mobile agent-oriented applications,' in *Mobile Intelligence: When Computational Intelligence Meets Mobile Paradigm*, L. Yang and A. Waluyo, Eds. John Wiley & Sons.

[5] T. Sander and C.F. Tschudin (1998) 'Protecting mobile agents against malicious hosts,' in *Mobile Agents and Security*, G. Vigna, Ed., LNCS vol. **1419**, Springer-Verlag, pp. 44–61.

[6] Y. Jiao and A. R. Hurson (2004) 'Modeling and performance evaluation of agent and client/server-based information retrieval systems – A case study,' in *Proceedings of Communication Networks and Distributed Systems Modeling and Simulation Conference*, San Diego, California, pp. 1–6.

[7] J. Claessens, B. Preneel, and J. Vandewalle (2003) '(How) can mobile agents do secure electronic transactions on untrusted hosts? A survey of the security issues and the current solutions,' *ACM Transactions on Internet Technology* **3**(1): 28–48.

[8] W. Jansen and T. Karygiannis (2000) 'NIST *Special Publication 800–19 – Mobile Agent Security*,' National Institute of Standards and Technology.

[9] E. Bierman and E. Cloete (2002) 'Classification of malicious host threats in mobile agent computing,' in Proc. of the Conference of the South African Institute of Computer Scientists and Information Technologists (SAICSIT), pp. 141–8.

[10] The Object Management Group, http://omg.org/, accessed Aug. 2009.

[11] M.S. Greenberg, J.C. Byington, T. Holding, and D.G. Harper (1998) 'Mobile agents and security,' *IEEE Communications Magazine*.

[12] Y. Jiao (2002) '*Multidatabase Information Retrieval Using Mobile Agents*,' Masters Thesis, Pennsylvania State University.

[13] J. Altmann, F. Gruber, L. Klug, W. Stockner, and E. Weippl (2001) 'Using mobile agents in real world: a survey and evaluation of agent platforms,' in *Proceedings of the 2nd Workshop on Infrastructure for Agents, MAS and Scalable MAS*, Montreal, Canada, May.

[14] IEEE Computer Society Foundation for Intelligent Agents (FIPA), http://www.fipa.org/, accessed Aug. 2009.

[15] Mobile Agent System Interoperability Facilities (MASIF) Specification, OMG TC Document orbos/97–10–05, http://www.omg.org/cgi-bin/doc?orbos/97–10–05.pdf, accessed Aug. 2009.

[16] C. Bäumer, M. Breugst, S. Choy, and T. Magedanz (1999) 'Grasshopper – a universal agent platform based on OMG MASIF and FIPA standards,' in *Proceedings of the First International Workshop on Mobile Agents for Telecommunication Applications (MATA)*, Ottawa, Canada, Oct., p. 1–8.

[17] K.E.B. Hickman (1995) '*Secure Socket Library*,' Netscape Communications Corp., Internet Draft RFC.

[18] Jumping Beans, Ad Astra Engineering Inc., http://www.JumpingBeans.com/, accessed Aug. 2009.

[19] B. Lange and M. Oshima (1998) 'Mobile Agents with JAVA: The Aglet API,' *World Wide Web* 1(3): 111–21.

[20] M. Kumar, B. Shirazi, S. Das, B. Sung, D. Levine, and M. Singhal (2003) 'PICO: a middleware framework for pervasive computing,' *IEEE Pervasive Computing* 2(3): 72–9.

[21] F. Akyildiz, W. Su, Y. Sankarasubramaniam, and E. Cayirci (2002) 'A survey on sensor networks,' *IEEE Communications Magazine*, Aug.

[22] J. Koch, S. Sedigh, and E. Atekwana (2007) 'Subsurface hydrological monitoring of a watershed with hybrid sensor networks,' in *Proc. of the Joint Assembly of the American Geophysical Union*, Acapulco, Mexico, May.

[23] T. Harms, B. Banks, S. Sedigh, and F. Bastianini (2009) 'Design and testing of a low-power wireless sensor network for structural health monitoring of bridges,' in *Proc. of the 16th International SPIE Symposium on Smart Structures and Materials and Non-destructive Evaluation and Health Monitoring*, San Diego, USA, Mar.

[24] A. Boulis, C. Han, and M. B. Srivastava (2003) 'Design and implementation of a framework for efficient and programmable sensor networks,' in *Proc. of the 1st International ACM Conference on Mobile Systems, Applications and Services*, New York, NY, USA, pp. 187–200.

[25] P. Levis and D. Culler (2002) 'Maté: A tiny virtual machine for sensor networks,' in Proc. of the 10th International Conference on Architectural Support for Programming Languages and Operating Systems (ASPLOS X).

[26] P. Levis, D. Gay and D. Culler (2004) '*Bridging the Gap: Programming Sensor Networks with Application Specific Virtual Machines*,' UC Berkeley Technical Report UCB//CSD-04–1343, August.

[27] P.K. Dutta, J.W. Hui, D.C. Chu, and D.E. Culler (2006) 'Securing the Deluge network programming system,' in *Proc. of the 5th International Conference on Information Processing in Sensor Networks (IPSN)*, pp. 326–33.

[28] J.W. Hui and D.E. Culler (2004) 'The dynamic behavior of a data dissemination protocol for network programming at scale,' in *Proc. of the 2nd ACM Conference on Embedded Networked Sensor Systems (SenSys)*.

[29] C. Fok, G. Roman, and C. Lu (2005) 'Rapid development and flexible deployment of adaptive wireless sensor network applications,' in *Proc. of the 24th International Conference on Distributed Computing Systems (ICDCS)*, June, pp. 653–62.

[30] L. Szumel, J. LeBrun, and J.D. Owens (2005) 'Towards a mobile agent framework for sensor networks,' in *Proc. of the Second IEEE Workshop on Embedded Networked Sensors*, Sydney, Australia, pp. 79–87.

[31] Y.K. Kwon, S. Sundresh, K. Mechitov, and G. Agha (2006) 'ActorNet: An actor platform for wireless sensor networks,' in *Proceedings of the 5th International Joint Conference on Autonomous Agents and Multi-Agent Systems (AAMAS)*, May.

[32] C. Han, K. R. Rengaswamy, R. Shea, E. Kohler, and M. Srivastava (2005) 'SOS: A dynamic operating system for sensor networks,' in *Proc. of the 3rd International Conference on Mobile Systems, Applications, and Services (Mobisys)*, Seattle, Washington, USA.

[33] P. Levis, N. Patel, D. Culler, and S. Shenker (2004) 'Trickle: A self-regulating algorithm for code maintenance and propagation in wireless sensor networks,' in *Proc. of the First USENIX/ACM Symposium on Network Systems Design and Implementation (NSDI)*.

[34] R. Garcia, A. Gordon-Ross, and A. George (2009) 'Exploiting partially reconfigurable FPGAs for situation-based reconfiguration in wireless sensor networks,' in *Proc. of the IEEE Symposium on Field-Programmable Custom Computing Machines (FCCM)*, Napa, CA, Apr.

[35] S. Commuri, V. Tadigotla, and M. Atiquzzaman (2008) 'Reconfigurable hardware-based dynamic data aggregation in wireless sensor networks,' *International Journal of Distributed Sensor Networks* 4(2) 194–212.

[36] Zigbee Alliance Zigbee Specification, ZigBee Alliance Technical Report, http://www.zigbee.org/Products/Technical DocumentsDownload/tabid/237/Default.aspx, accessed Aug. 2009.

[37] 'Wireless medium access control (MAC) and physical layer (PHY) specifications for low-rate wireless personal area networks (WPANs),' in *IEEE Standard for Information technology – Telecommunications and information exchange between systems – Local and metropolitan area networks – Specific requirements*, 2007, ch. 15.4., http://standards.ieee.org/getieee802/download/802.15.4a-2007.pdf, accessed Aug. 2009.

[38] K. Lee (2000) 'IEEE 1451: A standard in support of smart transducer networking,' in *Proceedings of the 17th IEEE Instrumentation and Measurement Technology Conference (IMTC)*, vol. 2, pp. 525–8.

[39] M. Botts, G. Percivall, C. Reed, and J. Davidson, 'OGC sensor web enablement: Overview and high level architecture (OGC 07–165),' Open Geospatial Consortium White Paper, http://portal.opengeospatial.org/files/?artifact_id=25562, accessed Aug. 2009.

[40] B.L. Gorman, M. Shankar, and C.M. Smith (2005) 'Advancing Sensor Web interoperability,' *Sensors Magazine* **22**(4), April 2005, http://www.sensorsmag.com/sensors/article/articleDetail.jsp?id=185897, accessed Aug. 2009.

[41] M. Shankar, B.L. Gorman, and C.M. Smith (2005) 'SensorNet operational prototypes: Building wide-area interoperable sensor networks,' in *Proceedings of the First IEEE International Conference in Distributed Computing in Sensor Systems (DCOSS)*, June, pp. 391–2.

[42] A. Sheth, C. Henson, and S.S. Sahoo (2008) 'Semantic sensor web,' *IEEE Internet Computing* **12**(4): 78–83.

[43] S. Rooney, D. Bauer, and P. Scotton (2006) 'Techniques for integrating sensors into the enterprise network,' *IEEE eTransactions on Network and Service Management* **2**(1), 2006, http://www.comsoc.org/etnsm/, accessed Aug. 2009.

[44] K. Aberer, M. Hauswirth, and A. Salehi (2007) 'Infrastructure for data processing in large-scale interconnected sensor networks,' in *Proceedings of the International Conference on Mobile Data Management*, May, pp. 198–205.

[45] P.B. Gibbons, B. Karp, Y. Ke, S. Nath, and S. Seshan (2003) 'IrisNet: an architecture for a Worldwide Sensor Web,' *IEEE Pervasive Computing* **2**(4): 22–33.

[46] F. Stann and J. Heidemann (2005) 'BARD: Bayesian-assisted resource discovery in sensor networks,' in *Proceedings of the 24th Annual Joint Conference of the IEEE Computer and Communications Societies (INFOCOM)*, vol. 2, March, pp. 866–77.

[47] S. Tilak, K. Chiu, N. Abu-Ghazaleh, and T. Fountain (2005) 'Dynamic resource discovery for sensor networks,' in *Proceedings of the IFIP International Symposium on Network-Centric Ubiquitous Systems (NCUS)*, pp. 785–96.

[48] M. Isomura, T. Riedel, C. Decker, M. Beigl, and H. Horiuchi (2006) 'Sharing sensor networks,' in *Proceedings of the 26th IEEE International Conference on Distributed Computing Systems Workshops (ICDCSW)*, Los Alamitos, CA, USA, p. 61.

[49] J. King, R. Bose, H.-I. Yang, S. Pickles, and A. Helal (2006) 'Atlas: A service-oriented sensor platform: Hardware and middleware to enable programmable pervasive spaces,' in *Proceedings of the 31st IEEE Conference on Local Computer Networks (LCN)*, pp. 630–8.

[50] F.C. Delicato, P.F. Pires, L. Pirmez, and L.F. Carmo (2005) 'A service approach for architecting application independent wireless sensor networks,' *Cluster Computing* **8**(2–3): 211–21.

[51] R. Marin-Perianu, H. Scholten, and P. Havinga (2007) 'Prototyping service discovery and usage in wireless sensor networks,' in *Proceedings of the 32nd IEEE Conference on Local Computer Networks (LCN)*, pp. 841–50.

[52] A.P. Jayasumana, Q. Han, and T.H. Illangasekare (2007) 'Virtual sensor networks – a resource-efficient approach for concurrent applications,' in *Proceedings of the 4th International Conference on Information Technology (ITNG)*, Apr., pp. 111–15.

[53] N. B. Priyantha, A. Kansal, M. Goraczko, and F. Zhao (2008) 'Tiny web services: design and implementation of interoperable and evolvable sensor networks,' in *Proceedings of the 6th ACM Conference on Embedded Network Sensor Systems (SenSys)*, Raleigh, NC, USA, Nov., pp. 253–66.

[54] N. Peterson, L. Anusuya-Rangappa, B.A. Shirazi, W. Song, R. Huang, D. Tran, S. Chien, and R. LaHusen (2008) 'Volcano monitoring: A case study in pervasive computing,' in *Pervasive Computing: Innovations in Intelligent Multimedia and Applications*, Eds. A. Hassanien, A. Abraham, and H. Hagras, Springer-Verlag.

[55] H. Alex, M. Kumar, and B. A. Shirazi (2005) 'Collaborating agent communities for information fusion and decision making,' in *Proceedings of the International Conference on Knowledge Integration and Multi-Agent Systems (KIMAS)*.

[56] T.L. Murray, J.W. Ewert, A.B. Lockhart, and R.G. LaHusen (1996) 'The integrated mobile volcano-monitoring system used by the Volcano Disaster Assistance Program (VDAP),' *Monitoring and Mitigation of Volcano Hazards*.

[57] A.R. Hurson and S. Sedigh (2009) 'A pervasive computing platform for individualized higher education,' In *Proceedings of the International Conference on Computational Intelligence and Software Engineering (CiSE '09)*, Wuhan, China.

3

Models for Service and Resource Discovery in Pervasive Computing

Mehdi Khouja,[1] Carlos Juiz,[1] Ramon Puigjaner,[1] and Farouk Kamoun[2]

[1]*Department of Mathematics and Computer Science, University of the Balearic Islands, Palma de Mallorca, Spain.*
[2]*National School of Computer Science, La Manouba University, La Manouba, Tunisia.*

3.1 Introduction

The omnipresence of mobile and embedded devices in various daily environments such as hospitals, schools, airports, etc aims to make people's lives easier. Thus, by being offered information and services users are experiencing a pervasive computing environment. The purpose of pervasive systems is to propose services within an heterogeneous environment and via devices with special requirements. A design approach for these systems is to consider them as a Service-Oriented Architecture. Interactions among the components include announcing, searching and invoking services. On the one hand, these interactions fall within the SOA concept. On the other, they have to stick to specific system requirements. The limitations of the devices and the context in which the components interact are two major constraints on a pervasive system. The expected result from a service orientated architecture (SOA) is the provision of an adequate service to fulfil a specific request. This process is simple if the requester and the service provider know each other at run time. But, owing to the heterogeneity of computers and especially pervasive systems, the task of finding the appropriate service is more complex. This mechanism is called service discovery (SD). Several models of service discovery have been specified for SOA. Some of them may be more suitable for a pervasive system than others. A study of the existing techniques for SD will help to establish the appropriateness of existing SD models for pervasive systems. Jini, SLP, Bluetooth discovery model and UPnP constitute a good set of models to investigate. Since these models are not oriented mainly to pervasive systems, various investigative works purport to enhance SD models with the characteristic features of pervasive systems, adding mainly context-awareness. In this chapter, we present our approach for SD in pervasive context-awareness systems. Our solution adds a semantic enhancement to a pervasive system at the thinking layer. This is done by organizing contextual information within an ontology. This approach aims to group the elements of a pervasive system according to contextual criteria. Hence, our group-based solution will ameliorate the service discovery process.

This chapter is structured as follows: first, we describe service oriented architecture principles. Then, existing model of service discovery is discussed in Section 3.2. In Section 3.3, a variety of work to do with enhancement of service discovery is listed. We also describe our approach to an SD model. Our conclusions and future directions for research directions are provided in the final section.

Pervasive Computing and Networking, First Edition.   Edited by Mohammad S. Obaidat, Mieso Denko, and Isaac Woungang.
© 2011 John Wiley & Sons, Ltd. Published 2011 by John Wiley & Sons, Ltd.

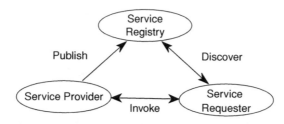

Figure 3.1 SOA core components.

3.2   Service Oriented Architecture

A SOA is a software architecture in which loosely-coupled services are defined using a description language and have evocable interfaces that are called upon to perform processes. From a conceptual level, a SOA is composed of three core parts:

- *The service provider*: The service provider defines the service description and publishes it to the service registry.
- *The service requester*: The service requester accesses the register directory to find the service description and its providers.
- *The service registry*: The service registry is an intermediary between the service provider and the service registry.

The activities that can be performed in a SOA, as described in Figure 3.1, are: publishing, discovering and invoking. First, the service provider has to publish the service description. Then, the service requester questions the registry in order to find the adequate service description. Finally, the requester invokes the desired service and interacts with the provider in order to perform the service.

In order to accomplish these tasks, several techniques have been specified. First, publishing a service requires its description. The available methods for this task vary according to the degree of expressiveness: key/value, template-based and semantic description. In the key/value approach, services are characterized by a set of key/value pairs. Requesting a service consists of specifying the exact value of an attribute. In the case where a query language is implemented, attributes values can be compared with the requested values using operators. The template-based description is done via semi-structured languages such as XML. The semantic description relies on the use of ontologies. They have higher and formal expressiveness. It is also possible to use ontology reasoning techniques. Knowledge sharing and reuse can be performed between heterogeneous context sources.

Once a description is submitted to the service registry, the requester may start looking for a service that is adequate for its needs. Therefore, the discovery process depends on the service registry architecture: centralized or distributed. The centralized-based model implies a dedicated directory that maintains service information and processes requests. The distributed model consists in that any components of the system maintain a local part of the registry. The service requests are distributed transparently to the clients across the other registry. There is a service-oriented model that does not include a registry. This is done through periodical announcements of the existing service via multicast protocols. Another aspect of service design is the communication mechanism for retrieving services. We may find two approaches: Query/pull and Notification/push. The former implies that clients query for services. In the latter approach, clients can register and wait to be notified if an adequate service is published.

3.3   Industry and Consortia Supported Models for Service Discovery

In this section, we present service discovery models supported by industry consortia. Then, we discuss their suitability for pervasive systems

3.3.1   Existing Models for Service Discovery

Surveys such as [1] and [2] (among others) have discussed the existing service discovery models.

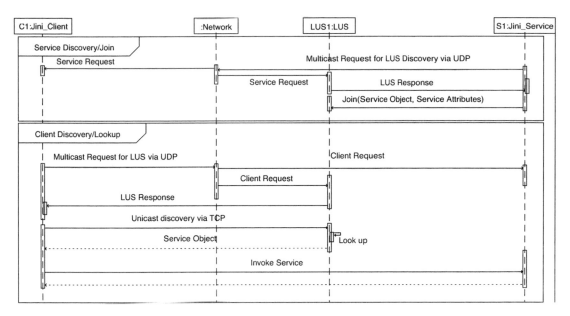

Figure 3.2   Service discovery interactions in Jini.

3.3.1.1   Jini

Jini [3], developed originally by Sun, offers a service oriented framework for constructing distributed systems. The goal of the Jini architecture is the federation of groups of clients/services within a dynamic computing system. Jini enables users to share services and resources over a network. The technology infrastructure is Java technology centered.

Since it is SOA-based, Jini's key concepts reflect the basic principles of service computing. The lookup service (LUS) plays the role of a registry. The service provider (Jini service) and service requester (Jini client) find each other via the LUS. The core of Jini communication is a trio of protocols, namely discovery, join and lookup. The interactions among the different Jini architecture components can be as follows:

• When a Jini service or client starts up; it sends a multicast request to search for a lookup service. Once one or more LUS respond to the request, the provider can start the join process. The Jini service registers a service object (proxy) and its attributes with the lookup service. The object consists of a Java interface for the service with the invocation methods for the service.
• The Jini client requests a service to the LUS. Then, the LUS sends a copy of the service object to the client. Finally, the requester uses the proxy to communicate directly with the service provider.
• The LUS sends at startup and periodically a multicast announcement via UDP to the network components.

Figure 3.2 illustrates the various communication scenarios that describe the Jini discovery process.

3.3.1.2   Service Location Protocol: SLP

The Service Location Protocol (SLP) [4] is being developed by the IETF. It provides a scalable framework for the discovery and selection of network services. SLP architecture includes three main components:

• The User Agent (UA) plays the role of a proxy for the client in the discovery process.
• The Service Agent (SA) announces the location and attributes of the service.
• The Directory Agent (DA) registers services published by the SA in their database and responds to service requests from UAs.

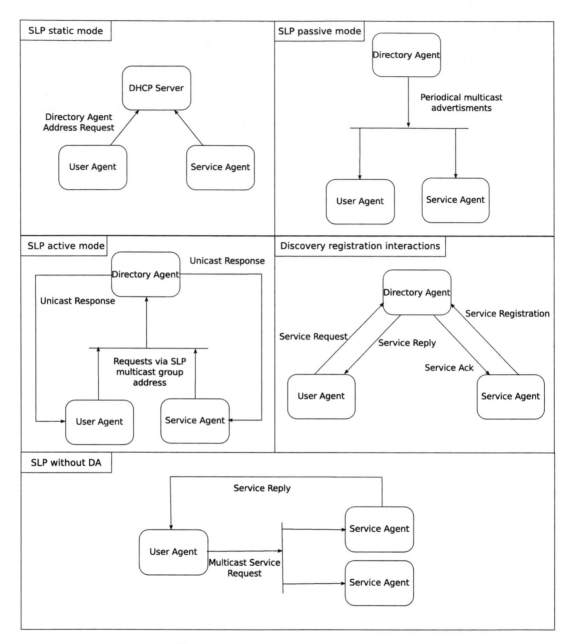

Figure 3.3   SLP Service Discovery scenarios.

To perform their respective roles UA and SA have to discover a DA. There are there methods for DA discovery: static, active and passive. In the static approach, SLP agents retrieve the address of DA via DHCP. The active method UAs and SAs use the SLP multicast group address to send service requests. Then, DAs that are listening on this address respond using via unicast to the agent. In passive discovery, DAs announce their service periodically via multicast. Hence, UAs and SAs are able to know the DA address and communicate directly with it. These discovery scenarios are better suited to large networks with many services. In a small network, the discovery process can be fulfilled without the DA. In this case, UAs send their service requests periodically to the SLP multicast address. The SAs announcing the service will send a unicast response to the UA. Moreover, SAs announce their presence via multicast. Figure 3.3 depicts the different communication scenarios in the SLP discovery process.

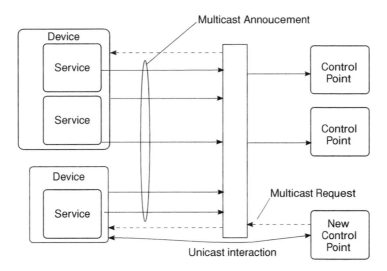

Figure 3.4   UPnP service discovery interactions.

3.3.1.3   UPnP

UPnP (Universal Plug and Play) [5] is maintained by the UPnP forum initiative. It aims to offer seamless connectivity to devices within a network. It comes with a set of specifications defining the addressing and discovery of resources as well as the description and control of the services within the network.

The UPnP architecture includes two main components: devices and control point. Devices offer services over the network whereas control point consumes these services.

The UPnP discovery process is based on the Simple Service Discovery Protocol (SSDP). The process is directory-less. When a device appears within a network, it announces its service to the control points. It sends HTTP requests over multicast via UDP, namely HTTPMulticast. These announcements contain information about the service types and links to the descriptions. SSDP permits a new control point to look for devices via multicast. The information received by the control point from the device during the discovery phase does not allow it to invoke a service. To do so, the control point must retrieve the device's description from the link provided previously. This description is XML-based and includes a list of the provided services and URLs for control, eventing and presentation. Once the description is retrieved the invocation process is initiated using SOAP (Simple Object Access Protocol) messages. Figure 3.4 depicts the interactions that take place during the discovery process.

3.3.1.4   Bluetooth SDP

Bluetooth comes with its own protocol stack. As part of it, it offers its proper service discovery method: Bluetooth Service Discover Protocol (Bluetooth SDP). The SDP specifies the actions required of a Bluetooth client application in order to discover the available services in the Bluetooth servers [6]. The protocol also establishes the searching mechanism for services. Service discovery in Bluetooth relies on a request/response model (Figure 3.5). This model allows devices to discover Bluetooth services offered in the vicinity via two modes: searching and browsing. Searching interaction consists of retrieving a specific service, while browsing is the process of looking at the services offered. These methods of inquiry are possible only if devices are first discovered and then linked. Therefore, in the discovery process, devices assume specific roles: Local Device (LocDev) and Remote device (RemDev). A LocDev initiates the service discovery mechanism. This device implements the client part of the Bluetooth SDP architecture. The service discovery application (SRVDscApp) is employed by a user to begin discovering other devices. The RemDev is the device that responds to the requests of the LocDev. It contains the server part of the Bluetooth SDP. Answering client inquiries is done by consulting the service records database. Figure 3.6 shows the architecture of the LocDev and the RemDev.

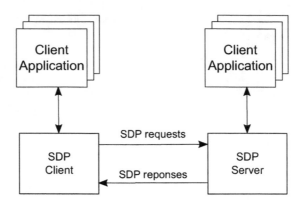

Figure 3.5   SDP client-server interaction [6, Figure 2.1].

3.3.2   Suitability of Existing Models for Pervasive Systems

Pervasive systems are heterogeneous and dynamic networks with limited devices. The industry supported models, with the exception of Bluetooth, are made for stable networks. Bluetooth has a peer-to-peer approach. A directory-based model, such as Jini and SLP, relies on a central registry. This design is not suitable for a pervasive system because it create dependency between the elements in the system. The models described above use multicast as an initial method of communication. This may cause an overhead in the network bandwidth and extra resource consumption caused by message forwarding. An enhancement of this communication mechanism would be controlled forwarding. In UPnP and SLP, devices are allowed to announce their services through the network. By storing these announcements for future use, devices may suffer from problems with memory.

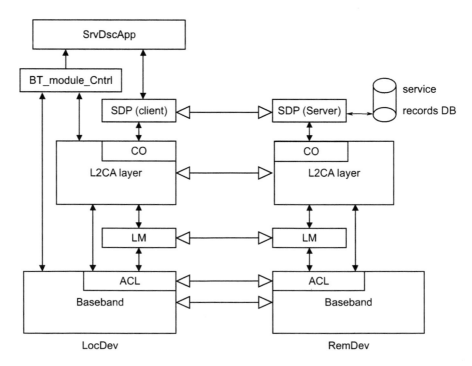

Figure 3.6   The Bluetooth protocol stack for the service discovery profile [7, Figure 2.1].

Services are described at a syntactic level. We can find Java interface description (Jini), Unique Universal identifier (UUID) for Bluetooth and XML template for UPnP. These descriptions may be limited for pervasive systems owing to the heterogeneity of service description.

Since Jini is Java-based, it assumes that devices support the Java Virtual Machine. Jini surrogate [8] consists of a solution for devices which do not fit Jini's specification in terms of a Java-based environment. It is an architecture that allows devices with limited resources to take part in the Jini network. A surrogate host represents a limited device in the Jini networks.

We can find a discussion about the suitability of these models in [9] and [10].

3.4   Research Initiatives In Service Discovery For Pervasive Systems

The models supported by industries and consortia are not specified initially for pervasive systems. In light of this, some research teams have described enhancements for these models. Others have proposed new solutions adapted for pervasive systems. In this section, we discuss existing research initiatives for service discovery. We will also present our solution for this issue.

3.4.1   Related Work

Avancha *et al.* [11] enhanced service discovery in Bluetooth with semantics. In regular Bluetooth SDP, services are associated with UUID (Unique Universal Identifier). The new matching mechanism is improved via the use of semantic information associated with services.

The pervasive discovery protocol (PDP) [12] is a fully distributed protocol. Services are published and discovered without a central registry. Devices in PDP are composed of a PDP User Agent (PDP_UA), a PDP Service Agent (PDP_SA) and PDP service cache. The role of PDP_UA is to find service information within the network. The PDP_SA announces the service offered by the device. It replies to the PDP_SA service requests in the case where it offers an adequate service. A PDP service cache maintains a list of services that are already known. PDP integrates both the push and pull mode for service discovery. Devices announce their services only if there is a service request in the network. These advertisements are stored in the cache of the devices that received the announcements. Once a PDP_SA receives a service request, it looks for an adequate service within its local services and then within its cache. Before replying to the request, the PDP_SA waits for a random period of time. During this time, it listens for service replies for the same request. The devices advertising the same services are added to the cache list of the PDP_SA.

The power problem is taken into consideration in PDP. Devices are classified in two categories: limited devices and less limited devices. This classification is based on memory size, processing capabilities, battery, etc. Limited devices have a higher probability of answering the requests. When a device roams to another networks or switches off, it sends a de-register notification to the other devices.

The Group-based Service discovery Protocol (GSD) [10] is a discovery protocol for Mobile ad-hoc networks. It describes a mechanism of request forwarding based on a group classification of services. It also includes a peer-to-peer caching process for service announcements. Services are described using the Web Ontology Language (OWL). The services are organized in a group following hierarchy.

Service Providers announce their service lists periodically. The nodes that receive these announcements forward the other nodes within radio range. Message forwarding is limited by the maximum number of hops. Each node stores the service advertisement in its service cache along with its local services. GSD allows nodes to announce group information of services seen in the vicinity. Apart from service advertisements, GSD specify the service requesting process. First, the service request is matched with the node's local cache. In the case of a cache miss, selective forwarding of the request is initiated. Thus, the message is transmitted to nodes associated with the group of the requested service.

In [13], Broens presents a context-aware and ontology-based service discovery approach. Ontologies are used to describe user requests, service properties and contextual information. These semantic descriptions are used in the matching process. It aims to classify the services according to the request into matching types.

Su *et al.* [14] surveyed the research approaches for services discovery for mobile ad-hoc networks.

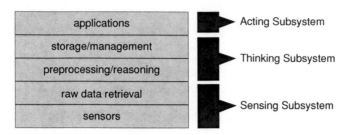

Figure 3.7 Abstract layered architecture for context-aware systems. Reproduced by permission of © TAYLOR & FRANCIS GROUP LLC and © Georgia Institute of Technology [15, 17].

3.4.2 Our Proposition: Using Semantic in Context Modeling to Enhance Service Discovery

Considering the layered architecture of a pervasive system [15] (Figure 3.7), our approach to SD falls within the thinking sub-system. We propose a semantic enhancement of service discovery. In our solution, the pervasive environment is viewed as a set of context elements, namely CoxEls. The CoxEls perform service discovery through a set of interactions [16] using a core ontology.

Figure 3.8 provides an overview of the CoxEl ontology. Since the system follows SOA principles, Task, Need and neighborhood concepts model the SOA main functionalities. The need concept allows CoxEl to specify service requests. Task represents the offered services. The neighborhood concept plays a key role in the service discovery process. It consists of a way of organizing the environment of CoxEls in order to facilitate the service search. The elements may form groups within the environment according to a specific criterion. Sensing information and type of available services are criteria of group creation.

Figure 3.9 specifies the group-based service discovery interactions.

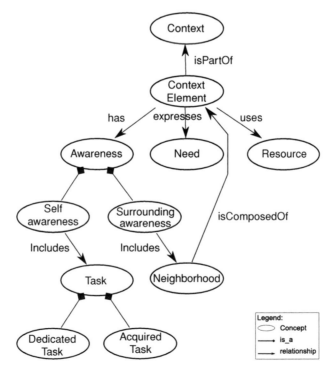

Figure 3.8 CoxEl core ontology.

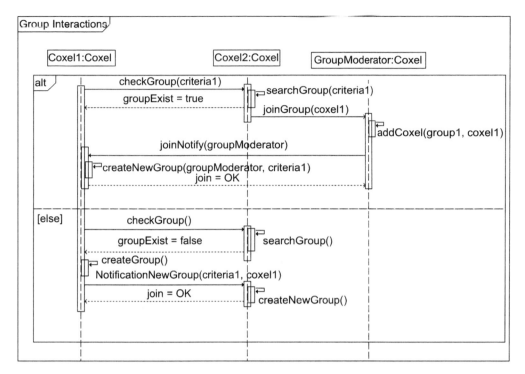

Figure 3.9   CoxEl service discovery interactions.

When a CoxEl (CoxEl1) starts up, it intends to form a group with other reachable CoxEls. It sends a request for group creation by specifying a criterion. The CoxEl (CoxEl2) that receives such a request checks the existence of such a group within its neighborhood. If so, the requester is notified about the group and a join request is sent on its behalf to the group moderator. Once the latter adds the CoxEl1 to the group members, a notification is sent. Therefore, the requester element updates its ontology by adding a new group and associates it with the corresponding moderator. If the requester receives negative notifications of group existence, it creates a new group and assumes the role of moderator. The CoxEls interested in joining the new group can express their needs to the new group moderator.

Once the group formed, CoxEls can search for services. Figure 3.10 shows the set of interactions for searching for a service within the environment. If the CoxEl knows the service provider, it starts the invoking process. If not, it contacts the group moderator corresponding to the service.

3.5   Conclusions

Pervasive systems do not only represent a set of mobile devices but also an environment that offers services to the users. In this chapter, we have considered the pervasive environment as a service oriented architecture. From this observation, we focused on a feature of SOA: service discovery. This process is crucial in SOA because the client and the service provider are not aware of each other at start up. This mechanism has interested industry consortia as well as research teams. First, we discussed the industry supported models for service discovery, such as Jini, UPnP, SLP and Bluetooth SD. Then, some research initiatives were presented. From these studies, we can conclude that any service discovery model for a pervasive system requires a decentralized design approach. Services should be designed with a structured language that gives a semantic add-on to the description. We also presented our solution for service discovery. It consists of using semantic information to organize services. The resources restriction issue has to be studied in a real test bed. A comparative study of the existing approaches focusing on consumption of resources must be carried out in next future. Another interesting issue would be how to integrate security constraints within the service discovery mechanism as this is also crucial for designing reliable pervasive systems.

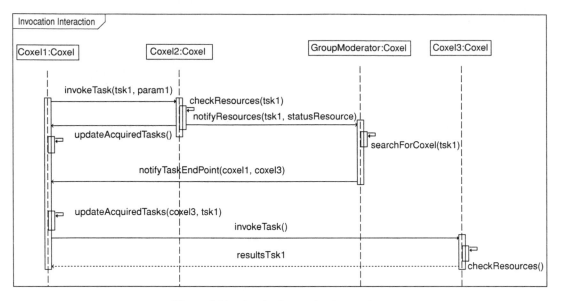

Figure 3.10   Service invocation interactions.

References

[1] G. Austaller (2007) 'Service discovery,' In *Handbook of Research on Ubiquitous Computing Technology for Real*, M. Mühlhäuser and I. Gurevych, eds. Hershey, PA, USA: IGI Global, pp. 107–27.

[2] F. Zhu, M.W. Mutka, and L.M. Ni (2005) 'Service discovery in pervasive computing environments,' *IEEE Pervasive Computing* **4**(4): 81–90.

[3] Jini Technology. [Online]. 'http://www.jini.org' http://www.jini.org.

[4] E. Guttman, C. Perkins, J. Veizades, and M. Day (1999) Service Location Protocol, Version 2, RFC 2608.

[5] UPnP Forum (2008) Universal Plug and Play Device Architecture Version 1.1, October 2.

[6] Bluetooth SIG (2009, August) Service Discovery Application Profile Version 1.1. http://www.bluetooth.com.

[7] Bluetooth SIG. (2009, August) Core Specification v3.0 + HS. http://www.bluetooth.com.

[8] (2009, August) Jini Technology Surrogate Architecture Specification, v 1.0. https://surrogate.dev.java.net/specs.html.

[9] D. Preuveneers and Y. Berbers (2004) 'Suitability of existing service discovery protocols for mobile users,' In *International Conference on Wireless Networks*, pp. 760–6.

[10] D. Chakraborty, A. Joshi, Y. Yesha, and T. Fini (2006) 'Toward distributed service discovery in pervasive computing environments,' *IEEE Transactions on Mobile Computing* **5**(2): 97–112.

[11] S. Avancha, A. Joshi, and T. Finin (2002) 'Enhanced service discovery in Bluetooth,' *Computer* **35**(2): 96–9.

[12] C. Campo, M. Munoz, J.C. Perea, A. Mann, and C. Garcia-Rubio (2005) 'PDP and GSDL: a new service discovery middleware to support spontaneous interactions in pervasive systems,' In *Third IEEE International Conference on Pervasive Computing and Communications Workshops, 2005. PerCom 2005 Workshops*, pp. 178–82.

[13] T. Broens, S. Pokraev, M. Van Sinderen, J. Koolwaaij, and P. Dockhorn Costa (2004) 'Context-aware, ontology-based service discovery,' In *Ambient Intelligence*.: Springer Berlin/Heidelberg, vol. 3295/2004, pp. 72–83.

[14] J. Su and W. Guo (2008) 'A survey of service discovery protocols for mobile ad hoc networks,' In *International Conference on Communications, Circuits and Systems, 2008. ICCCAS 2008*, pp. 398–404.

[15] S. Loke, *Context-Aware Pervasive Systems: Architectures for a New Breed of Applications*: Auerbach Publications.

[16] M. Khouja, C. Juiz, R. Puigjaner, and F. Kamoun (2008) 'Semantic interactions for context-aware and service-oriented architecture,' (2008) in *On the Move to Meaningful Internet Systems: OTM 2008 Workshops*. Berlin, Heidelberg: Springer Berlin Heidelberg, vol. **5333**, pp. 407–14.

[17] M. Baldauf, S. Dustdar, and F. Rosenberg (2007) 'A survey on context-aware systems,' *International Journal Ad Hoc Ubiquitous Computing* **2**(4): 263–77.

4

Pervasive Learning Tools and Technologies

Neil Y. Yen,[1] Qun Jin,[1] Hiroaki Ogata,[2] Timothy K. Shih,[3] and Y. Yano[2]

[1]*Department of Human Informatics and Cognitive Sciences, Faculty of Human Sciences, Waseda University, Japan.*
[2]*Department of Information Science and Intelligent Systems, Faculty of Engineering, Tokushima University, Japan.*
[3]*Department of Computer Science and Information Engineering, National Central University, Taiwan.*

4.1 Introduction

Through the centuries human beings have evolved greatly and nowadays, along with the development of technology, the provision of information and services have become more and more important. Indeed, fundamental knowledge is highly valued and has been considered as a quality that surpasses practical know-how and technical skill. Since the 1970s, information processing technology has brought fundamental changes in the way that societies work.

In the history of computing devices, it is not difficult to discover that the size of these devices is getting smaller and smaller while the capability of computing becomes more powerful. As shown in Figure 4.1, in the 1970s, users who wanted to perform certain operations, such as compiling code, had to access a main workstation located somewhere, and had to send commands to it for further processing. Of course, it took a long time to process each input from terminal users. In the 1980s, the personal computer changed that situation. It provided the capabilities for each user to work on one machine. In the 1990s, personal computers evolved into mobile devices such as the laptop, PDA (Personal Digital Assistant), or SmartPhone. It became easier for users to obtain information at any time and anywhere through such devices carried in the pocket. A few years ago, mobile devices could not only can track our location but had also replaced our payment procedures. Nowadays, they are intelligent assistants capable of anticipating many of our wishes and needs. With these characteristics, they can also be called ubiquitous devices. Recently, one is able to utilize these devices to arrange meetings, share information, or receive email messages, and one can listen to the music or the news through one's iPod at the same time. The use of ubiquitous devices has become an everyday routine. Ubiquitous devices have also changed our teaching and learning behaviour.

In the past, with mobile computing, much effort has been devoted to the design and the implementation of computing architectures that enable computing anywhere and any time. Pervasive computing takes another approach, aiming at enhancing the interaction between the user and the computer, or between users through computer(s), by taking the user's situation into consideration. Pervasive computing can also be considered as applications and services becoming accessible from Internet-enabled ubiquitous devices. It involves new technologies, services and business models.

Pervasive Computing and Networking, First Edition.   Edited by Mohammad S. Obaidat, Mieso Denko, and Isaac Woungang.
© 2011 John Wiley & Sons, Ltd. Published 2011 by John Wiley & Sons, Ltd.

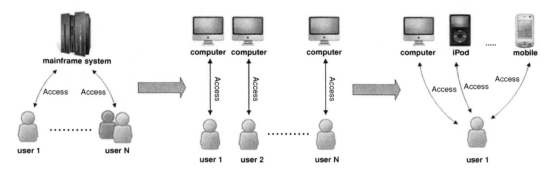

Figure 4.1   Development history of computer systems.

Pervasive computing technologies have already become the trend of the new century and consequently research into ubiquitous devices, such as wearable computers, has attracted significant attention from end users. The famous publicity slogan of Nokia Corporation – *Connecting People* (science and technology always stem from human nature) proclaims the interaction between scientific, technological products and people. Owing to the immature development of scientific and technological services in the past, there was a passive operational relationship between the computer and people. With the progressive improvement of the relevant technology in semiconductors, the trend in the development of computers was that they became smaller, cheaper and even more intelligent [42]. Steve Mann [29] considered that mobile computing: "belongs to the personal space, controlled by the users, and operated inter-dynamic continuation at the same time (constancy)". In other words, it is "always on and always accessible".

The technologies of pervasive computing can be applied in many research fields such as communication science [41], commerce [16, 55], artificial intelligence [20], human computer interaction [19, 24, 30], etc. In addition to these applications, the use of pervasive technology to facilitate the learning process has made a great impact on improving the feasibility of distance learning environments. The use of ubiquitous devices to support learning is not a new concept. Many innovative learning platforms [21, 45] and activities [56] have been proposed to promote the learning process, as well as exploiting the methods of e-learning. Although a variety of ubiquitous platforms provide a more flexible and extendable learning experience, hardware capabilities and restrictions have consequently become the challenges and barriers we need to overcome.

In this chapter, we focus on the applications of pervasive computing technologies and, especially on their integration with learning. We start with the historical development of the learning process and introduce pervasive technologies in learning. We then review the emerging technologies and systems for pervasive learning environments. We go further to introduce the current systems which support learning or teaching activities that are integrated with pervasive technologies. After that, we explore what the trend of pervasive technologies will be in the next few years.

4.2   Pervasive Learning: A Promising Innovative Paradigm

With regard to the term "Learning" from the viewpoint of psychology, the family can be regarded as the first educational environment we are in touch with. Far removed from home education, school instructors utilize different teaching styles, such as heuristic learning or collaborative learning, to make us understand what we have to learn at school age.

When Weiser and Preece proposed the concepts of ubiquitous computing and pervasive computing, it changed the way we utilize the Internet. It makes it possible for us to access the Internet without the limitations of the environment. With respect to the distance learning, ubiquitous devices have replaced the stand-alone computers of the past. The learning style has also moved from traditional face-to-face learning or web-based learning to mobile or ubiquitous learning. Distance learning has gained much benefit from ubiquitous devices. Of course, the way people acquire knowledge also becomes more and more diverse.

4.2.1   Historical Development of Computing and IT in Education

The development of distance learning can be traced back more than 20 years. We will take a brief look at the status and the future trends of distance learning by way of the Hype Cycle model proposed by Gartner. The vertical axis of the Hype Cycle model shows the development of visibility and the horizontal axis represents the approximate timeline, as shown in Figure 4.2.

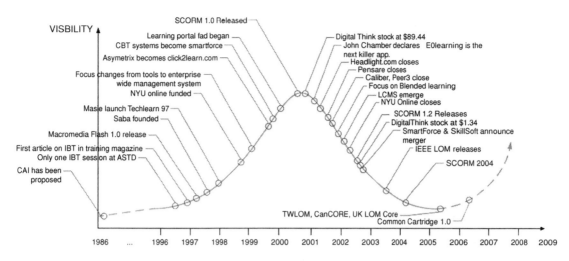

Figure 4.2   Hyper Cycle of e-learning.

Since CAI (Computer-Aided Instruction) was first proposed and utilized in the mid-1950s, more and more relevant technologies have appeared. With the popularity of personal computers, instructors can prepare teaching materials in advance and deliver them to learners in class. Learners can practice at home using computers. It is what we call CBT (Computer-Based Training/Teaching). After that, with the development of Internet technologies, CBT has evolved into a different presentation style – IBT (Internet-Based Training/Teaching). IBT makes it possible for instructors to teach their students at any time in an easy way. It allows students to learn in their spare time through the Internet. The instructor can also put relevant materials on a specific website at any time. When mobile devices became popular in the late 1990s, many researchers tried to transfer the original learning platforms into these novel devices. With its mobility and high performance, this kind of learning style, which is called m-learning (Mobile Learning) or u-learning (Ubiquitous Learning), has had a great impact on traditional learning.

Comparing Figures 4.1 and 4.3, it is not difficult to see that the original teaching or learning process has become deeply integrated with the technologies. From instructor's perspective, he or she can use the benefits of personal computers to create adaptive learning materials for each student instead of using the same materials for students who are at different levels. On the other hand, from the learner's perspective, he or she can not only obtain customized learning materials via his or her personal computer but can also acquire new information through ubiquitous devices at any time and any place. Teaching and learning behavior are no longer limited to

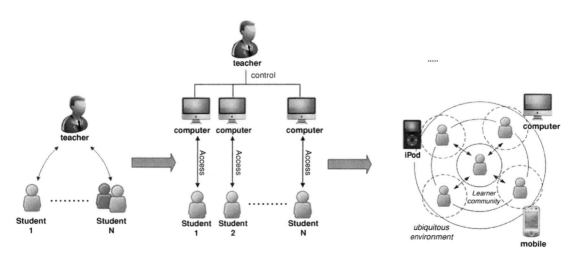

Figure 4.3   The historical development of the teaching/learning styles.

a specific place. Learning materials can also be retrieved through any device at any time. Furthermore, with the promising Web 2.0 technologies, people who have specific skills or experiences can be instructors within learning environments.

4.2.2 Past Experience and Issues

With regard to that discussed above, ubiquitous technologies should be taken into consideration. In the past, integration with GIS (Geographic Information System) and GPS (Global Positioning System) technologies were popular research topics.

For example, Cheok *et al.* [10] described the existing facilities, which involved ubiquitous multimedia space. It represented the interaction between humans and computers. However, it presented certain defects while supporting a large number of learners based in an electronically designed environment. In this research, Cheok *et al.* made use of three different systems including Virtual Kyoto Garden, Touchy Internet and Human Pacman. The main functionalities of these systems contained spatially-aware 3D navigation, tangible interaction and ubiquitous human media spaces. In the beginning, authors created a 3D virtual garden and a set of 12 tags, which had a specific function menu. Learners could perform a simple interaction such as picking up or dropping through wearing the monitor on their head. In the second system, the authors created a ubiquitous game environment by means of the Human Pacman system. This game emphasized collaboration and competition between learners. In this environment, one could monitor every learner and retrieve location-based information. The retrieved information could help in reconstructing the virtual 3D module. The third method was the Touchy Internet. It helped learners achieve a certain interaction with their pets through a mobile system, projecting an immediate image in a virtual garden.

Within the health care system, Lin *et al.* [5] proposed monitoring the status of elderly people in order to ensure their safety through GIS, RFID (Radio Frequency Identification) and GSM (Global System for Mobile Communications). The main modules of this system included (1) Indoor Residence Monitoring which monitored the patients who entered or left a specific area; (2) Outdoor Activity Area Monitoring which helped to detect the patients' location; (3) Emergency Rescue Module which provided a connection in an emergency and stored the traced procedure through TIPs. This system helped to reduce the accidents which might happen when the elderly were alone. It also helped trace the action path and send it directly to the family's portable devices.

Broadcasting is widely used in restaurants and other public locations. Nowadays, with the use of GPS, it enables us to access a smart guidance system and wireless environment. EDO (Evolution Data Optimization) has prompted the development of ubiquitous data distribution and location-based services. It will help reduce the cost of time when retrieving data through broadcasting for a large amount of users. Acharya *et al.* [2] proposed a broadcasting platform which allowed users to share or deliver location-dependent data such as weather, traffic and stocks. This research utilized Bluetooth to create a real-time broadcasting environment. To simplify the process, they created a wireless channel so as to connect the user's tuning channel to match the server and then send the data to the satellite for further delivering. This improved the retrieval time of location-dependent applications and data.

With the development of technologies, the amount of information has also become larger. It also became difficult for people who wanted to obtain information everywhere at any time. In [33], the authors made use of context-awareness techniques to assist people in retrieving and sharing information through portable devices such as mobiles and PDAs. The core mechanisms of this research depended on three RFID tags: Indoor Location ID, Recommended Information ID and User ID. They were managed by a central management server to control the requests from users and response messages to users. This research utilized a mobile with an active RFID reader and GPS receiver in order to achieve that goal. For example, we could know exactly where we were now and see if there were friends near us. We could also obtain relevant information when we went to tourist attractions and share it with others. This idea inspired Google to develop the service called "Google Latitude".

Morikawa *et al.* [34] took user modeling into consideration. The hardware requirements were almost the same as those in [1]. The authors emphasized the profile aggregator and service provider. This system would collect the following information: (1) the recommended information that users read through a browser; (2) the annotations or pictures of their favorite things; (3) the things that users bought; and (4) the people who were near to the users. This system utilized the RDF (Resource Description Framework) to model the user profile based on flexibility and interoperability. In addition, the users' privacy information was also considered in the profile aggregator. This research provided an ideal example designed to achieve mobile personalization.

With the maturity of GPS technologies, Liu *et al.* [27] developed a management system called ITEMS (Intelligent Travel Experience Management System) to assist users to manage a large number of photos taken at any place by comparing the annotation. The core mechanism was similar to optimal image similarity metrics and was also integrated into the search mechanism of the World Wide Web. The authors made use of SIFT matched points and illumination techniques so as to recognize the uploaded images. The system compared the uploaded images with those existing in a database through cluster matching. After that, the system generated the geographical information automatically for the images. For instance, the author utilized his own photos which were taken at notable places in Sydney, Paris and New York in order to evaluate the proposed system. The accuracy rate was as high as 90%. It was an interesting topic for the development of the ubiquitous environment.

In [12], Comelis *et al.* developed a real-time traffic forecast system within the context of city architecture. They also utilized 3D reconstruction techniques to restore the previous scene. They utilized two core techniques: the SfM (Structure-from-Motion) algorithm to compute the information of every target object captured by the camera; and Façade Reconstruction which computes the vector information in different dimensions and improves the efficiency of dynamic programming. This research could be applied to Topological Map Generation, Road Reconstruction and Texture Generation. Through GPS and CCD camera, it can construct a three-dimensional virtual model in order to provide real-time navigation information.

Moreover, Garner *et al.* [17] made use of RFID and NFC (Near Field Communication) techniques to achieve the goal of "SprayCan". It allowed users to paint public places through a mobile camera with authorized RFID tags. Users could put their graffiti on the web and share it with others. First, they utilized a mobile SprayCan site maker called MobSpray to generate graffiti. Second, the RFID tags would receive the data from MobSpray. With these processes, the graffiti could be shared not only on a website but also on the buildings or other real objects in the world.

According to the size of presentation of ubiquitous devices, there are many research issues that focus on how to provide the necessary information to the users instead of providing all of it. Lum and Lau [28] proposed a method designed to achieve content adaptation in mobile devices. This service was called "quality-of-service-aware". The process was managed by a decision engine. It negotiated automatically for the appropriate adaptation decision that could synthesize an optimal content version on mobile devices in different file types, and it also could minimize the downloading time of the content. Thus, users can utilize their mobile devices to browse the appropriate content which had already been adapted by the decision engine.

In [46], Sumita *et al.* proposed an automatic document indexing system. This system could connect each document through cross-indexing. By analysing the architecture of a document, users' browsers could summarize the representative parts of specific documents and create connections between them. The segmentation rules could be separated into two types: Restriction Rules for segmenting the text and Preference Rules for determining the local structures. These rules would help remove the superfluous parts of sentences and cluster the parts that were used most often. This structure was appropriate in full-length documents and utilized the double-layered document structure to delaminate them. In the extraction part, it included three processes: Document Hierarchy Analysis, Text Structure Analysis and Viewpoint Indexing. Through this procedure, they achieved a more optimal extraction rate than in previous research.

Lara *et al.* [13] proposed an object-based mobile computing system named "Puppeteer". This system could achieve interoperability between different applications without revising. The system could also ensure the security of applications when integrating them with each other. This system could be operated in the Windows NT environment and also supported other Windows-based applications such as MS PowerPoint and IE 5. "Puppeteer" had a four-layer main architecture. The applications and proxy belonged to the client-side in order to connect to the "Puppeteer" server and Data server. The connection between the client proxy and server proxy worked together to guarantee that there was no need to revise the applications and data in different mobile systems. The whole procedure was operated in DMI (Data Manipulation Interface) to ensure the interoperation policies, that is, to utilize this system to achieve the goal while interoperating between different mobile applications. However, system performance was also taken into consideration.

As stated, it is obvious that the use of ubiquitous technologies has originality and creativity. But it also has some important elements such as the portability and stability of devices, the retrieval methodologies of the information and the communication mechanism for these kinds of technologies.

4.2.3   Practice and Challenge at Waseda E-School

The classrooms for the distance learning environment can be regarded as any kinds of place. The Open University, founded in 1969, was the first to provide online training courses for students. The training courses are not limited

Table 4.1 Comparison of Waseda E-School and other online universities

	Waseda E-School	Open University	Phoenix Online
Foundation	2003	1969	1989
Number of Students	About 600	About 330 000	About 300 000
Number of student in one class	About 15	10–100	About 30
Class From	Video	Textbook, Video	Textbook
Quality of TA	Postgraduate	Domain Expert	Regular employee lecturer
Graduate Degree	Yes	No	No

to any specific level. In addition, it also provides well-organized teaching and assessment methods designed to evaluate the students. It has an organizational structure of entities such as Library, Administration Affairs, Student Affairs, etc. Students can decide whether to go to the school and attend classes or study at home through the Internet.

In Japan, in 2003, Waseda University established a new educational programme of Human Sciences, called "E-School". It is the first accredited undergraduate degree award programme by fully online e-learning in Japan. The basic concept of the Waseda "E-School" is how to support students who work full time and/or live away from the institute. The Waseda "E-School" is a typical example of the success of e-learning in higher education in Japan. The differences between the Waseda "E-School" and other online universities are shown in Table 4.1.

The success of the Waseda "E-School" can be perceived from the learners' motivation for active learning. Learners can ask questions on the website and the educational Coach (or Teaching Assistant) will provide the solutions within a few hours. Thus, the problems learners meet can be solved quickly. Learners realize that they can indeed learn something this way. Although the Waseda "E-Schoo'l" has already made steady progress, there are still some challenges to making that progress continual. It has to provide more courses to students in other departments and thus make it more comprehensive. According to the statistical data, most of the students in the Waseda "E-School" study while working full time. In this situation, management of the learning history and the personal profile of these students would help them achieve lifelong learning. In addition, the provision of learning materials to those students in their spare time, such as on the train, is also another issue for the "E-School".

4.3   Emerging Technologies and Systems for Pervasive Learning

The goal of distance learning is to provide a well-organized environment for learners to acquire knowledge. It can be regarded as an assistant for traditional learning. In order to achieve this goal, integration between distance learning and emerging technologies plays an essential role. In this section, we aim at drawing the outline of the current technologies or developing systems which can enhance the pervasive learning environment.

4.3.1   Emerging Computing Paradigms for Education

There are more and more computing technologies which enhance the education environment. For example, the concept of the social network makes it possible for a learner who has specific professional skills to become an instructor for others. The wireless communication mechanism for mobile devices allows learners to obtain extra services through their geographical information. The technologies have become more and more important for the current education environment and for the future. By summarizing the research issues from the IEEE and ACM digital library, the technologies for enhancing the e-learning environment can be categorized into the following groups.

- **Ubiquitous Computing in e-Learning**
  With the increasing development of wireless technologies, integration between these technologies and e-learning makes research organizations devote themselves into this environment. They assist learners in achieving their learning goals through PDA, smart phone, or the RFID technologies. It makes e-learning environments more plentiful.

- **Emergence with Web 2.0 Technologies**
  Learning resources on the Internet have become more plentiful. The costs of the Internet have also increased. In terms of research results, seamless interaction in the learning process can increase learning efficiency rapidly. Thus, utilizing the technologies of Web 2.0, such as AJAX, Wiki, or Blog, has become a more popular strategy in many e-learning systems.
- **Structured Ubiquitous Content Sharing**
  In early e-learning environments, there was no interactivity provided between learning resources, resources could not be shared between platforms and the costs of learning materials could not be reduced. Therefore, many researchers propose structured methodologies of learning resources sharing and discovery in both Internet and ubiquitous environments. They aim at providing a well-structured framework for sharing learning resources on the Internet.
- **Peer-to-Peer and Online Learning Community**
  In contrast to personal learning, collaborative learning can provide benign competition and high interaction so as to increase learning efficiency. For this purpose, many platforms pay more attention to interacting with the online community. They utilize P2P technologies to achieve an organizational learning process.
- **Intelligent Tutoring and Adaptive Testing**
  In e-learning environments, we aim at developing intelligent systems to provide learners with adaptive learning materials and assessments. This will reduce the costs of learning platforms and the efforts that instructors have to make. But this mechanism also presents some challenges such as the computing algorithm, or profile management, that researchers have to resolve.
- **Further Development in Distance Learning Standards**
  The integration of resources on the Internet is an important challenge that researchers aspire to overcome. To utilize a specific e-learning standard can help achieve this goal. It has yet to be supported by many research organizations like IMS, ADL, etc. E-learning standards can also assist us in providing pervasive information in the future.
- **Ubiquitous Game-based Learning**
  The development of a game-based learning style has become more and more popular since researchers found that it can increase the motivation of learners. The main reason is that it provides richly challenging and interesting components for traditional e-learning. Hence, the adoption of learning theories, such as problem-based learning theory and collaborative learning theory, is another important issue in this emerging learning style.

4.3.2   Pervasive Learning Support Systems and Technologies

Up to now, much effort with regard to the research issues discussed above has been expended in order to develop pervasive learning support systems and technologies. To construct an outdoor mobile-learning activity using up-to-date wireless technology, Chen *et al.* [8] developed a system for bird watching which shares the cognitive load of bird identification when a learner is engaged in outdoor bird-watching activities. The findings provided evidence that the system not only promoted bird-identification but also enhanced the independence of learners. With the Butterfly Watching System [9] the learners explore a natural area to locate and learn about butterflies. Once a butterfly's image is captured by the camera, the system supports students in order to identify the butterfly and provides information about it.

Language learning is one of the major domains of ubiquitous learning research because language is contextualized and the language thinking pattern is assimilated in the real world [22]. For example, there is the vocabulary teaching experiment by Miller and Gildea [31] in which they described how children acquire a vocabulary faster with the method used outside of school, by relating words to ordinary conversation, than with the traditional methods based upon abstract definitions and sentences taken from external contexts. Especially with overseas students who travel to another country to gather knowledge, language is one of the great barriers. The TANGO (Tag Added learNinG Objects) system detects objects around the learner using RFID (Radio Frequency Identification) tags, and provides the learner with the correct information in that context. This system helps learners, by sharing past experiences and presenting the right question at the right place and the right time, by detecting the context of the learning situation and solving the problem immediately, and by providing the correct information via a wireless network. The systems asks a question of the learner and then the learner replies with an answer. If the answer is correct, the learner will move to the next question. Again, if the answer is wrong, the learner will learn the correct answer from the system. In this way, TANGO helps learners to learn new vocabulary [36]. Yang *et al.* [50] developed a system for English oral practice and assessment using their handheld devices (PDAs) within a wireless connected environment in a classroom. With the support of handheld devices in the classroom with wireless communication capability, students have more of a chance to speak English than in a traditional classroom. There is much research in the field of language learning but very little about independent language learning and almost none about the use of technology in independent language learning. So Thornton and Sharples [47] proposed a new trend in this field which reports on an on-going, exploratory study to determine patterns of the use of technology in self-directed language learning. Again PALLAS [40] can provide support to a mobile language learner by providing personalized and contextualized access to learning resources. LOCH [37] also utilizes a mobile device, GPS, and PHS (Personal Handy-phone System) for outdoor language learning. It assists learners to apply the knowledge gained in the classroom for studying foreign languages, especially vocabularies.

Pervasive and ubiquitous technologies are often embedded in classrooms. A Reconfigurable Context-Sensitive Middleware (RCSM) has been developed by Yau *et al.* [52] for pervasive computing by providing development and run time support for the application software in a smart classroom. They focused on the characteristics of a smart classroom and how RCSM can be used to develop such an environment. Again, in order to provide basic support for classroom and field activities, the integration of a ubiquitous computing system in the classroom is needed. Ogata *et al.* [39] focused on the integration of a ubiquitous computing system in the classroom. By using the BSUL System, students interact amongst themselves and with the professor through an Internet-enabled PDA. To provide classroom help, a system called AirTransNote [32] has been developed which manages notes written by students on paper and enables the teacher to browse through the notes or show them to the students. AirTransNote enables real-time sharing of notes. When a student writes notes on a regular sheet of paper, the handwritten data are sent immediately to the teacher's computer. Lindquist *et al.* [26] described the design and use of a mobile phone extension for Ubiquitous Presenter, which allows students to activate learning exercises in the form of text or photo messages. Distance education has been around for a while and allows students and professors at different locations to share and learn from each other.

A number of technologies has been developed to support museum visitors. Museums have always seen themselves as having an educational role. Yet, more recently the need has been identified for a change in thinking about museums as places of education to places for learning, responding to the needs and interests of visitors. Mulholland developed a system that can support visitors in a museum. Bletchley Park Text is a novel approach to the exploration of an archive of resources, which allows the user to specify a set of interests, which can then be organized automatically into pathways and interconnected categories, explore the set, and then redefine it. The museum visit is usually an individual experience. Furthermore, electronic guides or interactive systems in museums are not designed to promote social interaction among visitors. Laurillau and Paternò [23] showed how to support collaborative learning in museum visits and presented an application based on mobile palmtop systems. Chou *et al.* [11] introduced a context-aware museum tour guide to adjust recommendations to the interests and contexts of individual visitors and to enable them to share their experience with others. Visitors enter the museum with a PDA. At first the PDA asks them some questions so as to identify their interests and relevant preferences. Based on this information the guide builds an initial tour. This tour is used to give directions to the visitor when needed. MyArtSpace [48] is a service that makes use of different technologies such as mobile phones and the Web to deliver a learning experience that traverses the museum and the classroom. By typing a two-letter code into multimedia mobile phones, students can collect an exhibit, which displays a multimedia presentation and sends an image and description of the exhibit automatically to their personal collection area in a Web portal.

Ogata *et al.* [38] propose LORAMS (Linking of RFID and Movie System) to provide learning materials to learners through scanning RFID tags. Learners can browse or review video segments to complete their learning activities. We also benefit from the development of ubiquitous devices such as mobile, smart phone and PDA. Chang *et al.* [25] develop a SCORM-Compliant (Sharable Content Object Reference Model) management system called "Pocket SCORM" to enhance the mobile learning environment. In the PERKAM (PERsonalized Knowledge Awareness Map) [15], the RFID tags have been utilized to assist students to learn how to solve problems and to share their knowledge and experience. In [6], a virtual learning environment named UPS (Ubiquitous Personal Study) has been proposed and discussed. This work integrates ubiquitous computing and Web 2.0 to provide retrieval, management, organization and recommendation functionalities for learners. In [4], many search services such as Google, Yahoo, etc have been surveyed, and a collaborative shared information retrieval model has been proposed in order to improve the performance of the traditional information retrieval model. In [53], weighting and ranking metrics have been proposed to evaluate the objects of learning and assist learners in finding related resources. Through this, learners do not have to spend time on filtering results. In [44], a reusability tree has been proposed to detect the derivation path for learning objects through their similarity and diversity. A u-learning model to enhance social technologies and interactivities through peer-to-peer technologies has been developed in [57]. And in [54], a transformation mechanism has been proposed and developed to solve the problem of interoperability between e-learning standards. Also, the fundamental system architecture is based on Web 2.0. For the ubiquitous game-based learning environment, we also take corresponding learning theories and assessment procedure into consideration [18, 51].

4.4   Integration of Real-World Practice and Experience with Pervasive Learning

Pervasive learning can be regarded as a kind of emerging learning strategy in the current learning environment. It aims at providing context awareness and location awareness materials for learners. In addition, learners can also benefit from ubiquitous devices with high intelligence to receive the materials provided by pervasive supported learning systems. In spite of this, there are many people who think that there is no difference between ubiquitous learning and pervasive learning. Thus, in this section, we will introduce briefly the ubiquitous learning environment and our UPS (Ubiquitous Personal Study) system.

4.4.1   Ubiquitous Learning

Educational technologies are growing rapidly in order to satisfy learners' needs. Learning at any time and anywhere is one of the trends researchers focus on. Ubiquitous computing technologies [1] assist us in organizing and mediating social interactions whenever and wherever the situations above might occur. It has been accelerated by improved Internet and wireless telecommunication capabilities, continued increases in computing power, improved battery technology and the emergence of flexible software architectures. With these technologies, the concept of the ubiquitous learning environment is realized, where individual and collaborative learning in our daily lives can be integrated seamlessly. The main characteristics of ubiquitous learning could be considered as the following.

- **Permanency**
  Learners never lose their work unless it is removed deliberately. In addition, all learning processes will be recorded continuously every day.
- **Accessibility**
  Learners have access to their learning contents, documents, multimedia data and personal profile information from anywhere. The information is provided based on learners' individual requests. Therefore, a self-directed learning style is involved in the ubiquitous learning environment.
- **Immediacy**
  No matter where the learners are, they can obtain any necessary information immediately. In this situation, learners can solve problems quickly. Otherwise, learners can also record the questions and find out the answers later.
- **Interactivity**
  Learners can interact with domain experts, instructors, or peers in the form of synchronous or asynchronous communication. Hence, experts are more reachable and knowledge becomes more available.
- **Situating of Instructional Activities**
  Learning could be embedded in our daily lives. The problems encountered as well as the knowledge required are all presented in their natural and authentic forms. This helps learners identify the features of problem situations that make particular actions relevant.

The concept of ubiquitous learning can be represented as various kinds of forms that focus on social knowledge building and sharing. The challenge in an information-rich world is not only to make information available to learners at anytime, at any place and in any form but also to be provided specifically with the right thing at the right time in the right way. The ubiquitous computing technologies make it possible for learners to learn at anytime and any place. But the fundamental issue is still how to provide learners with the right contents at the appropriate time and in the appropriate way.

4.4.2   UPS (Ubiquitous Personal Study)

We can take the benefits from the Internet technologies as stated above. In this situation, it is important for us to manage and share the personal content and knowledge. The concept of the Ubiquitous Personalized Study (UPS) environment was proposed in [7]. The framework of UPS includes the learners' profiles, personal activities and the interactivities between learners. We make use of the technologies of Web 2.0 and create personal data digitalization in this environment. We aim at supporting accessing, managing, organizing and sharing personal information.

The UPS includes two main functionalities: personal information collection and private information behaviour. The first functionality includes the set of personal information such as V-Book, V-Bookshelf, V-Desktop and V-Note. Learners can make use of these tools to manage their personal learning environment. The other functionality includes the learners' private information such as V-Card, user profile and V-Log. These tools will record all essential information when learners conduct learning activities. These private profiles can be exchanged with each other under the agreement.

Concerning system implementation, UPS integrates XSNS (Cross Social Network Services) to make information exchange between systems more flexible. The fundamental system infrastructure of UPS was developed based on OSS (Open Source Software) and the web services. The UPS emphasizes the mash up of web services. It utilizes the Google Toolbar and Google Web History to monitor and record the learning behaviour of learners. It also makes use of Twitter service as a discussion tool for learners to share what they have learned and utilize RFID to detect learners' location. The mash up of external services can also assist UPS in providing relevant information or services to learners.

4.5   Nature of Pervasive Learning and Provision of Well-Being in Education

4.5.1   Ubiquitous and Pervasive

The term "Ubiquitous Computing" means that computers will exist in everything that we may utilize in our daily life. We can obtain the computed information at any time and at anywhere. From the concepts proposed by Mar Weiser [49], we know:

1. The computer service style has been changed. In the future, there will be many computing devices to serve one person, and they will be exist around us invisibly.
2. Virtual Reality creates a computer-central virtual environment for people to access. On the contrary, Ubiquitous Computing has been integrated into a human-centred world.
3. Ubiquitous Computing aims at providing services to people and makes them use computers naturally and unconsciously, and to do everything they do by themselves.

"Pervasive Computing" is used to explain how "Ubiquitous Computing" develops and how we integrate it around us continually. Through the definition in the *Oxford Advanced Dictionary*, we can find that the word "Ubiquitous" means "seeming to be everywhere or in several places at the same time", and the word "Pervasive" means "existing in all parts of a place or thing" or "spreading gradually to affect all parts of a place".

"Ubiquitous computing" and "Pervasive computing" are two sides of the same coin. They are complementary. Ubiquitous computing creates convenient and ubiquitous networks while pervasive computing means the use of the networks in a ubiquitous manner, such as mobile, laptop, intelligent clothes, and so on. It is when we say that "Where ubiquitous computing would require information everywhere, pervasive computing would make information available everywhere" [14].

4.5.2   The Possible Trend of Pervasive Technology in Education

Pervasive Computing has had a great impact on the education environment and makes the teaching/learning process more convenient. For example, instructors do not have to spend time on collecting the necessary resources for creating specific learning materials. The resources could be gathered based on the instructors' profiles, such as teaching style and the teaching targets, and be retrieved through the ubiquitous devices at any time. When generating the learning materials, all instructors have to do is to point out which kinds of materials they would

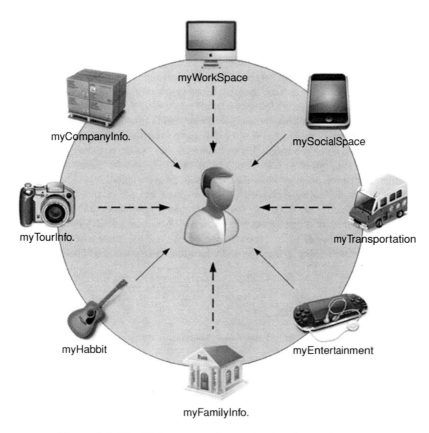

myWorkSpace

myCompanyInfo.

mySocialSpace

myTourInfo.

myTransportation

myHabbit

myEntertainment

myFamilyInfo.

Figure 4.4 The ideal concept of pervasive learning environment.

like to utilize. The learning materials will be generated automatically. On the other hand, in the pervasive learning environment, learners do not need to remember which classes they have to attend. The necessary information will be provided for them at the appropriate time and at the appropriate place. Furthermore, instructors and learners do not need to prepare any presentation files for a specific meeting. All the things they have to do will be finished based on their profile in advance. In summary, with the development of technologies, ubiquitous devices will become more and more intelligent. They know what we will do. They also know what we will need when we start the learning activities. The technologies and the learning environments are growing rapidly. With the support of the technologies and devices, the learning environments and educational services will become more and more amenable. Moreover, it will also best fit everyone's needs and learning styles.

The ideal pervasive learning environment should provide human-centric learning support. As shown in Figure 4.4, everything around the learners, not limited in the devices, will provide automatically the necessary information that learners need to utilize. The service is no longer limited to the process that "Learners request, devices response". The services become more and more integrated. The learners only have to think, and then the service will provide what learners want in various styles.

There is an old Chinese saying: "Life is finite, while knowledge is infinite." It shows that learning should not be limited to specific environments such as schools. The learning environment should be open and free. We consider future learning support environments to be ubiquitous, universal and unlimited. In current learning environments, with the support of ubiquitous devices, the learning process can be at any time and at any place. In the near future, the pervasive learning systems and technologies we introduced in the previous section will become more realizable and effective. The services can be designed specifically to meet everyone's needs and provided to users at the appropriate time and at appropriate places. Learners can benefit from the ubiquitous and pervasive technologies so as to achieve their lifelong learning. We believe that the provision of well-being in learning and education will come to pass in the future through the newly emerging ubiquitous and pervasive technologies.

4.6   Conclusion

In this chapter, we focused on the emerging technologies for pervasive learning environments. Through the historical development of learning styles, it is not difficult to ascertain that the technologies, especially the Internet and ubiquitous devices, play a very essential role in this domain. Ubiquitous learning can enable learners to learn at any time and anywhere. With pervasive technologies, the learning environment becomes more and more intelligent. Information and contents will be provided to learners according to their different needs through ubiquitous devices.

With the development of pervasive learning, we are always being asked: "Could this kind of learning styles replace the traditional learning?" The answer is obviously in the negative. However, this does not mean that a negative answer is not worthy of consideration. The technologies are considered to be a tool to enhance traditional learning. They aim at providing more a convenient learning environment for learners. As we discussed in the previous sections, there are more and more ideas that have been proposed to achieve this goal.

In the future, with highly developed technologies, the needs of schools should be taken into consideration. Providing different support for different life stages is also important for promoting life-long learning environments. In a pervasive learning environment, it is necessary to find a way to maintain motivation for learning in the future. The learning environment will change as time goes by. We believe that the concept of the human-centric learning process and the provision of well-being in learning and education environments will be a critical issue in the near future.

References

[1] Abowd, G.D., Mynatt, E.D. (2000) "Charting past, present, and future research in ubiquitous computing," *ACM Transaction on Computer-Human Interaction* **7**(1): 29–58.
[2] Acharya, D., Kumar, V., Yang, G.C. (2007) "DAYS Mobile: A Location based Data Broadcast Service for Mobile User," *ACM Symposium on Applied Computing*, pp. 901–5.
[3] Andreas, L., Ulrich, H. (2006) "Analysing processes of learning reading and writing in a computer-integrated classroom," *The 6th International Conference on Advanced Learning Technologies*, pp. 219–21.
[4] Chan, S., Jin, Q. (2006) "Collaboratively shared information retrieval model for e-learning," *The 5th International Conference on Web-based Learning*, pp. 123–33.
[5] Chang, H., Chang, W., Sie, Y., Lin, N., Huang, C., Shih, T. K., Jin, Q. (2005) "Ubiquitous learning on pocket SCORM," *ACM Symposium on User Interface Software & Technology*, pp. 171–9.
[6] Chen, H., Ikeuchi, N., Jin, Q. (2008) "Implementation of ubiquitous personal study using Web 2.0 mash-up and OSS technologies," *The 3rd International Symposium on Ubiquitous Application & Security Service*, pp. 1573–8.
[7] Chen, H., Jin, Q. (2009) "Ubiquitous personal study: a framework for supporting information access and sharing," *Journal of Personal and Ubiquitous Computing* **13**(7): 539–48.
[8] Chen, Y.S., Kao T.C., Sheu J.P. (2002) "A mobile learning system for scaffolding bird watching learning," *IEEE International Workshop on Wireless and Mobile Technologies in Education*, pp. 15–22.
[9] Chen, Y.S., Kao, T.C., Sheu, J.P. (2004) "A mobile butterfly-watching learning system for supporting independent learning," *The 2nd IEEE International Workshop on Wireless and Mobile Technologies in Education*, pp. 11–18.
[10] Cheok, A.D., Lee, S.P., Liu, W., James, T.K.S. (2005) "Combining the real and cyber worlds using mixed reality and human centered media," *International Conference on Cyberworlds*, pp. 14–26.
[11] Chou, S.C., Hsieh, W.T., Fabien, L., Norman, M. (2007) "Semantic web technologies for context-aware museum tour guide applications," *The 19th International Conference on Advanced Information Networking and Applications*, pp. 709–14.
[12] Cornelis, N., Cornelis, K., Gool, L.V. (2006) "Fast compact city modeling for navigation pre-visualization," *IEEE Computer Society Conference on Computer Vision & Pattern Recognition*, pp. 1339–44.
[13] de Lara, E., Wallach, D.S., Zwaenepoel, W. (2001) "Puppeteer: component-based adaptation for mobile computing," *The 3rd USENIX Symposium on Internet Technologies & Systems*, pp. 14–19.
[14] Dillon, T. (2006) "Pervasive and ubiquitous computing," *in digital Learning Weekly e-Newsletter* vol. I, Issue XX.
[15] El-Bishouty, M.M., Ogata, H., Yano, Y. (2007) "PERKAM: Personalized Knowledge Awareness Map for Computer Supported Ubiquitous Learning," *Educational Technology and Society Journal* **10**(3): 122–34.
[16] Gao, J.Z., Ji, A. (2008) "SmartMobile-AD: an intelligent mobile advertising system," *The 3rd International Conference on Grid and Pervasive Computing*, pp. 164–71.
[17] Garner, P., Rashid, O., Coulton, P., Edwards, R. (2006) "The mobile phone as a digital SprayCan," *ACM SIGCHI International Conference on Advances in Computer Entertainment Technology*, pp. 121–7.
[18] Huang, K.M., Chen, J.H., Chen, C.Y., Shih, T.K. (2009) "Developing the 3D adventure game-based assessment system with Wii remote interaction," *The 8th International Conference on Web-based Learning*, pp. 192–5.
[19] Jiang, L.F., Lu, G.Z., Xin, Y.W., Li, Z.L. (2006) "Design and implementation of speech interaction system in pervasive computing environment," *2006 International Conference on Machine Learning and Cybernetics*, pp. 4365–9.

[20] Kadouche, R., Mokhtari, M., Giroux, S., Abdulrazak, B. (2008) "Semantic matching framework for personalizing ubiquitous environment dedicated to people with special needs," *The Third International Conference on Pervasive Computing and Applications*, pp. 582–7.

[21] Korpipaa, P., Malm, E.J., Rantakokko, T., Kyllonen, V., Kela, J., Mantyjarvi, J., Hakkila, J., Kansala, I. (2006) "Customizing user interaction in smart phones," *IEEE Pervasive Computing* 5(3): 82–90.

[22] LaPointe, D., Barrett, A. (2005) "Language learning in a virtual classroom: synchronous methods, cultural exchanges," *Proceedings of Computer Supported Collaborative Learning*, pp. 368–72.

[23] Laurillau, Y., Paternò, F. (2004) "Supporting museum co-visits using mobile devices," *Proceedings of Mobile HCI*, pp. 451–5.

[24] Leichtenstern, K., Andre, E. (2008) "User-centred development of mobile interfaces to a pervasive computing environment," *The First International Conference on Advances in Computer-Human Interaction*, pp. 114–19.

[25] Lin, C.C., Chiu, M.J., Hsiao, C.C., Lee, R.G., Tsai, Y.S. (2006) "Wireless health care service system for elderly with dementia," *IEEE Transactions on Information Technology in Biomedicine* 10(4).

[26] Lindquist, D., Denning, T., Kelly, M., Malani, R., Griswold, W., Simon, B (2007) "Exploring the potential of mobile phones for active learning in the classroom," *ACM SIGCSE Bulletin Archive* 39(1): 384–8.

[27] Liu, C.C., Huang, C.H., Chu, W.T., Wu, J.L. (2007) "ITEMS: Intelligent travel experience management system," *The International Workshop on Multimedia Information Retrieval*, pp. 354–62.

[28] Lum, W.Y., Lau, F.C. "A context-aware decision engine for content adaptation," *IEEE Pervasive Computing* 1(3): 41–9.

[29] Mann, S. (1998) "Humanistic intelligence: WearComp as a new framework for intelligent signal processing," In *Proceedings of the IEEE* 86(11): 2123–51, Nov.

[30] McCrickard, D.S., Chewar, C.M. (2004) "Proselytizing pervasive computing education: a strategy and approach influenced by human-computer interaction," *The Second IEEE Annual Conference on Pervasive Computing and Communications*, pp. 257–62.

[31] Millar, G.A., Gileda, P.M. (1987) "How children learn words," *Scientific American*, pp. 94–9.

[32] Miura, M., Kunifuji, S., Sakamoto, Y. (2007) "AirTransNote: an instant note sharing and reproducing system to support students learning," *The 7th IEEE International Conference on Advanced Learning Technologies*, pp. 175–9.

[33] Morikawa, D., Honjo, M., Nishiyama, S, Ohashi, M. (2006) "Mobile web publishing and surfing based on environmental sensing data," *The 15th International Conference on World Wide Web*, pp. 995–6.

[34] Morikawa, D., Honjo, Yamaguchi, A., Nishiyama, S., Ohashi, M (2006) "Cell-phone based user activity recognition, management and utilization," *The 7th International Conference on Mobile Data Management*, pp. 51–51.

[35] Mulholland, P.; Collins, T.; Zdrahal, Z. (2005) "Bletchley Park Text: Using mobile and semantic web technologies to support the post-visit use of online museum resources" *Journal of Interactive Media in Education* 24: 1–21.

[36] Ogata, H., Akamatsu, R., Yano, Y. (2004) "Computer supported ubiquitous learning environment for vocabulary learning using RFID tags," *Technology Enhanced Learning*, pp. 121–30.

[37] Ogata, H., Hui, G., Yin, C., Ueda, T., Oishi, Y., Yano, Y. (2008) "LOCH: supporting mobile language learning outside classrooms," *International Journal of Mobile Learning and Organization* 2(3): 271–82.

[38] Ogata, H., Misumi, T., Matsuka, Y., El-Bishouty, M. M., Yano, Y. (2008) "A framework for capturing, sharing and comparing learning experiences in a ubiquitous learning environment," *International Journal of Research and Practice in Technology Enhanced Learning* 3(3): 297–312.

[39] Ogata, H., Saito, N.A., Paredes J.R.G., San Martin, G.A., Yano, Y. (2008) "Supporting classroom activities with the BSUL system," *Educational Technology & Society* 11(1): 1–16.

[40] Petersen, S.A., Markiewicz, J.K. (2008) "PALLAS: Personalized language learning on mobile devices," *The Fifth IEEE International Conference on Wireless, Mobile, and Ubiquitous Technology in Education*, pp. 52–9.

[41] Reddy, Y.V. (2006) "Pervasive computing: implications, opportunities and challenges for the society," *The 1st International Symposium on Pervasive Computing and Applications*, p. 5.

[42] Rhodes, B.J. (1997) "The wearable remembrance agent: a system for augmented memory," *The 1st International Symposium on Wearable Computers*, pp. 123–8.

[43] Riccardo, D., Paternò, F., Santoro, C. (2007) "An environment to support multi-user interaction and cooperation for improving museum visits through games," *The 9th International Conference on Human Computer Interaction with Mobile Devices and Services*, pp. 515–21.

[44] Shih, T.K., Lin, F.H., Du, Y.L., Chao, L.R., Kim, W. (2007) "Extending CORDRA for systematic reuse," *The 6th International Conference on Web-based Learning*, pp. 184–95.

[45] Storz, O., Friday, A., Davies, N., Finney, J., Sas, C., Sheridan, J.G. (2006) "Public ubiquitous computing systems: lessons from the e-Campus display deployments," *IEEE Pervasive Computing* 5(3): 40–7.

[46] Sumita, K., Ono, K., Miike, S. (1993) "Document structure extraction for interactive document retrieval systems," *The 11th Annual International Conference on Systems Documentation*, pp. 301–10.

[47] Thornton, P., Sharples, M. (2005) "Patterns of technology use in self-directed Japanese language learning projects and implications for new mobile support tools," *IEEE International Workshop on Wireless and Mobile Technologies in Education*, pp. 3–8.

[48] Vavoula, G., Meek, J., Sharples, M., Lonsdale, P., Rudman, P. (2006) "A lifecycle approach to evaluating MyArtSpace," *The Fourth IEEE International Workshop on Wireless, Mobile and Ubiquitous Technology in Education*, pp. 18–22.

[49] Weiser, M. (1993) "Some computer science issues in ubiquitous computing," in "Ubiquitous Computing", *Nikkei Electronics*, pp. 137–43.

[50] Yang, J., Lai, C., Chu Y. (2005) "Integrating speech technologies into a one-on-one digital English classroom," *IEEE International Workshop on Wireless and Mobile Technologies in Education*, pp. 159–63.

[51] Yang, M.J., Chen, J.H., Chao, L.R., Shih, T.K. (2009) "Developing the outdoor game-based learning environment by using ubiquitous technologies," *The 9th IEEE International Conference on Advanced Learning Technologies*, pp. 270–2.

[52] Yau, S.S., Gupta, E.K.S., Karim, F., Ahamed, S.I., Wang, Y., Wang. B. (2009) "Smart classroom: Enhancing collaborative learning using pervasive computing technology," *ASEE 2003 Annual Conference and Exposition*, available at http://shamir.eas.asu.edu/~mcn/publication/SmartClassroom-2003.pdf, retrieved Jul. 18th.

[53] Yen, N.Y., Hou, F.F., Chao, L.R., Shih, T.K. (2009) "Weighting & ranking the e-learning resources," *The 9th IEEE International Conference on Advanced Learning Technologies*, pp. 701–3.

[54] Yen, N.Y., Hou, F.F., Shih, T.K. (2008) "Extra-value for e-learning systems: an e-learning standard transformation mechanism," *IEEE International Symposium on IT in Medicine & Education*, pp. 121–5.

[55] Yih, J.S., Pinel, F., Liu, Y.H., Chieu, T. (2005) "Pervasive computing technologies for retail in-store shopping," *International Conference on Pervasive Services*, pp. 111–16.

[56] Yu, Z., Nakamura, Y., Zhang, D., Kajita, S., Mase, K. (2008) "Content provisioning for ubiquitous learning," *IEEE Pervasive Computing* 7(4): 62–70.

[57] Zhang, G., Jin, Q., Shih, T.K. (2005) "Peer-to-peer based social interaction tools in ubiquitous learning environment," *The First International Conference on Parallel and Distributed Systems*, pp. 230–6.

5

Service Management in Pervasive Computing Environments[1]

Jiannong Cao, Joanna Siebert, and Vaskar Raychoudhury
Department of Computing, Hong Kong Polytechnic University, Hong Kong.

5.1 Introduction

The cyber world has integrated with the physical world rapidly. Devices with communicating capabilities grow in number and become more powerful. The majority of devices in physical environments around us (home, means of transportation, office, shops, healthcare and entertainment facilities and so on) are embedded with computing capabilities and interconnected. How to coordinate these smart devices and make them serve people in a less obtrusive manner becomes one of the main research concerns in pervasive computing. In the vision of pervasive computing, the environment built on a physical world with embedded computing devices is a medium that provides the user with all the functionality he needs to satisfy his requirements. Such a functionality of a computational entity whose execution satisfies the requestor's requirement is called a service.

This chapter is concerned with service management in pervasive computing environments (PvCEs). Service management refers to mechanisms that help to find the best suited service providers in the environment, to identify and combine component functionalities to compose a higher level functionality and provide means to perform the requested functionality and make services better suited by adapting to changes in the request and in the environment. The motivation behind the research on service management in PvCEs derives from the huge gap between the high-level requirements from pervasive computing applications and the complexity of the PvCEs. The application requirements include high flexibility, reusability and reliability. The complexity of the PvCEs is characterized by hierarchical nature of services and service providers joining and leaving environments at run-time. Moreover, in PvCEs automatic delivery as well as returning the handle are both desired. Also, service duration must be taken into consideration. In order to allow the massive deployment of services in PvCEs, significant progress is needed along several dimensions. New architectures and techniques are needed in order to integrate seamlessly heterogeneous and changing devices and networks. For effective management of such systems, we need a way to capture the dynamic environment, compose available services and proactively recompose them depending upon specific contexts such as location, time and others.

In this chapter, we present the recent research on service management in pervasive computing. We describe a framework to analyze the functionalities of service management in PvCEs and survey the existing approaches and the corresponding techniques for implementing the key functionalities. The chapter also discusses the open problems and identifies the directions in future research.

[1] This work is supported in part by Nokia Research Center (Beijing) under Grant H-ZG19 and China National 973 Project under Grant 2009CB320702. Jiannong Cao, Joanna Siebert and Vaskar Raychoudhury ({csjcao, csjsiebert, csvray}@comp.polyu.edu.hk) are with the Department of Computing, The Hong Kong Polytechnic University, Hong Kong.

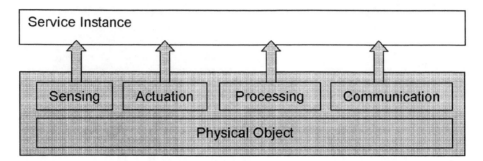

Figure 5.1 Physical object with extended capabilities.

The chapter is organized as follows. In Section 5.2, we present the background on services in PvCEs and analyse the functionalities of service management framework. In Section 5.3, we survey the approaches and the corresponding techniques for implementing the key functionalities with the evaluation of their advantages and disadvantages. Finally, in Section 5.4, we conclude the chapter with the discussion of the challenges, open problems and future directions of service management in PvCE research.

5.2 Service Management in Pervasive Computing Environments

In pervasive computing vision environments take care of user needs. Real life objects already existing in human lives are extended with capabilities that give them some level of intelligence. For this purpose objects can be augmented with sensing, processing, actuating or communication devices. These capabilities are implemented as services and exposed to the potential requestors. This concept is shown in Figure 5.1. Through services users can interact with PvCE. Therefore services are in the centre of the interests of the pervasive computing.

5.2.1 Introduction

In this section we describe the service-oriented architecture of PvCEs. We first provide the definitions necessary to describe an interaction between service requestors and service providers.

- **Service**
  A service is any hardware or software functionality (resources, data or computation) of a device that can be requested by other devices for usage. Examples of software services can be weather forecast, stock quotes and language translation. Services provided by the hardware devices can be, for example, controlling a temperature in the room or printing a document.
- **Service Implementation**
  A service implementation is software deployed on the service provider. It exists to be invoked by or to interact with a service requestor.
- **Service Description**
  The service description contains the details of implementation of the service, such as data types, operations, and location. It can also include categorization and other meta data. The service description may be advertised to a requestor or to a service manager.
- **Service Provider**
  Entities that expose their functionalities in form of services are called service providers. Services in an environment can be provided by hardware devices, software and other entities. Service providers are responsible for advertising their service description. Service providers in PvCE are characterized by their heterogeneity and dynamicity.
- **Service Requestor**
  Entities that need some functionality are called service requestors. Requestors of services can be users, other objects or middleware that facilitates interaction in PvCE. Requestors must be able to find service descriptions.

Table 5.1 Pervasive Computing Environments

	Properties	Issues	Examples
Infrastructured Pervasive Environment	1. Comprises enterprise environments, home and office settings 2. All devices in a single administrative domain 3. Generally resource-rich devices with low mobility and high bandwidth network support	Easy-to manage flexibility in design and reliability in operation	Jini [1], UPnP [2], SLP [3]
Ad hoc Network	1. High mobility, dynamic topology, resource-constraints 2. Extensive use of broadcast and multicast mechanisms	Energy efficiency Achieving scalability Achieving reliability	GSD [5], Allia [6], Konark [7], Service Rings [8], LANES [9], DEAPspace [10], Bluetooth SDP [11]

- **Service Requirement**
  In service requirement, the service requestor specifies what service descriptions it needs to find. Service requestor may retrieve a service description directly from service provider or from the service manager. Often, in order to fulfill service requirement, several services have to be integrated (several service descriptions retrieved).
- **Service Advertisement**
  In order to be accessible, a service needs to advertise its description such that the requestor can subsequently find it. Service description can be advertised to the service manager or directly to the service requestors.
- **Service Manager**
  This is an entity which fills the gap between service advertisement and requirement. The service manager manages the service registry, to which service providers advertise their service descriptions and service requestors retrieve required service descriptions. It can also support service composition and other service management functions.

5.2.2   Pervasive Computing Environments

PvCE consists of service providers and service requestors. It can be classified into infrastructure-based and infrastructure-less or ad-hoc environments.

Infrastructure-based pervasive environments have access to computing elements in the wired infrastructure through gateways, proxies and base stations. Examples of such an environment are wireless home or office networks based on Wi-Fi. These environments are able to leverage from the high bandwidth resource-rich wired environments.

Infrastructure-less pervasive environments, on the other hand, refer to environments where computing elements (mobile devices) do not have any access to the wired infrastructure and network connections with peer devices are created and broken down on-the-fly and on real-time on-demand basis. Examples of such environments are warfront activities, ad-hoc sensor networks, walkways in cities, futuristic malls etc.

Service management in the above-mentioned PvCEs comes with associated challenges as listed in Table 5.1.

5.2.3   Service Management Framework

Service management mechanisms facilitate interaction between entities in PvCEs and free users from tedious and redundant administrative and configuration works. Therefore, service management research is critical to the success of pervasive computing. Figure 5.2 presents a framework of service management in PvCE. A self-contained service management framework for pervasive computing must provide support for certain basic functionalities along with some management supports. Basic functions include the means to specify the preferences by requestors and service descriptions by service providers, finding service providers in the environment, matching requestors

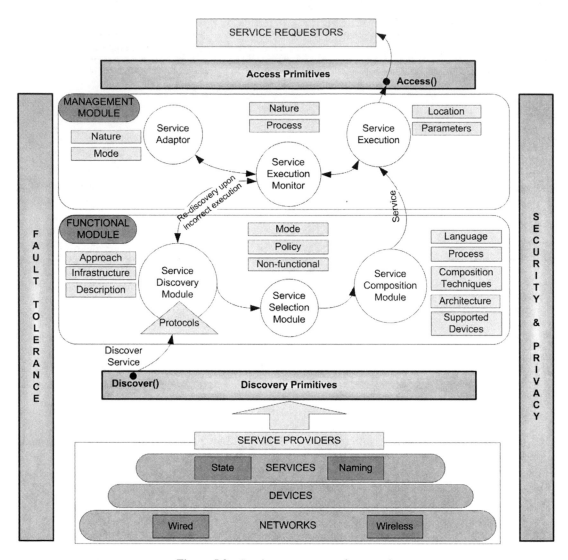

Figure 5.2   Service management framework.

preferences, selecting the best suited service and combining elementary services to compose a higher level service. The management functionalities include service execution monitoring and required adaptation policies to suit the user need under dynamic environment changes. Also, there are fault tolerance and reliability supports.

5.2.4   General Components of a Service Management System

In this section we shall discuss the major components of a service management system. As described in Figure 5.2, the service discovery framework has two operational modules – the functional module and the management module. The lowest layer of a service management framework consists of the underlying network structure. Above that we have a pool of services provided by multiple devices in the network. Issues related to service information management are listed in Table 5.2. Service management is facilitated by the use of various protocols designed for different network structures and management models. Users, human or devices alike, can discover and access services using different primitives provided by the protocols. Service management systems also require fault tolerance supports as well as measures to protect security and privacy of the users. Due to the dynamic nature

Table 5.2 Service information management

	Categories	Examples
Service Naming	Template-based	Jini, UPnP, Salutation and Bluetooth
	Predefined	SLP
Service State	Soft	Jini, UPnP, SLP, Bluetooth SDP, DEAPspace
	Hard	Salutation

of pervasive environments, service management systems should also provide support to handle user and device mobility. We consider all the horizontal boxes as core components of a generic service management system and the vertical boxes as essential system support service components. Different protocols may choose to implement the system support modules depending on the application and user requirements.

- **Network Issues**
  Based on the nature of the underlying network, service management mechanisms adopt different designs. Wired networks consist of resource-rich and static computing devices connected through high-bandwidth network cables. Examples of these types of systems are enterprise networks. Wireless networks, on the other hand, can be employed either in static or in mobile settings. Static wireless networks mostly contain high bandwidth network backbone infrastructure, which supports the networking need of the participating computing entities. Mobile wireless networks, however, lack any such infrastructure support. They create an ad hoc composition of multiple resource-constrained portable and handheld devices connected by unreliable and intermittent wireless connectivity. Wired networks and static wireless networks, which have some common properties, are together called as infrastructure-based networks. Mobile wireless environments, on the other hand, are called infrastructure-less or ad hoc and are more difficult to handle. Pervasive computing systems, though can adopt either infrastructure-based or ad hoc network environment, or a hybrid of them, they are characteristically more close to the ad hoc wireless environment due to their limited resource availability, extreme dynamism and unreliable network connections.
- **Service Naming Issues**
  Services are identified by their names and discovered by matching their attributes. So, it is very important to have easily discoverable names and attributes for services. Some mechanisms choose user friendly service naming (e.g. Jini [16]), whereas others used some standard naming template (e.g. Rendezvous [10]).
- **Service Management Architecture**
  Service management mechanisms adopt either a centralized or decentralized management model. For the centralized model there is a dedicated manager along with the service providers and service requestors. The manager node maintains service information and processes service requests and announcements. Some managers provide additional functionality, such as support of secure announcements and requests in Ninja SDS [15]. The decentralized model, on the other hand, has no dedicated manager. When a service request arrives, every service provider processes it. If the service matches the requirement, it replies. When hearing a service announcement, service requestor can record service information for future use.

  Based on the number of services and the size of the network, centralized systems can use a single manager or multiple manager nodes distributed across the network in strategically important locations. Depending on the organization of the manager nodes, centralized models can be distinguished either as flat or as hierarchical. In a flat structure, managers have peer-to-peer relationships. For example, within an INS sub-domain, managers have a mesh structure: a manager exchanges information with all other managers. Salutation [31] and Jini [16] can also adopt flat structure. On the other hand, a hierarchical structure follows the DNS model in which information, advertisements and requests are propagated up and down through the hierarchy. Parents store information of their children and thus in this type of system, the root node may possibly become a bottleneck. Examples include Rendezvous [10] and Ninja SDS [15], both of which have a tree-like hierarchy of managers.

5.2.5　System Support Components

After we have discussed the prime issues associated with the core components of service management in general, we also discuss the supportive operations that are important for service management in real pervasive environments.

The first one of this is fault-tolerance, the second one is mobility management while the third and final one is support for security and privacy.

- **Fault Tolerance and Mobility Support**
  We divide the faults in pervasive environments into two types – infrastructure-related and software and service-related. While the infrastructure-related faults cover hardware failures, such as, devices and network failures, software and service-related faults are concerned with failures of pervasive software and networked services. Existing service management mechanisms mostly consider crash of service providers or manager nodes which gives rise to service unavailability. This type of fault requires redundancy support to keep the operation ongoing if a manager node fails. Service unavailability can also arise due to user or device mobility and network disconnection. Mobility is a crucial issue for infrastructure-less pervasive environments and requires special attention to cope with the challenges.

  When a service becomes unavailable it must be noted by the manager in order to keep the service information up-to-date and freeing the memory of stale service information. This is achieved by maintaining a soft service state which is required to be renewed at specified intervals by the service provider. In case the service is unavailable, the service state cannot be renewed, which leads to the deletion of the service from the registry. A hard service state, on the other hand, does not require to be renewed by the service providers. However, the managers must poll the service providers to find out whether the service information is up-to-date.

  Another way of doing away with accumulated service usage information in service providers is by providing lease-based service. In lease-based service usage, service users must renew their lease with the service providers before the lease gets expired or the user fails or leaves the environment. Otherwise, the service providers simply delete the service states associated with the user and free the memory space. The alternative way is to explicitly release a service for a user and maintain the service execution information.
- **Security and Privacy Support**
  Security is another important issue required to be addressed for service management in pervasive computing. Since users might need to interact with possibly unknown devices acting as managers and service providers in different environments, none of the parties involved is keen to take the first move for the fear of breach of privacy. So, it is important to maintain proper security and privacy at all times.

5.2.6   Service Management Challenges

We list some of the major issues (Table 5.3) and concerns that are associated with different modules of a service management system and are required to be addressed while designing such a system.

Table 5.3   General issues for service management

	Primitives	Discover and Access
CORE COMPONENTS	**Service Discovery Protocols**	*Discovery*: Approach (Pull/Push), Service Information State (soft / hard), Scope, Selection Policy (Automatic / manual),
		Access: Invocation Policy, Usage Policy (Lease-based/completely released)
	Services	Naming, Attributes
	Management Architecture	Centralized/Decentralized
	Network	*Connection Type*: Wired/Wireless (single / multi hop) *Dynamics*: Static or Mobile
SYSTEM SUPPORT COMPONENTS	**Fault Tolerance**	*Types of Faults*: Hardware-related (Device & Network), Software-related (Software & Service)
	Mobility Management	Network Partition, Unreachability of devices
	Security & Privacy	Securing Privacy of service provider and user

5.3   Techniques for Service Management in PvCE

5.3.1   Introduction

As mentioned in the previous section, service management in PvCE comprises of finding the best suited service providers in the environment, identifying and combining component functionalities to compose a higher level functionality and adapting to changes in the request and in the environment, whereby a service becomes better suited. Services in PvCEs are different than in traditional environments. They are characterized by their hierarchical nature with service providers joining and leaving environments at run-time and automatic delivery of service as well as returning the handle being both desired. Moreover, service duration must be taken into consideration. In PvCEs service requestors require services instantaneously to accomplish their goals. They have no prior knowledge about the available services, and the state of the environment changes frequently due to service providers joining and leaving environments at run-time. Also, the demand of service requestors changes together with changes in context. Therefore there are several challenging issues, such as: detection and management of changes in the environment, requirement specification and dynamic detection of changes and the best service selection with regards to dynamic ranking and state of the execution of the composite service [74].

Detection and management of changes in the environment – How to discover services? Which is the best suited service discovery mechanism among of many proposed, such as JINI, UPnP, and Bluetooth proximity detection? Can we propose a generic discovery mechanism? How information about available services should be maintained and how is it updated? How do different layers of a system access and make use of it? If the substitution of a service during recomposition is necessary how to resume service in the same point as it was stopped?

Requirement specification and dynamic detection of changes – How is the requirement specified and represented internally? How rich must this information be in order to be useful? When and how it should be updated? Can changes in requirement be inferred, or does it have to be explicitly provided? In the latter case, is it statically specified or obtained on demand through dynamic interactions? Will the attempt to obtain requirement place an undue burden on the user? Will it hurt usability and performance unacceptably? Is the benefit worth the cost? How to quantify this benefit? How do different layers of a system access and make use of the requirement specification? Is incomplete or imprecise requirement specification still useful? At what level of uncertainty is it better to ignore such knowledge in making decisions?

The best service selection – How to quantify 'the best' service instance? What factors should be taken into account for a good decision about service selection? How should different factors of service ranking be weighted? How to support dynamic changes in ranking? Should the user play any role in making this decision? If it is not possible to find all requested service instances to satisfy the requirements of the composed service, is incomplete service composition still useful?

5.3.2   Classification of Service Discovery Protocols

Service discovery protocols provide users the means to automatically discover networked services. Existing service discovery protocols can be grossly classified based on their underlying network structure and the discovery infrastructure built over that. In Figure 5.3 we give a classification of existing protocols.

Over the past few years, many organizations have designed and developed service discovery protocols. Examples in academia include the Massachusetts Institute of Technology's Intentional Naming System (INS) [1], University of California at Berkeley's Ninja Service Discovery Service (SDS) [15], and IBM Research's DEAPspace [24]. Major software vendors ship their service discovery protocols with their current operating systems – for example, Sun Microsystems' Jini Network Technology [16], Microsoft's Universal Plug and Play (UPnP) [36], and Apple's Rendezvous [10] (currently known as 'Bonjour'). Other organizations have also proposed discovery protocols standards, including Salutation Consortium's Salutation protocol [31], Internet Engineering Task Force's Service Location Protocol (SLP) [13], and Bluetooth Special Interest Group's Bluetooth SDP [4].

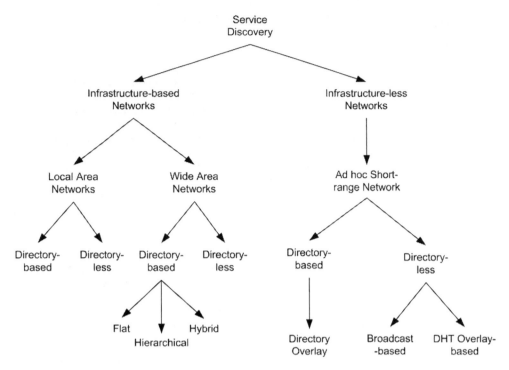

Figure 5.3   Classification of existing service discovery protocols.

5.3.3   *Service Discovery in Infrastructure-Based Networks*

Infrastructure-based networks are characteristically either wired networks or wireless networks with network backbone support. Service discovery protocols (SDP) for infrastructure-based networks have been developed either for limited area networks (LAN) or for wide area networks (WAN).

- **SDPs for Local Area Networks**
  As a LAN is covered by a single administrative domain, DHCP services can be used for them. Also they contain resource-rich devices which are mostly static and provide high-bandwidth network support. Enterprise environments and smart office and home environments can be considered as ideal examples of these types of systems. Some of the well-known protocols for this environment are Jini [16], UPnP [36], SLP [13], and FRODO [33].

  Jini is a Java-based service discovery protocol introduced by Sun Microsystems where services are represented as Java objects. Jini adopts the directory-based discovery model where all the service information is centrally maintained in the registry or lookup servers. Universal Plug and Play or UPnP, on the other hand, is a Microsoft-initiated standard that extends the Microsoft Plug-and-Play peripheral model. UPnP uses a directory-less model and enables peer-to-peer network connectivity of intelligent appliances, wireless devices, and PCs of all form factors. The Service Location Protocol or SLP has been designed for TCP/IP networks and can choose to operate using a directory-based or a directory-less model. FRODO is a directory-based service discovery protocol for home networks where directory nodes are elected based on available resources.

  Another directory-based protocol, named Salutation [31], has been developed by IBM. This can be used for any network and can perform either in P2P mode or in centralized manner. Salutation is useful to solve the problems of service discovery and utilization among a broad set of appliances and equipment and in an environment of widespread connectivity and mobility. The Salutation architecture defines an entity called the Salutation Manager (SLM) that functions as a directory of applications, services and devices, generically called 'Networked Entities'. The SLM allows networked entities to discover and to use the capabilities of the other networked entities.

- **SDPs for Wide Area Networks**

  Service discovery protocols for wide area networks have separate design concerns. These types of networks usually contain large number of devices and services that need to be managed and so, the developed service discovery protocols must be highly scalable. Also WANs do not have support for broadcast or multicast mechanisms. Moreover, to ensure network-wide service availability, they must have multiple replicas for each service. So, there must be a trade-off between the need to maintain consistency among service replicas and the generated network traffic. Also, the requested service should be available to the user within optimal time and by using minimum number of messages. Some of the well-known protocols for infrastructure-based wide area networks are SSDS [15], CSP [23], INS/Twine [3], Superstring [28] and GloServ [2]. All these protocols use a structured distribution of directory servers which stores the service information. The directory structure can be well classified into three modes – flat, hierarchical or hybrid.

  INS/Twine forms a flat directory structure which is a peer-to-peer overlay network of resolvers constructed by means of distributed hash tables (DHT). INS/Twine uses the Chord [32] DHT system. Another example is One Ring Rules Them All [6] which uses structured P2P overlays as a platform for service discovery and implements over Pastry [29] DHT system. The infrastructure proposed relies on a universal ring that all participating nodes are expected to join. The major advantage of using DHT based protocols is the efficient lookup which usually takes $O(log(N))$ hops, where N is the number of nodes in the overlay network. However, DHT systems do not take into account the actual physical distance between the overlay nodes and also they are costly in terms of resources. Due to this, using DHT based protocols for pervasive environments can incur higher costs. Contrary to the peer-to-peer directory overlay structure, SSDS, CSP and GloServ form a hierarchical structure of directory nodes.

  With a view to tackle the disadvantages associated with either type, Superstring, combines both the peer-to-peer and hierarchical topologies and provide a hybrid model. It utilizes a flat topology to discover top-level nodes that specialize in a particular kind of service. From this top-level node, a hierarchy is created which reflects the hierarchical structure of service descriptions and helps to resolve user queries. Another research based on hybrid directory structure is the Project JXTA protocols [34] which establish a virtual network overlay on top of existing physical network infrastructure. They store the service information in rendezvous peers. This approach combines DHT methods with a random walker that checks for non-synchronized indices. In order to avoid expensive network traffic, the resolver nodes are not required to maintain a consistent distributed hash index. Instead, a limited-range walker is used to walk the rendezvous from the initial DHT target.

- **Fault Tolerance and Mobility Management Mechanisms**

  In infrastructured networks, the existing service discovery protocols are mostly directory-based. So, fault tolerance supports are required to safeguard the system against possible failures of directory nodes. Usual approach to achieve this is by redundancy. For central storage, such as, FRODO, there is a backup of the directory node which takes over in case the central directory fails. For distributed storage, e.g., INS/Twine, JXTA, and One Ring Rule Them All, multiple copies of service information are maintained in different servers. When one or more fails, the rest can still answer the queries.

  Service information maintenance on the face of mobility is tackled by many protocols in these types of networks. Jini uses a lease-based mechanism for service access by users. SLP and UPnP, however, enforce expiry time for service registrations and advertisements, respectively, and GloServ requires periodic renewal of service registrations, all in order to remove outdated service information. SSDS maintains service information state as soft states.

5.3.4 Service Discovery in Infrastructure-Less Networks

Infrastructure-less networks are highly dynamic and consist of multiple resource-constrained and possibly mobile devices. They are connected through weak wireless connectivity and form a mobile ad hoc network. In these environments, we assume that the participating devices provide some services to their peer devices which can be discovered and accessed through carefully designed service discovery protocols. Depending on where the service information is maintained, the existing service discovery protocols for ad hoc environment have been classified into directory-based and directory-less models. We first discuss the existing directory-less service discovery protocols and analyze their features. Next, we shall elaborate on the available directory-based protocols.

- **Directory-less SDPs**

  In the directory-less service discovery protocols, the service information is stored with the service providers themselves. There are two distinctly identified methods for directory-less service discovery in ad hoc networks.

The first method works by broadcasting service information as well as service requests and the second approach works by building DHT-based P2P overlay [18, 38, 26, 15, 12, [5] for mobile ad hoc networks (MANET). We first describe the broadcast based method followed by the DHT overlay approach.

Broadcast-based service discovery can adopt either push or pull model. In a push-based discovery model, service advertisements are distributed by the service providers to all the nodes in the network. A pull-based discovery model, however, necessitates a service requestor to broadcast their service request to other nodes until a matching service is found. The broadcasting nature of this model is grossly unsuitable for the mobile ad hoc networks due to their high demand of bandwidth and energy. So, these protocols can only be used in small scale networks. Some of the protocols which use broadcast policy are Bluetooth [4], DEAPspace [24], Allia [27], GSD [7], DSD [8] and Konark [14].

Among all the protocols mentioned above, Bluetooth and DEAPspace are designed for single hop ad hoc networks where the rests are for multi-hop networks. Bluetooth has been developed following a client server model whereas, DEAPspace follows a peer-to-peer architecture. DEAPspace is a push-based decentralized discovery protocol in which each node maintains information about its known services and periodically exchanges its known service list with the neighbors via broadcast. Similar policy for broadcast is also used in Konark which allows each node to store service information from itself and other known services. However, Konark is multihop in nature and supports both pull and push based advertisement and discovery using multicast. The broadcast in DEAPspace significantly increases the message overhead and can lead to multicast storm. To cope with this problem, Konark proposes a service gossip algorithm which suppresses repeated message delivery by caching the service information and also the multicast is performed at random intervals. In GSD, every node stores service advertisement from any other node within a maximum of N hop distance, known as advertisement diameter. Allia and DSD, on the other hand, allow nodes to advertise their services only within their transmission range. In Allia, nodes which cache service advertisements form an alliance with the advertising node. Difference between Allia and DSD is that, in DSD, nodes which stores advertisements from their neighbours, can also forward the advertisement to other nodes, unlike in Allia, based on the forwarding policy.

Some other protocols in this category are Wu and Zitterbart [38], Cheng and Marsic [9] and Varshavsky et al. [37]. These protocols support cross-layer types of service discovery. Wu and Zitterbart is a directory-less P2P protocol based on DSR routing protocol in which every node caches service advertisements and performs both pull and push based discovery. Cheng and Marsic, on the other hand, is based on on-demand multicast routing protocol (ODMRP) in which nodes cache service advertisements depending on their interest. Varshavsky et al. proposes a protocol which has two main components – a routing protocol independent Service Discovery Library (SDL) and a Routing Layer Driver (RLD). SDL function is to store information about the service providers. RLD, which is closely coupled with the MANET routing mechanism, is used to disseminate service discovery requests and advertisements. Each node has the stack containing SDL and RLD and form a P2P networking with other nodes.

There are several MANET-oriented DHT systems [26, 40, 12] which integrate DHT with different ad hoc routing protocols to provide indirect routing in MANET. Ekta [26], MADPastry [40], and CrossROAD [12], each integrates Pastry [29] with DSR [17], AODV [25], and OLSR [11], respectively, to share routing information between network layer and application layer. An opposite approach is adopted by virtual ring routing (VRR) [5] which is a network-layer routing protocol inspired by overlay routing on DHTs and can significantly reduce traffic compared with broadcast-based schemes. Ekta implements a service discovery protocol and there are other DHT overlay based service discovery protocols [18, 38] for MANET. DHT-based systems for service discovery, however, have certain drawbacks. Ekta and VRR construct a DHT substrate without taking into account the actual physical distance between nodes. This can cause undesirably long search latency and deterioration of success ratio of service discovery with the growing network scale. On the other hand, MADPastry proposed a clustering method which groups the overlay nodes according to their physical distance. But, MADPastry routes information using location-dependent addresses. These identifiers can change with mobility and node failures, so, it always requires mechanisms to look up the location of a node given a fixed identifier which is costly.

- **Directory-based SDPs**
Given the resource-poor nature of the devices and their mobility, it is difficult to choose single centralized directory nodes. To cope with this limitation, directories are dynamically selected from mobile nodes considering their available resources, such as, processing power, memory size, battery life, or node coverage, etc. It is true that, dynamic directory assignment incurs extra overhead to the network because, directories should be selected and their identities should be informed to the rest of the network nodes. Moreover, directory nodes must be constantly available on the face of node failures and dynamic topology changes and network partitions. Even with

these difficulties, a directory-based system proves to be more scalable and fault tolerant than a directory-less one in the infrastructure-less environments. Moreover, using directories will most certainly decrease the discovery delay and will enhance load balancing among service providers, so as to reduce the load on individual services and enhance the overall service discovery performance. Examples of directory-based service discovery protocols for ad hoc wireless networks are – Service Rings [20], Lanes [21], DSDP [22], Tyan *et al.* [35], Sailhan *et al.* [30], Kim *et al.* [19], and Splendor [41].

The basic approach followed by these protocols is the same. They select some nodes as directories based on some parameters and then form an overlay or backbone network connecting those nodes. Here we give a brief account of the protocols named above. Service Rings forms an overlay of nodes which are physically close and offer similar services. This structure, built over the transport layer, is called service ring. Each service ring provides a service access point (SAP), which acts like a directory and through which services provided by any of the members of a ring can be accessed. SAPs of different rings connect with each other to form a hierarchical structure. The protocol Lanes is inspired by the content addressable network (CAN) protocol, for wired P2P networks. In Lanes, nodes are grouped together to form a linear structure, called lanes. Each node in a lane contains same service information and share the same anycast address. Multiple lanes are loosely coupled together. DSDP selects certain directory nodes in the network, based on available resources, to form a dominating set, or a virtual backbone. The backbone of directory nodes is then used for both service discovery and routing. Tyan *et al.* proposes a protocol in which the network is divided into hexagonal grids, each having a gateway. The gateway nodes are used for routing and work as directory nodes. The connected overlay of gateways forms a virtual backbone. Kim *et al.* proposed a volunteer node based protocol where volunteers are relatively stable and resource rich nodes and form an overlay structure with other volunteers. The volunteers in fact act as directory nodes.

Sailhan *et al.* proposed a protocol for large-scale mobile ad hoc networks in which multiple directory nodes distributed across the network interconnect to form a backbone. The directory nodes are so deployed that at least one directory is reachable in at most a fixed number of hops, H, known as the vicinity of the directory. Directories store service information available in their vicinity. Their protocol builds a hybrid network bridging mobile ad hoc networks and infrastructure-based networks, where some nodes have the same network interface, where others hold several network interfaces, and act as gateways with other networks.

- **Fault Tolerance and Mobility Management Mechanisms**
  Fault tolerance in infrastructure-less environments is more difficult than in infrastructured networks. The dynamic nature of the environment, resource-constrain of the participating devices, and the unreliability of wireless connection makes it hard to design robust fault tolerant mechanisms for them compared to the traditional distributed environment. Fault tolerance requires withstanding failure of directory nodes. INS [1] and LANES cope with directory failure by maintaining multiple copies of service information at different nodes. Most of the directory-based protocols replace failed directory nodes with newly selected ones.

  Mobility management, however, is very challenging in ad hoc environments. Frequent node mobility renders the topology unstable and disconnections give rise to inconsistency in service information. In order to maintain consistency of service and route to service information, either a proactive or a reactive method has been adopted by existing protocols. In a proactive approach, the participating nodes periodically exchange messages to update information. Example is, when a service provider periodically sends advertisements to update its location. A reactive method, however, updates information based on triggering of certain events. So, if a user finds out that a previously cached service is unreachable, then it seeks for new service information.

  Directory-less service discovery protocols cope with node mobility by adjusting the service advertisement rate and the diameter of announcements. For example, GSD implements a small advertisement time interval for highly dynamic environments, opposed to a larger value for comparatively stable networks. The advertisement diameter (in number of hops) is also regulated depending on different mobility situations. Similarly, Allia controls the frequency of advertisements and the diameter of the alliance considering the mobility of nodes.

  Most of the directory-based protocols for ad hoc service discovery require special mechanisms to maintain the directory structure – backbone or overlay. The job of these algorithms is to ensure smooth operation by handling node joining or leaving scenarios, broken connections, network partition and partition merges. Service Rings, Lanes, DSDP, Tyan *et al.* and Sailhan *et al.* all propose similar mechanisms.

  We have proposed an efficient and fault tolerant service discovery protocol [43] for MANETs. A virtual backbone of directory nodes, called directory community, is first constructed selecting top K nodes considering higher resource and lower mobility, to ensure that they are more reliable and less fault-prone. We model the directory community formation problem as top-K weighted leader election in mobile ad hoc networks [42], and develop a distributed algorithm to achieve the objective. Here, the weight indicates available node resources in terms of memory, processing power or energy. Using the afore-mentioned directory community, we propose a quorum-based fault-tolerant service discovery protocol. The elected directory nodes are divided into multiple quorums. Services registered with a directory are replicated among its quorum members, so that, upon the failure of a directory, services can still be available. This approach guarantees network-wide service availability using the quorum intersection property and reduces replication and update costs by minimizing the quorum size.

5.3.5   Multiprotocol Service Discovery

Along with the advances in pervasive computing technologies, users are no longer constrained to only a single computing environment but can work across different environments, meeting with a diversity of software, hardware devices, and network infrastructures. Heterogeneity of the environments brings significant problems that application developers must cope with. Many service discovery protocols have been proposed. Existing service discovery protocols differ in the way service discovery is performed, and no single protocol is suitable for all the environments. Due to the differences in the service discovery approaches implemented, a user working in one environment may not be able to search for the services available in another environment that suits the user's need. To support service discovery across different environments, new techniques are needed to integrate or bridge the service discovery protocols used in a diversity of environments. Works can be found on providing interoperability of service discovery protocols [64–69].

All the existing service discovery mechanisms realize the concept of client/server application. Clients are entities that need some functionality (service) and servers are entities existing in the environment and offering this functionality. Service discovery is the framework for connecting clients with services. Existing works towards achieving multi-protocol service discovery use a middleware approach but differ in where the interoperability support is provided. The middleware can be on the service side, client side, or intermediate entity. An example of providing the support on the service side is INDISS [67] designed for home networks. INDISS provides parsers and composers, which decompose a request from the source service discovery protocol into a set of events and then compose them into message understood by target service discovery protocol. INDISS can also be deployed on the client side. Another example of middleware on the client side is ReMMoC [66]. In ReMMoC, the client side framework provides the mappings between all the supported service discovery mechanisms. Abstraction of discovery protocols to the generic service discovery is achieved by using a generic API or doing discovery transparently to the client. In this approach, all possible mappings need to be provided at the client side. Service side support can also be based on service proxies [69]. When a new service appears or disappears in one of the environments, the framework detects it and creates or removes its proxies from other environments. However, this approach requires dynamic service proxies to be implemented for each environment, which increases developer's workload. Another way of providing interoperability support is to provide an intermediate entity [65, 64]. In Open Service Gateway Initiative OSGi [65] all the connections and communications between the devices are brokered by a Java-based platform. An internal service registry exists in the framework. Interoperability is achieved by providing an API to map the given service discovery protocol to OSGi and vice versa. Supporting new service discovery protocols requires defining new APIs for them. Gateway functionality is also utilized by protocol adapters in the FuegoCore Service broker [64] designed for mobile computing environments. The service broker registers the mappings between its internal template and the external templates used by different service discovery protocols. Extending the FuegoCore service broker to new service discovery protocols requires creating and deploying additional service discovery protocol adapters. INDISS [66], mentioned above, can be deployed as an intermediate entity as well. The intermediate entity approach requires broker to integrate all the adapters into one system. In a network with a large number of service discovery protocols the framework may not be scalable. Another approach is providing service that discovers services across different environments [68]. In MUSDAC, it registers itself in all the environments, so clients can use whatever protocol to discover it. However, clients must have the knowledge about MUSDAC and the process of discovering the service has high processing requirements.

- **Universal Adaptor**

  We have studied how to provide service discovery across different environments supported by different service discovery systems, which may use standard protocols such as SLP and Jini, as well as tailor-made mechanisms that support multiple protocols within an environment, such as ReMMoC [65]. Our work [71] is based on the analysis of the following requirements on the interoperability system for pervasive computing: (i) no change

Table 5.4 Service discovery approaches

	Categories	Examples
Discovery Approach	Pull-based/Query-based	Jini, UPnP, SLP, GSD, Konark
	Push-based/Announcement-based	Jini, UPnP, SLP, Bluetooth SDP, Konark

should be imposed on the existing service discovery mechanisms; (ii) no change should be imposed on the services registered in domains; (iii) no functionality of the environment should be compromised; (iv) the system should be lightweight, scalable, and extendable; (v) support both standard and tailor-made service discovery mechanisms. Our approach addresses all of these requirements. We proposed the Universal Adaptor (UA) approach, which consists of two major components: the Universal Adaptor Primitives (UAP) and the Universal Adaptor Mapping (UAM). UAP is the universal set of primitives used by the user to discover the services across different environments, while UAM provides the mapping between the Universal Adaptor Primitives to the primitives used in various service discovery systems. The Uniform Adaptor can be implemented in any environment, independent of the service discovery system used in that environment. This approach enables users to discover available services with no knowledge of the service discovery system adopted in an environment. Universal Adaptor provides a simple and flexible solution. It provides only a single set of APIs and supports not only all existing but also future service discovery systems. It is lightweight and easy to implement in diverse infrastructures and to use by users.

5.3.6   Service Discovery Approaches

The methods of exchanging service discovery and registration information among clients, services, and directories are basically of two types – active/pull-based/query-based and passive/lazy/push-based/announcement-based. They are listed in Table 5.4.

In the query-based approach, a party receives an immediate response to a query and does not need to process unrelated announcements. Multiple queries asking for the same information are answered separately. In the announcement-based approach, interested parties listen on a channel. When a service announces its availability and information, all parties hear the information. So in this approach, a client might learn that the service exists and a directory might register the service's information. Many protocols support both approaches.

5.4   Service Composition

Services provide different functionalities. Taking services as building blocks, service composition is a process of identifying and combining component functionalities to compose a higher level functionality and provide means to perform the requested functionality. Figure 5.4 shows service composition application in PvCE. Consider that user specifies the requirement of watching the movie, for which following services must be satisfied: file with the movie, movie player, device to output sound of the movie and device to output its display. During runtime usually multiple instances of requested services can be found. For example, one instance of movie player is a DVD player in a work room and a second instance of it is a video player in a sitting room. Similarly, the computer monitor and the TV monitor are two instances of display service. Apart from the services specified in the requirement, other services exist in the environment as well. According to requirements we need to select one instance for each service type.

Service composition mechanisms facilitate interaction between entities in PvCEs and free users from tedious and redundant administrative and configuration works. Therefore, service composition research is critical to the success of pervasive computing. Existing works divide the service composition process problem into several fundamental problems.

5.4.1   Service Composition Functions

The basic functions of service composition are shown in Figure 5.5 and include the means to specify the preferences by users and service descriptions by service providers, finding service providers in the environment matching user

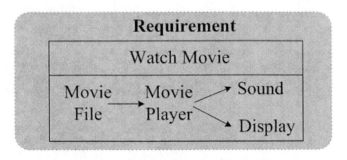

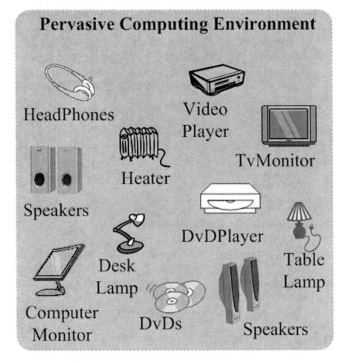

Figure 5.4   Service composition application.

preferences, selecting the best suited service and combining elementary services to compose a higher level service. The composition functionalities also include adaptation policies to suit the user need under dynamic environment changes.

Service composition system interacts with the application layer by receiving functionality requests from users or applications. It needs to respond to the functionality requests by providing services that fulfill the demand. These services can be atomic or composite. The composition system has two kinds of participants, service provider and service requestor. The service providers propose services for use. The service requestors consume information or services offered by service providers.

The process of service composition includes the following phases:

Describing services. Firstly, the service providers will provide description of their atomic services. In the meantime, the service requestor can also express the requirement in a service specification language.

Specifying composition plan. Next, service composition system tries to provide the needed functionalities by composing the available service technologies, and hence composing their functionalities.

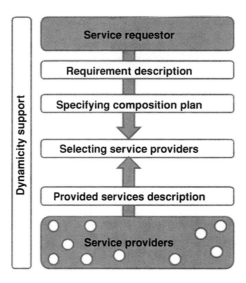

Figure 5.5　Service composition functions.

It tries to generate one or several composition plans with the same or different technology services available in the environment. It is quite common to have several ways to execute the same requirement, as the number of available functionalities in pervasive environments is in expansion.

Selecting service providers. It is quite common that many services have the same or similar functionalities. In that case, the services are evaluated by their overall utilities using the information provided from the non-functional attributes. In pervasive environments, this evaluation depends strongly on many criteria like the application context, the service technology model, the quality of the network, the non functional service QoS properties, and so on.

Dynamicity support. Service provision needs to be dynamic and adaptable as changes may occur unpredictably and at any time.

In the next section, we describe existing methods that fulfill above mentioned functions of service composition process.

5.4.2　Survey of Methods in Service Composition Process

We outlined in the previous sections that service composition is an important and challenging topic in the area of pervasive computing. We have identified basic functions of service composition process. Some research results have been already reported in this field. In this section we summarize this works and conduct a comprehensive review of various types of approaches to service composition in PvCEs.

- **Describing Services**
  To describe functional and nonfunctional attributes of services, many service description languages have been developed. Existing service composition approaches are characterized by different expressiveness of the language used to describe the provided as well as requested services. Some of them allow expressing semantically the functional and nonfunctional attributes of the services for a later use at composition process [6–18]. Regarding to the way that requested services are defined, we observe two approaches: low-level description and high-level description. In low-level description [45– 48, 50–53, 56–60] requested service is specified as a workflow, given the set of atomic services to be composed. In highlevel description, requested service is specified as a goal to be achieved [44, 45, 62, 63]. The description of provided services, they are mainly described as atomic

functionalities. In another approach [51], they may have enough knowledge to specify workflows in which they can take part.

Regarding the way that requested services are defined, we observe two approaches: low-level description and high-level description. In low-level description, requested service is specified as a workflow, given the set of atomic services to be composed. In high-level description, requested service is specified as a goal to be achieved. For the description of provided services, they are mainly described as atomic functionalities. In another approach, they may have enough knowledge to specify workflows in which they can take part.

- **Specifying Composition Plan**
  In order to create the composition plan diverse techniques can be used. One of these approaches is AI Planning [44, 48, 50, 54–56, 62], where the composition of atomic services into composite service is viewed as planning. Atomic services are mapped into planning operators, planning algorithm links them, and generated plan constitutes a composite service. In workflow approach [45–47, 51–53, 55, 58, 59–61, 63] a composite service is broken down into sequence of interactions between atomic services. The system generates a customized workflow that describes how various services should interact with one another as well as with the requestor. Data Mining [49] refers to extracting knowledge from service usage historical data and is used to describe patterns discovered through mining in order to help with service composition.

- **Selecting Service Providers**
  In the majority of the solutions selection is done by the service requestor [24, 13, 36, 16, 14]. After receiving a full or partial list of available services, the requestor chooses a service based on service information and additional information, such as context. In some cases, protocols may select services for a user [24]. The advantage of protocol selection is that it simplifies client programs or little user involvement is needed. On the other hand, protocol selection may not reflect the actual user's will. Predefined selection criteria may not apply to all cases. Alternatively, too much user involvement causes inconvenience. For example, it may be tedious for a user to examine many printers and compare them. A balance between protocol selection and user selection is preferred. Context information is useful in selecting services [24, 45, 47–51, 53–55, 57–63, 16]. Providing users with better services is nice feature for service composition mechanisms. For better service matching, service requests may be directed to services with higher QoS [46, 51, 59–61].

- **Dynamicity Support**
  The majority of proposed solution to service composition problem assume static service composition. In this approach, if one of the services fails, service composition needs to start over again. Dynamic service composition approaches [44, 52, 59, 61, 62] support replanning of the composed service during the execution of the composition. Services can be replaced, added or removed if necessary without starting over the service composition processes. Dynamic service composition is more difficult to implement than static service integration since every service component of the dynamic service is being monitored and should be replaced immediately in case of failure.

5.4.3   Service Composition Approaches

As described above, the service composition system executes the selected composition plan and produces an implementation corresponding to the required composite service. In PvCEs the service composition system should find and choose the services taking part into the composition process and to choose contextually the most suitable ones if many are available.

Many works [47, 51, 53, 55, 58] for PvCE propose centralized solutions to service composition. This category of works depends on the existence of a centralized directory of services. Service requestors submit requirements to the directory and the directory makes a decision on the list of services that should be returned to the requestors. This approach works well for environments that are well managed and relatively stable. However, the assumption on centralized control becomes impractical in scenarios with dynamic arrivals and departures of service providers, which requires frequent updates of the central entities, resulting in large system overhead. Moreover, relying on central entities to maintain global knowledge leads to inability of the system to serve the request when the central manager is compromised.

In order to address deficiencies of centralized approaches, some works exploiting distributed approach have already been proposed. As the first step towards decentralization of service composition they introduce distributed directory of services. For example, in [72] a hierarchical directory has been proposed. The resource-poor devices depend on resource-rich devices to support service discovery and composition.

The next category of distributed service composition approaches removes the need for a service directory and provides a fully distributed search for needed services. In [73] a hierarchical task-graph based approach to do service composition in ad hoc environments has been proposed. A composite service is represented as a task-graph and sub trees of the graph are computed in a distributed manner. This approach assumes that service requestor is one of the services and relies on this node to coordinate service composition. The coordinator uses global search across the whole network to do composition. The same domain of the problem was studied in [52]. It is different from [73] in terms of the way of electing coordinator. For each composite request, a coordinator is selected from within a set of nodes. The service requestor delegates the responsibility of composition to the elected coordinator. The elected coordinator utilizes distributed service discovery architecture, and subsequently integrates and executes services needed for a composite request.

- **Utilizing Localized Interactions for Service Composition**
  We have proposed a fully decentralized approach to the service composition process [70]. Our work is based on the analysis of the requirements of the fully decentralized service composition: (i) there should be no special entity to manage the service composition process; (ii) service providers can communicate only with their local neighbors, not all other service providers, (iii) no service provider knows the full global information or gathers it.

  We assume that all service providers in the environment can access to a whiteboard showing the user-specified service description, which is a set of component functionalities together with composition relationships. Our problem is to construct the composite service in a distributed manner, through localized interactions between service providers. We transform this problem to subgraph isomorphism problem, which is proven to be NP-complete.

  Unlike previous algorithms that rely on global knowledge, in our algorithm each device only maintains local state information about its physical neighbors. As a result, service providers will build an overlay network graph which satisfies the requirement with the quality of service comparable to centralized approach. We propose an algorithm for the devices to cooperatively construct the requested services through localized interactions. Device candidates decide with whom to cooperate only based on the information of the devices within its physical neighborhood. Our idea is to grow sections of composed service from available services. First, service provider identifies what service type it needs to interlock with through his output and then searches for appropriate service provider with matching input type. It is possible that service provider forms a section with another section. By merging pieces with each other, eventually global solution emerges.

  For the purpose of reducing redundant broadcast we propose the service composition backbone (SCB) built in a fully localized way. SCB is an overlay communication infrastructure which the property that each node is in the SCB or 1-hop away from the SCB.

5.5   Conclusions

As discussed in this chapter, service management concerns several functionalities blended seamlessly to provide users with the finest experience of service access and use without any interruption. We have developed several protocols that benefit such a system. We are working to develop a distributed protocol which will support seamless service mobility for mobile users in a mobile ad hoc network through localized interactions of the participating entities.

Several issues are still worth investigating. One core challenge that remains more or less unexplored is how we can utilize localized algorithms for service management in PvCEs. Developing localized algorithms can be the key to address the challenges posed by the dynamic and ad hoc nature of pervasive systems. Pervasive devices must coordinate locally with peer devices to achieve some global objective with respect to service discovery, or composition. Also the fault tolerance issue has grossly been overlooked by the researchers. Pervasive computing is mainly application-specific and many of the applications are safety-critical, e.g. health-care or elderly-care applications. This type of systems certainly needs highest reliability support that can be provided. Another important issue is about maintaining security and privacy of users in widely heterogeneous pervasive environments. Services in pervasive computing systems can be provided by software modules that are available in markets as COTS. These software modules are not properly tested for reliable performance as well as not guaranteed to provide necessary security supports. So, more research works must be carried out in order to develop secured and fault tolerant software for using in pervasive applications.

References

[1] W. Adjie-Winoto, E. Schwartz, H. Balakrishnan, and J. Lilley (1999) 'The design and implementation of an intentional naming system,' In *Proceedings of the Seventeenth ACM Symposium on Operating Systems Principles (SOSP'99)*, ACM Press, pp. 186–201, Charleston, SC, December.

[2] K. Arabshian and H. Schulzrinne (2004) 'GloServ: Global service discovery architecture,' In *Proceedings of MobiQuitous*, pp. 319–25, June.

[3] M. Balazinska, H. Balakrishnan, and D. Karger (2002) 'INS/Twine: a scalable peer-to-peer architecture for intentional resource discovery,' In *Proceedings of the International Conference on Pervasive Computing*, August.

[4] Bluetooth SIG. Specification. http://bluetooth.com/.

[5] M. Caesar, M. Castro, *et al.* (2006) 'Virtual ring routing: network routing inspired by DHTs,' In *Proceedings of ACM SIGCOMM*, pp. 351–62.

[6] M. Castro, P. Druschel, A.-M. Kermarrec, and M. Rowstron (2002) 'One ring to rule them all: Service discovery and binding in structured peer-to-peer overlay networks,' In *Proceedings of the SIGOPS European Workshop*, Saint-Emilion, France, September.

[7] D. Chakraborty, A. Joshi, T. Finin, and Y. Yesha (2002) 'GSD: A novel group-based service discovery protocol for MANETs,' In *Proceedings of the Fourth IEEE Conference on Mobile and Wireless Communications Networks (MWCN)*.

[8] D. Chakraborty, A. Joshi, Y. Yesha, and T. Finin (2006) 'Toward distributed service discovery in pervasive computing environments,' *IEEE Transactions on Mobile Computing*, February.

[9] L. Cheng and I. Marsic (2000) 'Service discovery and invocation for mobile ad hoc networked appliances,' In *Proceedings of the Second International Workshop Networked Appliances (IWNA 00)*, December.

[10] S. Cheshire and M. Krochmal (2008) 'DNS-Based Service Discovery,' IETF Internet draft, September, http://files.dns-sd.org/draft-cheshire-dnsext-dns-sd.txt.

[11] T. Clausen and P. Jacquet, (2003) 'Optimized link state routing protocol (OLSR),' RFC 3626, October.

[12] F. Delmastro (2005) 'From Pastry to CrossROAD: Cross-layer ring overlay for ad hoc networks,' In *Proceedings of the 3rd IEEE International Conference on Pervasive Computing and Communication Workshops*, pp. 60–64, March.

[13] E. Guttman and C. Perkins (1999) 'Service location protocol,' version 2, June.

[14] S. Helal, N. Desai, V. Verma, and C. Lee (2003) 'Konark – a service discovery and delivery protocol for ad hoc networks,' In *Proceedings of the Third IEEE Conference on Wireless Communication Networks WCNC*, March.

[15] T.D. Hodes, S.E. Czerwinski, B.Y. Zhao, A.D. Joseph, and R.H. Katz (2002) 'An architecture for secure wide-area service discovery,' *ACM Wireless Networks Journal*, vol. **8**, nos 2/3, pp. 213–230.

[16] Jini Technology Core Platform Specification, v. 2.0, Sun Microsystems, June 2003; www.sun.com/software/jini/specs/core2_0.pdf.

[17] D. Johnson, D. Maltz, and J. Broch (2001) *DSR: The dynamic source routing protocol for multihop wireless ad hoc networks*, Chapter 5, pp. 139–72, Addison-Wesley.

[18] E. Kang, M.J. Kim, E. Lee, and U. Kim (2008) 'DHT-based mobile service discovery protocol for mobile ad hoc networks,' In *Proceedings of the Fourth International Conference on Intelligent Computing: Advanced Intelligent Computing Theories and Applications – with Aspects of Theoretical and Methodological Issues (ICIC '08)*, September.

[19] M.J. Kim, M. Kumara, and B.A. Shirazi (2006) 'Service discovery using volunteer nodes in heterogeneous pervasive computing environments' *Journal of Pervasive and Mobile Computing* **2**: 313–43.

[20] M. Klein, B. Konig-Ries, and P. Obreiter (2003) 'Service rings – a semantic overlay for service discovery in ad hoc networks,' In *DEXA Workshops*, pp. 180–5.

[21] M. Klein, B. Konig-Ries, and P. Obreiter (2003) 'Lanes – a light weight overlay for service discovery in mobile ad hoc networks,' *Technical Report 2003–6*, University of Karlsruhe, May.

[22] U.C. Kozat and L. Tassiulas (2004) 'Service discovery in mobile ad hoc networks: an overall perspective on architectural choices and network layer support issues,' *Ad Hoc Networks* **2**(1): 23–44.

[23] C. Lee and S. Helal (2003) 'A multi-tier ubiquitous service discovery protocol for mobile clients,' In *Proceedings of the International Symposium on Performance Evaluation of Computer and Telecommunication Systems (SPECTS'03)*, Montréal, Canada.

[24] M. Nidd (2001) 'Service discovery in DEAPspace,' *IEEE Personal Communications*, pp. 39–45.

[25] C.E. Perkins and E.M. Royer (1999) 'Ad-hoc on-demand distance vector routing,' In *Proceedings of the Second IEEE Workshop on Mobile Computer Systems and Applications*, pp. 90–100, IEEE Computer Society, February.

[26] H. Pucha, S. Das, and Y. Hu (2004) 'Ekta: An efficient DHT substrate for distributed applications in mobile ad hoc networks,' In *Proceedings of the Sixth IEEE Workshop on Mobile Computing Systems and Applications (WMCSA)*.

[27] O.V. Ratsimor, D. Chakraborty, A. Joshi, T. Finin (2002) 'Allia: alliance-based service discovery for ad hoc environments,' In *Proceedings of the ACM Workshop on Mobile Commerce (WMC'02)*, September.

[28] R. Robinson and J. Indulska (2003) 'Superstring: a scalable service discovery protocol for the wide area pervasive environment,' In *Proceedings of the Eleventh IEEE International Conference on Networks*, Sydney, September.

[29] A. Rowstron and P. Druschel (2001) 'Pastry: scalable, decentralized object location, and routing for large-scale peer-to-peer systems,' In *Proceedings of theIFIP/ACM International Conference on Distributed Systems Platforms (Middleware), Lecture Notes in Computer Science (LNCS)*, Vol. 2218, pp. 329–50, Heidelberg, Germany, November.

[30] F. Sailhan and V. Issarny (2005) 'Scalable Service Discovery for MANET,' In *Proceedings of IEEE PerCom'05*, pp. 235–46.

[31] The Salutation Consortium. Salutation architecture specification version 2.0c, June, 1999, available online at http://www. salutation.org/.

[32] I. Stoica, R. Morris, D. R. Karger, M.F. Kaashoek, and H. Balakrishnan (2001) 'Chord: A scalable peer-to-peer lookup service for internet applications,' In *Proceedings of the 2001 ACM SIGCOMM Conference*, pp. 149–60, August.

[33] V. Sundramoorthy, J. Scholten, P.G. Jansen, and P.H. Hartel (2003) 'Service discovery at home,' In *Proceedings of the Fourth International Conference on Information, Communications & Signal Processing and Fourth IEEE Pacific-Rim Conference on Multimedia (ICICS/PCM)*. IEEE Computer Society Press, Singapore. pp. 1929–33, December.

[34] B. Traversat, M. Abdelaziz, and E. Pouyoul (2003) 'Project JXTA: a loosely-consistent DHT rendezvous walker,' Sun Microsystems Inc. http://www.jxta.org/project/www/docs/jxta-dht.pdf, May.

[35] J. Tyan and Q.H Mahmoud (2005) 'A Comprehensive Service Discovery Solution for Mobile Ad Hoc Networks', In *ACM/Kluwer Journal of Mobile Networks and Applications (MONET)* **10**(8): 423–34, August.

[36] UPnP Device Architecture 1.0, UPnP Forum, December, 2003, www.upnp.org/resources/documents/CleanUPnPDA 10120031202s.pdf.

[37] A. Varshavsky, B. Reid, and E. de Lara (2005) 'A cross layer approach to service discovery and selection in Manets,' In *Proceedings of the Second International Conference on Mobile Ad-Hoc and Sensor Systems (MASS'05)*, IEEE Press, Washington DC, USA, November.

[38] J. Wu and M. Zitterbart (2001) 'Service awareness in mobile ad hoc networks,' *Paper Digest of the 11th IEEE Workshop on Local and Metropolitan Area Networks (LANMAN)*, Boulder, Colorado, USA, March,.

[39] H.J. Yoon, E.J. Lee, H. Jeong, and J.S. Kim (2004) 'Proximity-based overlay routing for service discovery in mobile ad hoc networks,' In *Proceedings of the Nineteenth International Symposium on Computer and Information Sciences (ISCIS)*.

[40] T. Zahn and J. Schiller (2005) 'MADPastry: a DHT substrate for practicably sized MANETs,' In *Proceedings of the Fifth Workshop on Applications and Services in Wireless Networks (ASWN)*, June.

[41] F. Zhu, M. Mutka, and L. Ni (2003) 'Splendor: a secure, private and location-aware service discovery protocol supporting mobile services,' In *Proceedings of the First International Conference on Pervasive Computing and Communication PerCom'03*, pp. 235–42.

[42] V. Raychoudhury, J. Cao, W. Wu (2008) 'Top K-leader election in wireless ad hoc networks,' *Proc. of the 17th International Conference on Computer Communication Networks (ICCCN08)*, St. Thomas, US Virgin Islands, August 4–7.

[43] V. Raychoudhury, J. Cao, W. Wu, and S. Lai (2010) 'K-directory community: reliable service discovery in MANET,' In *Proceedings of the 11th International Conference on Distributed Computing and Networking (ICDCN2010)*, January 3–6, Kolkata, India.

[44] K. Carey, D. Lewis, S. Higel, and V. Wade (2004) 'Adaptive composite service plans for ubiquitous computing,' In *2nd International Workshop on Managing Ubiquitous Communications and Services (MUCS 2004)*, December.

[45] C. Hesselman, A. Tokmakoff, P. Pawar, and S. Iacob (2006) 'Discovery and composition of services for context-aware aystems,' *1st European Conference on Smart Sensing and Context (EuroSCC'06)*, Enschede, The Netherlands

[46] A. Mingkhwan, P. Fergus, O. Abuelma'Atti, M. Merabti, B. Askwith, and M.B. Hanneghan (2006) 'Dynamic service composition in home appliance networks. *Multimedia Tools and Applications*' **29**(3): 257–84.

[47] H. Pourreza and P. Graham (2006) 'On the fly service composition for local interaction environments,' In *IEEE International Conference on Pervasive Computing and Communications Workshops*, p. 393. IEEE Computer Society.

[48] A. Qasem, J. Heflin and H. Munoz-Avila (2004) 'Efficient source discovery and service composition for ubiquitous computing environments," In *Workshop on Semantic Web Technology for Mobile and Ubiquitous Applications, ISWC.*

[49] S.Y. Lee, J.Y. Lee, and B.I. Lee (2006) 'Service composition techniques using data mining for ubiquitous computing environments,' *International Journal of Computer Science and Network Security* **6**(9): 110–17.

[50] M. Sheshagiri, N. M. Sadeh, and F. Gandon (2004) 'Using semantic web services for context-aware mobile applications,' Second International Conference on Mobile Systems (MobiSys 2004), Applications, and Services – Workshop on Context Awareness.

[51] S.B. Mokhtar, N. Georgantas, and V. Issarny (2006) 'Cocoa: Conversation-based service composition in pervasive computing environments,' In *Proceedings of the IEEE International Conference on Pervasive Services*.

[52] D. Chakraborty, A. Joshi, T. Finin, and Y. Yesha (2005) 'Service composition for mobile environments,' *Journal on Mobile Networking and Applications*, Special Issue on Mobile Services, **10**(4): 435–51, January.

[53] Z. Song, Y. Labrou, and R. Masuoka (2004) 'Dynamic service discovery and management in task computing," In *Proceedings of the 1st Ann. Int'l Conf. Mobile and Ubiquitous Systems*, IEEE Press, pp. 310–18.

[54] Q. Ni (2005) 'Service composition in ontology enabled service oriented architecture for pervasive computing,' In *Workshop on Ubiquitous Computing and e-Research.*

[55] M. Vallee, F. Ramparany, and L. Vercouter (2005) 'Flexible composition of smart device services,' In *International Conference on Pervasive Systems and Computing (PSC05)*, pp. 165–71. CSREA Press.

[56] D.N. Kalofonos and F.D. Reynolds (2006) 'Task-driven end-user programming of smart spaces using mobile devices,' *Technical Report NRC-TR-2006–001*, Nokia.

[57] T. Cottenier and T. Elrad (2005) 'Adaptive embedded services for pervasive computing,' In *Workshop on Building Software for Pervasive Computing – ACM SIGPLAN conf. on Object-Oriented Programming, Systems, Languages, and Applications.*

[58] A. Bottaro, J. Bourcier, C. Escoer, and P. Lalanda (2007) 'Autonomic context-aware service composition,' *2nd IEEE International Conference on Pervasive Services*.

[59] W.L.C. Lee. S. Ko, S. Lee and A. Helal (2007) 'Context-aware service composition for mobile network environments,' In *4th International Conference on Ubiquitous Intelligence and Computing (UIC2007)*.

[60] G. Kaefer, R. Schmid, G. Prochart, and R. Weiss (2006) 'Framework for dynamic resource-constrained service composition for mobile ad hoc networks,' *UBICOMP, Workshop on System Support for Ubiquitous Computing*.

[61] U. Bellur, N.C. Narendra, I.I.T. Kresit, and I. Mumbai (2005) 'Towards service orientation in pervasive computing systems,' *International Conference on Information Technology: Coding and Computing* **2**: 289–95.

[62] A. Ranganathan and R.H. Campbell (2004) 'Autonomic pervasive computing based on planning,' *International Conference on Autonomic Computing*, pp. 80–7.

[63] A. Ranganathan and S. McFaddin (2004) 'Using workflows to coordinate web services in pervasive computing environments,' In *Proceedings of the IEEE International Conference on Web Services*, pp. 288–95.

[64] T. Koponen, T. Virtanen (2004) 'A service discovery: a service broker approach,' In *Proceedings of the 37th Annual Hawaii International Conference on System Sciences (HICSS 2004)* – Track 9, January 05–08, 2004, vol. 9, IEEE Computer Society, Washington, DC.

[65] P. Dobrev, D. Famolari, C. Kurzke, B.A. Miller (2002) 'Device and service discovery in home networks with OSGi,' *IEEE Communications Magazine* **40**(8): 86–92.

[66] P. Grace, G.S. Blair, S. Samuel (2005) 'A reflective framework for discovery and interaction in heterogeneous mobile environments,' *SIGMOBILE Mob. Comput. Commun. Rev.* **9**(1).

[67] Y.D. Bromberg, V. Issarny (2005) 'INDISS: Interoperable Discovery System for Networked Services,' In: Alonso, G. (ed.) *Middleware 2005*. LNCS, vol. 3790, Springer, Heidelberg.

[68] P.G. Raverdy, V. Issarny, R. Chibout, A. de La Chapelle (2006) 'A multi-protocol approach to service discovery and access in pervasive environments,' In *Proceedings of MOBIQUITOUS – The 3rd Annual International Conference on Mobile and Ubiquitous Systems: Networks and Services*, San Jose, CA, USA (July).

[69] Kang, S., Ryu, S., Kim, N., Lee, Y., Lee, D., Moon, K. (2005) 'An architecture for interoperability of service discovery protocols using dynamic service proxies,' In Kim, C. (ed.) *ICOIN 2005*. LNCS, vol. 3391, pp. 786–795. Springer, Heidelberg.

[70] J. Siebert, J.N. Cao, L. Cheng, E. Wei, C. Chen, J. Ma (2010) 'Decentralized service composition in pervasive computing environments,' In *International Wireless Communications and Mobile Computing Conference* (IWCMC 2010).

[71] J. Siebert, J.N. Cao, Y. Zhou, M.M. Wang, V. Raychoudhury (2007) 'A novel approach to supporting multiprotocol service discovery in pervasive computing,' *The 2007 IFIP International Conference on Embedded and Ubiquitous Computing* (EUC 2007).

[72] S. Kalasapur, M. Kumar, B. Z. Shirazi (2007) 'Dynamic service composition in pervasive computing,' In *IEEE Transaction on Parallel and Distributed Systems*.

[73] T. D. C. Little, B. Prithwish, K. Wang (2002) 'A novel approach for execution of distributed tasks on mobile ad hoc networks,' In *IEEE WCNC*. Orlando. Florida.

[74] M. Satyanarayanan (2001) 'Pervasive computing vision and challenges,' *IEEE Personal Comm.* **6**(8): Aug., pp. 10–17.

6

Wireless Sensor Cooperation for a Sustainable Quality of Information[1]

Abdelmajid Khelil,[1] Christian Reinl,[2] Brahim Ayari,[1] Faisal Karim Shaikh,[1] Piotr Szczytowski,[1] Azad Ali,[1] and Neeraj Suri[1]
[1]*DEEDS group of the Department of Computer Science and Engineering, Technische Universität Darmstadt, Hochschulstr, 10, 64289 Darmstadt, Germany.*
[2]*SIM group of the Department of Computer Science and Engineering, Technische Universität Darmstadt, Hochschulstr, 10, 64289 Darmstadt, Germany.*

6.1 Introduction

Increasingly, the notion of a widely inter-connected, adaptive and dynamic ubiquitous computing environment is being proposed for virtually all application domains. Consequently, the underlying Wireless Sensor Networks (WSN) represent a key enabling technique for the emerging ambient/ubiquitous/pervasive computing. The drivers facilitating its actual utility are often touted as the self-*attributes such as self-organizing, self-repairing, etc. These result from the spontaneous cooperation between radio-enabled sensors. This sensor cooperation also provides for sustained on-demand functionality in the presence of both network and sensor level changes and perturbations. The key service provided by a WSN is to characterize the physical world as required by the user(s). Hereby, the main interest of users is to receive specified information (events, real world snapshots) with a certain desired quality level that may include precision, freshness of sampled data or data coverage range among other relevant data quality attributes – collectively termed as the Quality of Information (QoI). Accordingly, the main operations of a WSN are (1) to create a minimal suitable sensor data stream, (2) to extract useful information from these data streams, and (3) to deliver the extracted information to the users. The main design objectives are to sustain the desired Quality of Information (QoI).

Unfortunately, the world of static battery-powered WSN, by definition, does not provide for a long-lived system providing the required full functionality. Hence, the 2-tier architecture has been proposed, where mobile assist sensor nodes and static sensor nodes cooperate side by side. This mobile sensor cooperation is driven by mobility that can be easily injected to the WSN and provided by cheap commercial-off-the shelf platforms [1]. This cooperation is mainly twofold: (1) the mobile assist nodes move so as to enhance functionality, i.e. to sustain QoI by improving the sensing coverage, increasing the sampling rate in certain regions, etc., or (2) the mobile assist nodes move for dependability, mainly to enhance the network lifetime, e.g. to reconnect a partitioned WSN or to delay coverage drops in energy holes. One of the main research trends in the scope of mobile sensor cooperation is QoI-aware delay-tolerant information transport [1, 3, 4]. Here, mobile assist nodes exploit the long time-relevance

[1] This work was supported in part by DFG GRK 1362 (TUD GKMM), EC INSPIRE, EC CoMiFin, LOEWE CASED, MUET and HEC.

of some data to save network resources while maintaining the required QoI. In [5], we have presented gMAP, an extremely efficient mobility-assisted approach to collecting sensor data. gMAP opportunistically exploits node mobility to collect data of interest, keeping sensor nodes transmitting only their own readings on-demand to a mobile node in their transmission area. In [6] we have extended the gMAP approach [5] by a path planning algorithm for one single mobile assist node.

For a clear coordination schema facilitating and enhancing the vital self* properties, QoI-aware cooperation across mobile entities is needed. Mobile assist nodes need to plan their paths cooperatively for optimized data collection, network maintenance, etc. In this chapter, we extend our single node path planning algorithm [6] to support multiple mobile nodes. In order to allow for a traceable coordination of autonomous mobile elements, it is necessary to record coordination data. Furthermore, coordination operates on distributed data and assumes a distributed/autonomous decision/agreement. Transaction processing allows for consistent data and agreed coordination, rendering them suitable for cooperation across mobile assist nodes [7] along with sensor cooperation.

The result of efficient cooperation across mobile entities and sensor nodes surrounding them is a long-lived WSN delivering the desired information to users at the required QoI level. WSNs are usually mission-oriented and are tailored to specified information and also specified users. In order to provide for actionable knowledge for varied users the information obtained from the different WSNs should be composed. Consequently, inter-WSN cooperation is needed. The sensor map [8] or sensor web enablement [9] approaches are two contemporary examples. Overall, the goal is to integrate information collected from varied WSNs to information infrastructures (e.g. WWW). Hereby, the integration of information into visual maps is an intuitive way to present such spatial-temporal information. We utilize such a map approach and present an integrated framework for map-based information retrieval and presentation in WSN [10].

In this chapter, we describe techniques for intra-WSN sensor and mobile sensor cooperation, as well as techniques for coordination across mobile entities and WSNs. Our contribution is twofold: (1) we survey the existing approaches and (2) present cooperation approaches based on our ongoing research [6, 5, 7, 10]. It is noteworthy that we assume cooperative WSN entities and do not consider non-cooperative behaviors such as selfishness and deliberate attacks. We also extend our system model progressively and cover the broad spectrum of popular WSN system classes.

The remainder of this chapter is structured as follows. Section 6.2 illustrates how a WSN instruments the real world for users and provides for core primitives within the ubiquitous and pervasive computing paradigm. In Section 6.3, we investigate the inter-sensor cooperation for provisioning the desired QoI in static networks. Section 6.4 considers the deployment of mobile assist nodes that cooperate with the multiple static sensor nodes in order to enhance the functionality and dependability of WSN systems. In Section 6.5, we present work that shows that additional cooperation between the mobile assist nodes improves WSN performance while allowing for novel transaction-based functionalities. Cooperation between different WSN systems is discussed in Section 6.6. We conclude by presenting relevant future research directions in Section 6.7.

6.2   Sensing the Real World

Typically, a user is interested in observing and possibly controlling a certain physical phenomenon or generally the physical world through a WSN that represents a dashboard level abstraction of the attributes of that physical world. The WSN delivers the required high-level user information with the specified QoI [15]. Accordingly, a WSN has to create an appropriate model of the physical world of interest, collect the required raw data, synthesize this data and provide the required information to the user. Similarly, an administrator requires a dashboard for the 'network world', e.g. to show where the energy distribution is low. We refer to both physical and network worlds as the 'world'. We also denote by the 'user' all users of both world models. Therefore, we model a WSN as a (physical/network) World Model. The user observes the world through this world model.

Usually, the user (represented by the WSN sink) and the WSN agree on one or more (information) models that should be kept consistent during the deployment. Common examples of these models include the ambient temperature map and the notification of a certain event. Upon deployment, these models are initialized. The WSN should report changes in the agreed model. The user can query the implemented model or trigger a model replacement if necessary. An adaptive negotiable model is desirable to allow for evolvable WSN systems. The model update should be incremental in order to minimize the consequent communication overhead. Prediction models target predictability of desired data/event trends while carefully reducing the number of communication messages.

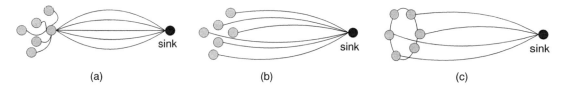

Figure 6.1   Information categorization. (a) Atomic information. (b) Replicated atomic information. (c) Composite information.

The information exists in the physical world and the WSN monitors it and delivers it to the digital world, where the user can access it. Being captured in digitalized form, the information can then be used reproducibly, analyzed and disseminated on demand. Subsequently, the WSN is considered as a tool for observing and controlling the states of the real physical world [28] or as a bridge to the physical world [29]. Thus, WSNs deliver basic primitives for the ambient, ubiquitous or pervasive computing world.

Each sensor node measures the physical signals of interest at a given sampling rate (with a certain noise level). The sampling rate should be tuned as a function of physical world dynamics (which impacts the dynamics of world model data) and the QoI required by the user. Subsequently, a sensor node produces a time series of the signal (temporal sampling of the world). Spatial sampling of the world is completed by the set of sensor nodes. Both temporal and spatial samplings play an important role in the accuracy and consistency of the world model, and specifically the achievable QoI. The physical world can be controlled through deployed actuators and the network world through network maintenance or reconfiguration.

We refer to an information entity either as the raw sensor data or its transformation as required by the application/user. Usually, a transformation results from applying data processing techniques on raw data (e.g. aggregation, filtering and compression). Information entities can be generated centrally on a single node (e.g. a designated node in a sensor cluster termed as a cluster head) or in a distributed manner by some nodes (e.g. a few cluster heads). In the latter case we say the information entity is replicated. The information entities can further be grouped/composed for a higher semantic such as grouping the location of the nodes that detected the same event and define a new information, i.e. the event/region perimeter. Accordingly, we classify the information required by the applications into two broader classes (Figure 6.1): *Atomic information and Composite information*. Atomic information is composed of a single information entity whereas composite information is composed of more than one information entity. When the atomic information is generated at many sensor nodes we refer to it as replicated atomic information. For example, atomic information can be generated by a sensor node, e.g. after aggregation or by many sensor nodes, e.g. in case of the fire. On the other hand, composite information is divided into a set of information from different sensor nodes (no redundancy). Composite information can be viewed as a set of atomic information (spatial composition) and if it can be aggregated centrally (temporal composition), then it is transformed into a single atomic information.

From the literature, three main system-level design paradigms arise, namely, considering a WSN either as a network, database or an event service. The categorization is the following:

(i) *WSN as a network*: WSN can be viewed as a self-organized communication platform. The standard WSN communication architecture is layered in a similar way to the OSI layer model. However, a modification of this architecture called a cross-layer model is commonly adopted. The cross-layer design is an envelope of optimizations that benefit from the cooperation of non-adjacent layers [16]. Furthermore, new communication patterns such as data-centric communication have been proposed [17].

(ii) *WSN as a database*: Major research issues are in designing efficient query dissemination and in-network selection, projection, join and aggregation. An example is tinyDB [18] that treats sensor data as a single table (sensors) with one column per sensor type. Research focuses here on data-centric communication [19–23].

(iii) *WSN as an event service*: A WSN is usually deployed for missions providing services such as tracking targets or detecting events. The mission determines the overall operation of the WSN including data generation, processing, filtering and transport. To support multiple missions and augment the service flexibility, the publish/subscribe (pub/sub) service architecture has been advocated for WSNs [24–27].

These design paradigms are the main existing approaches to realizing the world model and its update or query. In order to provide for more comprehensible abstractions, we provide in [10] a Map-based World Model (MWM). In Sections 6.3–6.5, we focus mainly on the network and event service view on the WSN. We highlight how cooperation across static and/or mobile nodes is a perquisite for sensing the real-world by a QoI-aware information extraction and delivery. In Section 6.6, we focus on the map-based view on the WSN and show that this approach allows for inter-WSN cooperation.

6.3 Inter-Sensor Cooperation

With diversity as a hallmark, WSNs often comprise computing nodes with similar communication, sensing, processing and storage capabilities. WSNs can be embedded in varied environments with the desired goals of sensing, monitoring and predicting phenomena of interest in the physical world. In this section, we consider the established static WSN model. The main functionality of the WSN is implemented by a large number of stationary resource-limited sensor nodes that are deployed following either an arbitrary or structured spatial distribution in the area of interest. Also, one dedicated stationary sink is selected as the interface to the user. We use a CSMA/CA based MAC layer, where communication links are symmetric and bidirectional, and collisions may occur.

Self-organization is a key aspect of WSN as it allows for high scalability as well as adaptability to changing conditions. Without inter-sensor cooperation this key feature is hard if not impossible to realize. Therefore, we will briefly review existing works that allow for self* properties through sensor to sensor node cooperation. Next, we briefly survey how the inter-sensor copperation maintains the desired QoI. Subsequently, we present our work on QoI-aware information transport.

Following most of the existing work, our WSN perspective in this section is mainly network driven. The cooperation between sensor nodes should be driven by the main objective of observing the world and extracting and delivering relevant information to the user/sink. Therefore, nodes should cooperate to create the desired information as close as possible to their sources. In addition, sensor nodes should cooperate to maintain a suitable path to the sink to transport required information to the user and receive commands for control, maintenance, etc. from the sink.

6.3.1 Sensor Cooperation for Self-Organization

The basis of cooperation in WSN is to provide for the popular self-organization property. This begins with deploying the sink and the sensor nodes in the desired field. The deployment may range from manual to unstructured by randomly dispersing the sensor nodes from the air. The deployment conditions do not only vary from application to application but also evolve with the time for the same WSN. The research challenge over years has been to design techniques that allow the deployed autonomous WSN elements to self-organize into a meaningful network that provides for basic communication primitives such as routing, dissemination, flooding, etc. The limited capabilities of sensor nodes with respect to communication, processing and especially energy, renders the task particularly crucial. Most of developed techniques rely on nodes' local cooperation with the nodes in their communication range. The local cooperation ranges from sharing proprietary information (e.g. node ID, node position and node energy level) to participating in executing distributed algorithms (e.g. forwarding the traffic of neighbours and aggregation of sensor values received from neighbors).

Sharing of proprietary data allows one to aggregate global information such as the list of neighbours. Similarly, sharing local data leads to the generation of more global, e.g. regional information (e.g. two hop neighborhood, minimal connected dominating sets and map isolines). There is always a trade-off between the cooperation range (geographic spread) and the cooperation data update efficiency (communication and energy overhead). This tradeoff should be balanced depending on the application requirement on data consistency and the data change frequency. Generally, the trend is to allow localized inter-sensor cooperation, i.e. between nodes in each others communication range. The acquisition of localized knowledge is the prerequisite for self-organization.

The autonomous sensor nodes and the sink should self-organize to form a meaningful network that provides for the required network functionalities (e.g. routing, convergecast) with the desired dependability. To this end, cooperation to execute distributed algorithms on different nodes is essential. Usually, the sink starts with a discovery phase to establish routing paths to and from all sensor nodes. Here nodes cooperate to realize the flooding of the route request message. After the establishment of routes, the relay nodes support the flow of information from their sources to the sink. Without the cooperation of all nodes along the path, information may not reach the user. In order to maintain the desired functionality and dependability beyond operational and design failures (node failures and link disruptions, etc.), nodes should observe their proprieties and shared information and if needed reshape their behavior with respect to distributed algorithms.

We refer to [30] for an excellent classification of well-known protocols allowing for self-organization in ad hoc and sensor networks, primarily, on the medium access control and network layers. In Subsection 6.3.2, we investigate QoI in more detail and briefly review how cooperation impacts the reachable QoI level during the information lifetime cycle. In Subsection 6.3.3, we will focus on dependable information transport and present our own contributions to the state-of-the-art.

6.3.2 Cooperation for Quality of Information Provisioning

The main functionality of a WSN consists of (1) the sampling of the physical phenomenon, (2) the in-network processing of raw data to generate user information and (3) the actual information transport to the user. These fundamental sequential functional building blocks present the lifetime cycle of the information. Accordingly, these blocks play a major role in determining the QoI when the information reaches the user.

The notion of QoI is not new as it was already advocated in the context of databases and web searches. However, QoI in WSN presents new challenges such as sensor resolution, communication unreliability and in-network processing inaccuracy. QoI extends existing metrics such Quality of Service (QoS, i.e. end-to-end latency, packet loss rate), which is commonly considered in standard communication networks.

As mentioned before, the WSN digitalizes the information of interest from the physical world and makes it available to the users. Usually, the quality of information suffers from this 'digitalization' resulting from data sampling losses. However, users do require a certain level of quality relevant for the application being considered. The main goal of sensor node cooperation is to generate sufficient amount of raw data that allows extracting correct information from the physical world. The required operations on raw data in order to create and deliver the information to the user should be carefully accomplished in order to maintain a QoI level acceptable by the user.

In the literature, there exist several examples that focus on separate functional blocks. In addition, we identify a few recent projects that combine blocks and target computation of the QoI [33–35]. Obviously, considering all blocks is crucial but will allow for a holistic approach, where the QoI provisioning is well shared between all blocks.

The spatial and temporal sampling of the physical world is achieved by the sensor nodes' spread in the area of the physical world. This sampling is usually not perfect due to the limited number of sensor nodes and due to the limited resources on these nodes. Therefore, the quality of raw data does not reflect perfectly the information in the real world, but is only an approximation of it. The cooperation of all sensor nodes here is mandatory as missing samples will decrease the achievable QoI. Usually, the quality of sampling depends on the nature of the physical signal and on the performance of the sampling nodes. Consequently, there exist some approaches that target adaptive sampling to maximize QoI [31, 32]. The drop of QoI caused by sampling should be considered in the subsequent functional blocks.

Given the limited resources of sensor nodes and their wireless links, in-network processing of raw data and the formation of the information is a standard design paradigm in WSN. In-network processing allows for an adaptive balance between computation and communication. The popular in-network processing operations are data aggregation, filtering and compression.

Sensor nodes cooperate spontaneously or on-demand in order to share sensor data and compute locally the desired aggregates and take decisions concerning event detection or query result. Most of aggregation techniques in WSN operate on less accurate sensor data and may introduce more uncertainties in the aggregation result due to perturbations that lead to incomplete data. Usually, sensor nodes coordinate in distributed way to fix a subset of nodes that aggregate the raw data.

Only a subset of nodes is needed to increase efficiency. This is a kind of filtering, which may decrease the reachable QoI level at the in-network-processing and information extraction stage. In [11–13], the authors investigated the performance of data dissemination depending on sensor cooperation for data compression. They considered dense static WSNs, where sensor readings show high spatial correlation. Varied cooperation levels were studied including the extreme cases of no cooperation and network-wide cooperation. Their main result is that localizing the cooperation within certain regions of the network allows for the best trade-off between the generated traffic and latency.

The consequent functional block to information extraction is the information transport towards the sink, which is the last block. In Section 6.3.3, we will detail the information transport under QoI requirements.

6.3.3   Sensor Node Cooperation for Information Transport

The basis behind assured QoI requires responsive information transport. Responsiveness in these environments corresponds to the reliability and timeliness of the information, required by the user. The users require different types of information from a WSN along with a set of requirements from them. In order to provide QoI-aware information transport in WSN, the sensor nodes have to cooperate with each other to fulfill the global task of achieving user requirements. The application requirements impose consequent responsiveness requirements for the base information transport in a WSN. Being an ad-hoc and volatile environment, the WSN is obviously subject to a wide range of operational perturbations that lead to deviation between the attained and desired QoI complicating the design of information transport in WSN.

The general information transport process involves the flow of the *raw data* from source nodes towards the sink. However, the utility of a WSN based application arises from delivering responsive information, necessitating the cooperation of sensor nodes. One possible solution is to involve all nodes in transporting the information, i.e. flooding. Such cooperation where all nodes forward information is not very practical for WSN, since it is expensive (in terms of energy and communication bandwidth) and can lead to more perturbations such as collisions, contention and power depletion and subsequently, timeliness cannot be assured. The problem becomes of how to enable cooperation among the sensor nodes to adapt to both the evolving application requirements and network properties. It is not only the forwarding of sensor information, but also the assurance of the reliability and timeliness of the information that is a necessary feature of the cooperation within WSNs. Cooperation between sensor nodes should increase the network robustness and lifetime, and such a mechanism is to be designed. The highly dynamic and possibly unstructured characteristics of the environment itself introduce additional complexity; e.g. if we consider that disconnection is the norm rather than the exception, the information transport can only be loosely coupled. However, despite this loose coupling, the forwarding of information to neighboring sensor nodes or a sink needs to be responsive to satisfy the application requirements.

There has been intensive research to design suitable transport protocols for WSN. To achieve QoI, existing information transport protocols exploit temporal cooperation (e.g. retransmissions [36], erasure codes [37]), spatial cooperation (e.g. number of sources [38] or paths [39,41]) or some combination of them [40]. These mitigate perturbations to some extent, however, they are not able to cope with evolving network conditions as they are designed for specific classes of applications. Our comparative study [42] has confirmed that the current approaches perform well only for carefully selected deployment scenarios. Furthermore, they are not designed to consider explicitly the variable application requirements as their main design driver is to maximize efficiently the attained responsiveness. However, it is obvious that a WSN runs different applications (network management, event detection, event perimeter tracking, etc.) that require different information types with varied QoI requirements. It is also expected that the same WSN application may tune its requirements over time. The existing approaches over-utilize precious network resources even when the application does not require higher responsiveness. Therefore, a tunable information transport responsiveness is required which fulfills the application requirements and also adapts according to network conditions.

6.3.3.1   QoI Awareness for Information Transport

The challenging environment of WSN can be envisioned as a two dimensional problem space for QoI aware information transport, i.e. reliability and timeliness. Maximizing efficiency is the primary goal of WSN due to limited energy resources. Therefore, we have to reduce the message overhead as much as possible. Thus, WSN should transport critical information with high reliability potentially using more transmissions, while transporting less important information at a lower reliability using fewer transmissions. Furthermore, composite information has fine-grained requirements on reliability, i.e. depending on the shape and size of phenomenon the application requirements are changing and may require all or a subset of nodes to report the target's location. This necessitate to provide $x\%$ (probabilistically-guaranteed) QoI-aware information transport instead of best effort or always transporting information. Based on the application requirements we define atomic information transport reliability as the degree of tolerating the information loss over time. For composite information it is defined as how much loss of information entities can be tolerated by the application without losing the meaning of the composite information. Therefore, we transform the application requirement of $x\%$ reliability to provide statistically $(100 - x)\%$ successful transport of atomic information over time. To categorize the reliability of composite information we introduce a k-of-m reliability model, where m is the total amount of information entities required and k is desired amount of information entities to be transported by the WSN. We express $x\%$ and k-of-m by a probability with which the WSN transports information entity towards the sink.

Some of the real-time applications include industrial monitoring, oil/gas/water pipeline monitoring and border surveillance. In such applications missing messages or messages not received by a specified time limit implies unusable or erroneous information. Accordingly, there is a great need for sensor nodes to cooperate with each other and assure information transport in a reliable and timely manner. To achieve the required timeliness in such application scenarios, we follow primarily the best effort approach. In future, we aim at providing tunable timeliness.

6.3.3.2   Classification of Cooperation Schemes for Information Transport

The main objective of information transport is to achieve efficiently the desired QoI level. To this end, nodes have to cooperate despite the operational and design perturbations. The cooperation should result in a QoI level close to the desired level and avoid over or under provisioning. Such cooperation prohibits that the precious WSN resources are over utilized.

Figure 6.2 depicts the horizontal and vertical dimensions for the cooperation of sensor nodes inside WSN. Reading the figure from the left, the required cooperation is increased and requires more state information to be available. Also shown in this figure is a mapping of the basic cooperation mechanisms required for responsive information transport. Solutions developed for WSN need to avoid global state information in order to increase the scalability of the cooperation.

Node Level Cooperation: At node level local state is always available and can be utilized for information transport. For example, if the acknowledgment of the message is not received and the retransmission timer expires, a node can locally conclude the loss of the message and can retransmit the message. Also, sensor nodes can monitor the congestion locally and cooperate to store or discard the data. For example a sensor node monitors its neighbor's congestion and link losses and cooperates accordingly with the neighbor nodes, i.e. adaptive retransmissions.

Neighborhood Level Cooperation: As we increase the cooperation from node level to neighbor level the state is increased to keep track of neighbors. In this case, only neighborhood information is available to perform the necessary decisions for information transport. Usually, HELLO messages are exchanged at regular time periods. This keeps the neighborhood information up-to-date and allows the exchange of performance measures such as the current buffer status and link reliability across the neighbors.

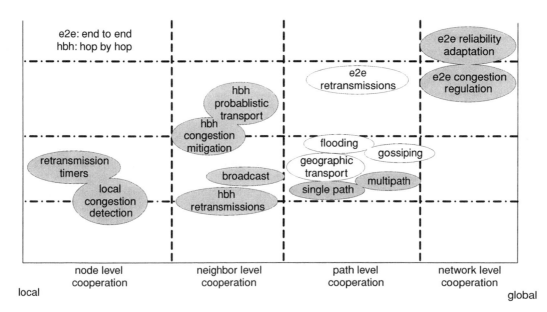

Figure 6.2   Cooperation schemes for information transport.

Path and Network Level Cooperation: Both path and network level cooperation requires partial or global state to be available, for example, to create paths in advance, to identify the network wide congestion or to have overall path responsiveness. In traditional networks this may be feasible but in WSNs to maintain the global state is not viable due to the resource constraint nature of sensor nodes. In general, node and neighborhood level cooperation is exploited for QoI aware transport.

6.3.3.3 Mechanisms and Strategies Enabling Cooperation

To enable cooperation among the sensor nodes inside the WSN different methods and approaches are utilized at different levels of cooperation as depicted in Figure 6.2. For example a sensor node monitors its neighbor's congestion and link losses.

Message Loss Detection Strategies: Retransmissions are required to overcome message loss, i.e. if the message does not reach the destination (usually the sink). To enable retransmissions it is necessary to detect the message loss by observing the local state, using neighbor cooperation [40] or network wide cooperation [38]. Several *message loss detection (MLD)* techniques can be adopted by transport protocols such as Acknowledgment (ACK), Negative ACK (NACK), Implicit ACK (IACK), Selective ACK (SACK), Selective NACK (SNACK) and timers. In comparison to wired networks, where only the source caches messages and retransmissions are done end-to-end, in WSN hop-by-hop retransmissions are more feasible [43]. If only the source caches and retransmits, the retransmission strategy is termed *end to-end cooperation*. If the intermediate nodes also cache and retransmit, the strategy is termed *hop-by-hop cooperation*. This poses the problem of where to cache the packet on the way from sources to the sink, either all intermediate nodes on the path or a subset of them should cache. Recently, we developed an adaptive retransmission strategy using hybrid ACK and local retransmission timers providing reliable information transport [44]. In [45], we utilize spatial cooperation to tune the maximum number of retransmissions and provide reliable information transport.

Congestion Control Mechanisms: These comprise schemes to detect congestion, and alternatively to avoid or mitigate it. In the WSN literature, we identify congestion control using node level cooperation [44], neighbour level cooperation [48] and network level cooperation [47]. The first approach is to monitor the channel utilization, e.g. through observing the collision rate [46]. The second approach is to monitor the buffer utilization, e.g. by observing the buffer length [38] or the average message queuing time. Upon congestion detection, nodes trigger congestion notification by disseminating the appropriate information to relay nodes and sources. Source nodes realize congestion avoidance by adjusting their data rate dynamically. The common approach for the adjustment is Additive-Increase and Multiplicative-Decrease [38, 46]. Some approaches propose conducting the adjustment in a discriminative manner depending on the fidelity of the source. Recently, we presented an on demand split path strategy [44] to mitigate the congestion by enabling cooperation at node and neighbor level. This technique is shown to be more robust and to provide tunable reliability.

6.3.3.4 Cooperation for QoI Aware Reliable Information Transport

In order to increase reliability different ACK schemes are widely adapted. [50] uses neighbor cooperation and sends the sequence of packets to the next hop with explicit acknowledgement (EACK) to ensure reliability. Reliable Multi-Segment Transport (RMST) [36] and Asymmetric Reliable Transport (ART) [51] utilize timer driven retransmissions for loss detection and notification and use path level cooperation for ensuring reliability. Distributed Transport for Sensor Networks (DTSN) [52] and Sensor Transmission Control Protocol (STCP) [53] provide differentiated reliability using end-to-end retransmissions. DTSN uses forward error codes (FEC) beside retransmissions to enhance reliability. These works rely on network level cooperation. For high information rates, Reliable Bursty Convergecast (RBC) [40] provides a reliability design based on a windowless block ACK and IACK along with fixed number of retransmissions. RBC lacks tunability and always provides high reliability. In RBC nodes cooperate with each other to provide responsiveness. Another approach to increase the responsiveness is to utilize multiple paths. In [41] MMSPEED protocol is proposed for probabilistic QoS guarantee in WSNs supported by multipath forwarding. MMSPEED comprises two modules, i.e. reliability and timeliness. The messages are forwarded according to their reliability cum timeliness requirements. MMSPEED exploit neighbor cooperation for transporting the information. Majority of approaches in multipath [54, 55] utilize multiple paths from the source node to the sink in order to load balance and enhance the lifetime of the network. These solutions use path level cooperation. Event to Sink Reliable Transport (ESRT) protocol [38] achieves reliability by adjusting the reporting rate of sensor nodes depending upon current network load. Upon congestion detection nodes inform the sink for appropriate action. [47] also works on the principle of ESRT and provides end to end best effort reliability. These approaches utilize both neighbor and network wide cooperation. In [49], two classes of reliability are considered: high priority and low priority with multiple sinks. Each sink receives either low or high priority information based on its role. The proposed algorithm in [49] degrades the low priority data by suppressing them to forward high priority data.

6.3.3.5 Our Contributions to QoI Aware Reliable Information Transport

Our solution in [45] employs a reliable algorithm using hop-by-hop IACK for information loss recovery. To ensure end-to-end reliability, the adaptive retransmissions mechanism is introduced and used at each sensor node. In addition, to increase robustness the sensor nodes adapt the number of retransmissions according to the number of sources. Furthermore, [45] provides tunable reliability which is very promising for utilizing resources efficiently inside the WSN. [45] also relies on neighbor level cooperation to achieve tunable reliability.

In [44], we present Tunable Reliability with Congestion Control for Information Transport (TRCCIT) comprised of two components: a hop by hop reliability component and a congestion control component. The first component ensures the desired application reliability by controlling the number of retransmissions dynamically at each intermediate node based on channel quality. TRCCIT uses HACK comprising IACK and EACK for information loss recovery. Determining how long the node should wait for an HACK, TRCCIT uses an adaptive local retransmission timer at each node by observing the information flow across it. The second component ensures tunable reliability in the face of congestion by splitting the information flow on multiple paths. Congestion is detected proactively by observing the input and out put information flow across the node. By combining the tunable reliability and congestion control, TRCCIT offers desired reliability to the application users with efficiency. TRCCIT utilizes both node and neighbor level cooperation.

6.4 Mobile Sensor Cooperation

Here we consider the established mobile Wireless Sensor Network (mWSN) model. This model is used in a variety of WSN deployments, in particular in emergency and military scenarios. In addition to the multiple static nodes and the dedicated sink, a few mobile assist nodes are deployed with generalized cooperation objectives such as (1) application support (e.g. additional interface to users), (2) functionality support (e.g. delay-tolerant data transport) and (3) network support (e.g. maintenance). The mobile nodes cover functional capability spanning robots, Unmanned Air Vehicle (UAV), etc. Hereby, we assume that an assist node is able to move to and stop at any position in the sensor field. The assist node possesses high processing, storage and energy capabilities compared to sensor nodes. Furthermore, it has no energy limitations because it can recharge its batteries by means of on-board renewable energy resources or through moving to recharging energy-stations. For simplicity, we consider all nodes (assist nodes and sensor nodes) are equipped with a conformal level of communication technology and are able to communicate if they are in each other's transmission range R. Furthermore, we assume that network can get partitioned, i.e. some sensor nodes may not be able to communicate with the sink.

The injection of a few powerful (in terms of processing, storage, energy and communication) and/or mobile nodes into an existing homogeneous and static WSN is shown to improve the overall system performance such as extending the network lifetime. This is achieved through the cooperation of assist nodes with the static sensor nodes supplemental to inter-sensor cooperation. Overall, the benefits from adding a few mobile nodes are twofold in either enhancing network functionality or network dependability. We survey the literature systematically and present our own work for enhancing functionality and dependability through mobility.

6.4.1 Mobility to Enhance Functionality (gMAP)

Mobility-assisted data collection achieves high bandwidth and energy efficiency at the cost of a high end-to-end delay. In the following we survey the techniques to collect data from the WSN in a mobility-aided fashion. Next, we review the existing approaches to control mobility.

6.4.1.1 Mobility-Aided Spatial Sampling

Sampling resolution is a key aspect of QoI for WSNs. The introduction of mobility offers new possibilities for enhancing and adjusting spatial sampling resolution.

One basic approach is to augment the existing static network with mobile nodes equipped with additional sensing capabilities. Static nodes, upon detecting a specified event, request the mobile nodes to provide advanced analysis of the detected event. If the number of simultaneous events is lower than the number of mobile nodes, mobile nodes closest to the event are directed there. In case there are less mobile nodes than events, then events are clustered such that the number of clusters corresponds to the number of mobiles nodes and each mobile node is responsible for providing sensing for a separate cluster [56].

Mobility can also be employed for sampling the regions where the accuracy falls below the given threshold. Existing mobile nodes move to measure the value of sensed phenomena in the locations that will decrease the measurement variance. [57] presents an algorithm based on the divide and conquer strategy which is used to reduce the number of needed sampling performed by mobile node.

The innovative strategy for adapting sampling resolution to the intensity of monitored phenomena is achieved using the virtual forces approach. Nodes are attracted by the event and move to in turns. The distance travelled depends on the location of the event. The mobile node can only get closer to the event region but cannot go through its area [58].

An alternative approach for matching the mobile nodes distribution to the distribution of monitored phenomena is to create local clusters, where the cluster heads set the position of mobile nodes [59].

6.4.1.2 Mobility-Aided Data Collection

There exist different mobility-assisted data collection techniques in the literature. These techniques have been developed to collect user and control data. In the following we briefly review the popular techniques. For further readings we refer to an excellent survey [60].

Data Mule [65–69] is a mobile node that collects data from sensor nodes as it passes by. In [2], a basic theoretical analysis has been developed. Hereby, simplistic deployments and communication protocols have been considered. Sensor nodes are deployed randomly on a grid and multiple data mules move according to the random walk model. The sensor nodes buffer their data until the mule can receive it through direct communication. The mule buffers the data until it can deliver it to the sink. In [65–68] the movement paths are fixed and the coverage is not guaranteed. Therefore, some sensor nodes may need to form a multihop path to reach the mule. Nodes not on the movement path establish a local routing tree to send data to sensor nodes on the path, which collect the data till the data mule passes by.

Message Ferrying [61] is similar to data mules, but is designed mainly to overcome network partitioning. Mobile nodes act like ferries to reach disconnected sensor nodes. [61] uses controlled movement to deliver data. Ferries can either follow a predefined movement path or change their movement path on-demand. To 'order' the ferry, nodes can increase their communication range spontaneously to reach a message ferry near by and inform it about communication need so that the ferry can change its path. In order to allow for differentiated end-to-end latencies in [62] the authors suggest prioritizing messages. They present a forwarding strategy for the ferry route and discuss the buffer requirements to deal with this proposal. In [63,64], the authors improve message ferrying by introducing a power management framework. If nodes know when they are going to encounter the ferry they can sleep to conserve power. [72] analyses two simple data delivery schemes, namely, the direct transmission and flooding, based on (1) the likelihood that a sensor node can deliver data messages to the sink, and (2) the message fault tolerance, i.e. the probability that at least one copy of the message is delivered to the sink by other sensor nodes.

Compared to data collection in static WSN, mobility-aided data collection provides for a comparable high accuracy while outperforming them with respect to load balancing on sensor nodes and communication/energy efficiency. Furthermore, the delay-tolerant data collection approaches charge all sensor nodes similarly and contribute to the desired energy balancing in WSNs. These approaches are resilient to network partitioning, which increases the dependability of the WSN, since data collection continues even if critical failures/situations occur.

6.4.1.3 The gMAP Approach

In [5], we proposed a technique to collect data from all static sensor nodes. One of the main applications for such data collection is to construct a map of the physical world (e.g. the temperature map) or the network world (e.g. the residual energy map). Our approach is called gMAP, referring to constructing global maps of interest. Obviously, using mobility to collect data is suitable for data with high time validity, i.e. during collection the data remains valid/consistent compared to the physical/network world.

Compared to existing map-driven continuous data collection techniques, gMAP uses a minimal number of messages without sacrificing the completeness of sensor information. This provides for high efficiency with respect to both energy and bandwidth consumption. In gMAP we decouple the collection of the sensor values from the construction of the map, which results in minimal processing on sensor nodes, thus reducing the energy consumption on them.

The main design principle for the gMAP approach is to exploit the mobility of nodes to transport messages and collect information in a delay-tolerant way, thus reducing the communication overhead. Data collection in gMAP is similar to that of data mule and message ferrying approaches. However, most of existing approaches have been developed for specific scenarios such as sparsely deployed and structured WSNs (data mule) or partitioned networks (message ferrying). We focus on WSN and provide techniques that are for generalized scenarios (from structured to unstructured). We consider two major classes of mobility: Structured mobility, i.e. predictable & controllable, and unstructured mobility, i.e. unpredictable & uncontrollable.

We briefly present our gMAP approach comprising algorithms to collect samples in a mobility-assisted manner. We consider that the assist node knows its position and the position of all other sensor nodes. We let the mobile assist node scan the sensor field and collect the energy information from each node it encounters. We consider a single mobile assist node for simplicity of communicating the idea whereas we extend the gMAP approach to consider multiple cooperative nodes in Section 0. The assist node sends a short beacon, on which nodes reply with their sensor reading value and optionally their positions. We proceed progressively by first considering a structured scenario, then a semi-structured one and finally an unstructured one. For scenarios with controllable node mobility we design an integrated path planning algorithm. For all scenarios we design appropriate algorithms to collect energy information.

Scenario Classification

In the gMAP approach, we focus on three important types of scenarios that provide basic features to build more detailed realistic scenarios.

1. In a *structured* scenario we assume that the spatial deployment of sensor nodes is known *a priori* and that the mobility of the assist node is controllable (Figure 6.3(a)).
2. In a *semi-structured* scenario with an *a priori* known (or reliably estimated) spatial deployment of sensor nodes we assume that the mobility of the assist node to be controllable (Figure 6.3(a)).
3. In an *unstructured* scenario the topology is unknown (e.g. random spatial deployment) and the mobility of the assist node is assumed to be unpredictable and uncontrollable (Figure 6.3(b)).

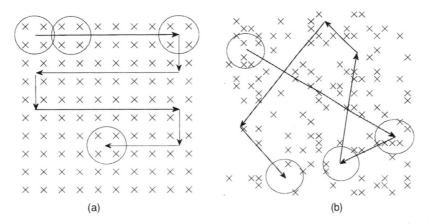

(a)   (b)

Figure 6.3   Scenario classification (All sensor nodes (x) are covered with the transmission ranges that depicted as (o) for some examples). (a) Structured scenario with a known spatial deployment of nodes and a controllable assist node. (b) Unstructured scenario with unknown node positions and unpredictable movement node.

Our main driver for the scenario selection is the proof of the concept in extreme scenarios.

Furthermore, in a realistic scenario, the spatial deployment of sensor nodes can be structured or known only partially. The mobility of the assist node can be either controllable or uncontrollable and may follow varied patterns.

Path Planning of Assist Nodes

Path planning is required for structured and semi-structured scenarios. The assist node plans its movement according to the positions of sensor nodes. The problem we are addressing here is planning an optimal tour to collect the data with a minimal number of messages. The tour is optimal if it consists of a minimal number of break-points with a minimal overlap of their corresponding transmission areas. Further optimizations of the tour are possible, e.g. concerning tour time, length and the energy overhead for an assist node. As our main goal is the proof-of-concept, we plan to consider these optimizations in future work.

Path planning for coverage problems are well studied and many centralized and distributed approaches, for various indoor and outdoor scenarios (e.g. [76, 77]), have been proposed. Many of these approaches are built basically on studying vehicle routing and scheduling problems [78], especially travelling salesman problems (TSP) and vehicle routing problems (VRP, cf., e.g. [71]). These approaches are well suited to the case when certain waypoints (e.g. node positions in our application) are known or reliably estimated. Many different instances of TSP and VRP have been proposed. The large number of proposed algorithms for solving and approximating these problems shows impressively its relevance to many applications in research and industry. For the case of unknown or only roughly estimated node positions, that have to be covered, distributed planning on the mobile node based on its local information is more adequate [74].

The basic problem of planning a single shortest path that covers certain fixed positions without considering a certain locomotion dynamic (known as the traveling salesman problem with neighborhoods [83]) is NP-hard. Thus many approximating algorithms for the TSPN under mild assumptions were proposed in recent years [79, 81]. For small scale WSN scenarios (a few dozens of nodes) a solution for multiple assist nodes based on TSP-path-planning has recently been tested and presented in [70]. The TSPN-problem is only briefly discussed there without presenting a solution for larger scenarios with a dense setting of nodes. For larger WSN scenarios a problem similar to TSPN is solved in [75], but disregarding overlaps within the transmission areas of subsequent breakpoints. Data collection for sparsely distributed nodes on basis of a TSP-solution is presented in [72]. For a further survey, we refer the reader to [73].

In structured mWSN scenarios, the sensor nodes are deployed according to a specific uniform scheme (e.g. on a grid). The knowledge of the uniformly structured spatial distribution of the sensor nodes can be used to simplify the complex problem of optimal assist node path planning. For the grid topology, the work [77] suggests a zigzag movement of nodes. Accordingly, the assist node crosses the full length of the sensor field in a straight line, turns around, and then traces a new straight line path adjacent to the previous one and $2R$ far from it. The assist node repeats this till the entire WSN field is covered.

In semi-structured mWSN scenarios, nodes deployment is not structured, which complicates an intuitive path planning of the assist node. In the following, we investigate the approaches to plan the movement of one controllable assist node to efficiently collect sensor data.

For clarity we elaborate upon the assumptions for our path planning strategy:

1. In our scenarios, a mobile entity knows the (reliably estimated or accurate) positions of sensor nodes and plans its movements accordingly.
2. A mobile node has to stop to communicate with stationary sensors that are within its circular neighborhood of radius R.
3. The mobile node's movements are fully controllable.
4. A set of breakpoints is optimal if useless overlaps of the mobile nodes' communication ranges are reduced as well as the number of points itself. The reasoning behind this is that sensor nodes in coverage overlap will waste energy to listen to redundant beacons.

Our approach consists of stepwise decomposition of the problem into different and less complex sub-problems: (1) finding suitable breakpoints, (2) reducing overlaps in the communication range and (3) planning a shortest-path-tour. Subsequently, we integrate the sub-solutions into single path planning algorithm. We are looking for a set of breakpoints, significantly fewer than the number of sensor nodes, where the assist node stops and communicates with the sensor nodes within the assist node's communication range. To find the really smallest number of these breakpoints, one may in general solve a mixed-integer nonlinear optimization problem, which may become impractical for larger settings. On the other hand, algorithmic approaches were shown to work well in determining a sufficiently small set of candidates. In [6], we proposed solving a constraint based nonlinear optimization problem (NLP) and subsequently removing the unnecessary breakpoints.

In a second step, we repeat the solving of a mixed-integer linear program (MILP) to reduce overlaps within the transmission areas. Therefore, pairwise distances and local adjacencies are approximated with linear expressions and binary variables. By minimizing a sum of binary variables that indicates the coverage of each node with different communication ranges, we gain an optimized set of positions for an assist node to stop.

Interrupting MILP optimization and then checking for unnecessary break-points in the best solution so far may reduce the combinatorial character of the MILP before the minimization starts again. Because of the MILP's structure with many minima of equal value, this shown to be more efficient than waiting until the solver gives back a proven global optimum.

Assuming that the mobile nodes have to stop to communicate, the costs to go from breakpoint i_1 to breakpoint i_2 can be seen as constant and independent from the sequence of breakpoints visited before i_1 and after i_2. Accordingly, the resulting path planning problem falls into the class of classical TSP or VRP. TSPs can be solved efficiently for hundreds of positions using existing solvers like [112].

Thus we are solving a TSP to get an optimal path for the mobile assist node. As an example in [6] we proposed the integration of a scalable path planning algorithm based on solving the sub-problems above. Depending on R, the number of nodes and on their spatial distribution, the steps can be scaled by time constants and by some parameters in the implementation such as the size of a regarded local adjacency. Figure 6.4(a) depicts a solution resulting from this algorithm.

Data Collection

For the structured and semi-structured scenarios, we present the following data collection algorithm. The assist node performs a first snapshot by sending a REQ-beacon to all sensor nodes in its transmission area using a MAC broadcast. A sensor node replies by sending a message containing its node-ID, location (loc) and sensor reading S_{val}. In order to reduce collisions, nodes schedule their reply for a random time t_{rand} between 0 and a maximum value T_{max}. The assist node performs the subsequent snapshot after visiting the next breakpoint according to the path planning algorithm. The optimal result of the collection operation is a tuple from each sensor node with the following structure: $\langle node\text{-}ID, loc, S_{val}\rangle$.

For unstructured scenarios the movement of the assist node is neither controllable nor predictable. If the assist node performs a snapshot, moves $2R$ away without changing the direction, and performs a second snapshot, then both snapshots are covering disjoint areas. Subsequently, we let the assist node perform a second snapshot, only after moving $2R$ from the location of the previous snapshot. The data collection completes, when the total WSN area is covered by all snapshots. We note that if the assist node changes its movement direction, then the snapshots overlap and some nodes may receive redundant REQ beacons. The major concern for sensor nodes is to minimize the number of messages to be sent or received. The assist node is powerful enough to send REQ beacons frequently. However, the REQ beacons are received by energy precious sensor nodes. Therefore, we have to minimize the number of unnecessary REQ messages sent by AN. To avoid unnecessary snapshots, the assist node maintains a history of snapshots $\langle snapshot_{id}, snapshot_{loc}\rangle$. After moving $2R$ from the location of the previous snapshot and before performing a second snapshot, the assist node uses the history to calculate if the second snapshot has an additional coverage higher than a fixed threshold coverage $COV_{th}\%$. Only in this case the assist node performs a snapshot. The value of $COV_{th}\%$ allows investigating the trade-off between the number of redundant REQ beacons and the sampling latency. Once the assist node scans the whole sensor field, the history of snapshots will be flushed and a new round will be initiated by the assist node. To avoid unnecessary transmissions, sensor nodes send information only once in a round.

6.4.2 Mobility to Enhance Dependability

Mobility has been used by various researchers for many dependability-driven purposes such as deployment [84–87], ensuring coverage [88–92], repairing the network [94–97] or extending lifetime [98–100].

Mobility opened up new strategies for WSN deployment by delivering qualities such as autonomicity and flexibility. Autonomicity means the self-deployment of the mobile node, according to a set objective function. The nodes have the capability to cooperate with static sensor nodes in order to fulfill the mission without additional interaction with the user. The user has high flexibility in setting the deployment details as the user is not constrained to low level designs, but concentrates on the functionality of the network.

The intuitive approach is to deploy the mobile nodes progressively one after another. After deploying each node, the feedback (current topology, coverage, etc.) from the network is collected and evaluated. As a result the decision is taken on placing consequent sensors (position, movement patterns) [84]. The more sophisticated techniques for WSN deployment use the concept of virtual forces. In opposite to previous approach, it allows one to deploy all of the nodes at once. The mobile nodes calculate the relative attraction and repulsion forces from obstacles and other (mobile) sensor nodes. The calculated force determines the direction and distance of the movement. The goal is to achieve uniform distribution of the nodes within the network [87]. The approach can be further refined to assure that each node has at least k-neighbors [86]. The mobility can also be used with special deployment requirements. An example of such a requirement is nodes density distribution allowing one to combat the sink centered energy hole problem. The nodes, after deployment, negotiate their position among themselves and move to reflect a given distribution [102].

WSNs are often deployed with the task of monitoring certain attributes. To assure the high quality of monitoring the deployment should assure, sufficient for the functional requirement, coverage of the area. Coverage means that each point within area is in the sensing range of at least one node. Therefore, coverage is a key aspect of WSN. In traditional static WSNs, the network lacks the potential to continue ensuring the coverage despite the frequent perturbations. However, mobility allows for a flexibility that provides for sustained sensing coverage.

There are two classes of coverage algorithms. The first aims at preserving uniform convergance. In a hybrid network of static and mobile sensor nodes with limited mobility it can be ensured that the maximum distance to be moved by a mobile node is bounded. Moreover k-coverage is available with a constant sensor node density [90]. In a more relaxed scenario, where all nodes are mobile, sensor nodes can move to the least covered area only within their neighborhood. They choose direction and velocity randomly according to the local information about sensor node density [90]. The goal of the second approach is to deliver coverage with respect to a given profile. Certain quantitative metrics are defined based on medium and phenomenon characteristics learned by each node. Desirable network configuration is calculated based on the optimization of these metrics [91]. An alternative is an algorithm designed in the iterative manner. In each iteration, the space is partitioned randomly into small neighborhoods creating node clusters. Each cluster head directs the redistribution process locally [88].

Node mobility is highly beneficial in the case of repairing the network. As WSN is based on resource constrained nodes, their depletion leads to common failures. As a result of these failures WSN may suffer coverage and connectivity issues. Repositioning the nodes allows for compensating for these failures.

The small number of specialized mobile nodes is utilized to aid the network. They cover the holes created by the failure of the static nodes. The static nodes detect the holes and send a bidding request to mobile nodes to provide coverage for them [103]. The nodes can also use fuzzy-logic based rules for choosing the proper actions based on nodes' requests [94]. Nodes are injected iteratively into the network to bridge the connectivity gap. They travel in a given direction until the edge of the network is detected by measuring the drop in connectivity degree [105]. Sensor nodes can also control assist nodes [93] and relocate them to the holes. Robots take their own initiative based on the failure notices received from sensors [97]. The function of the assist nodes may be taken over by flying assist nodes. It monitors the connectivity of the network and if necessary drops additional sensors to restore connectivity [104].

Even more flexibility is provided when each node is mobile. Redundant nodes are identified and relocated to deal with node failures or respond to events [92]. Neighbors of failed nodes are used to restore connectivity [96]. Nodes evaluate their coverage using Voronoi diagrams. They optimize coverage by adjusting their positions [85]. Mobile nodes move to replace failed parent nodes linking it to sink [101]. Nodes are injected iteratively into the network to bridge the connectivity gap. They travel in given direction until the edge of the network is detected by measuring the drop in the degree of connectivity [105]. The alternative approach is to split the network in a virtual grid. Each cluster head monitors the occupancy of the neighboring cells and moves its spare nodes when an empty cell is detected [107]; [108]. The energy efficiency of movement is improved by use of cascading movement [106].

The nodes also take a proactive approach and measure the coverage they provide for the network. Before depleting their energy they inform their neighbors about the possible loss of coverage so that neighbours can adjust their position to alleviate the problem [109].

Some research has also been done to extend the lifetime of WSN using mobility [98–100]. In a sink mobility scheme a mixed integer linear programming analytical model is used to determine the movement pattern for the sink that maximizes the network lifetime [98]. A heuristic is also proposed to move the sink to a new location with the highest residual energy. In the presence of a few resource rich mobile nodes, the lifetime of the network is extended by moving to the area heavily burdened by high network traffic 0. Scalable model and centralized heuristic are used for concurrent and coordinated movement of multiple sinks to provide a benchmark for evaluating controlled sink mobility schemes [100].

6.4.2.1 Predictive Approach to Improve Cooperation

Mobility can be used to enhance the dependability of WSN. However, it is triggered typically by different events in real-time. Such a reactive approach may be not sufficient or efficient. For example, in the case of fire it could be too late to react, signifying insufficiency. Another example is when nodes move to restore the connectivity in the case of network partitioning. However, it may not be the optimal movement of the nodes that has taken place, highlighting the inefficiency of the reactive approach.

These shortcomings of the reactive approach can be addressed with a proactive approach that predicts a containing event that might happen in the future by reacting to it within a sufficient margin of time and in the most efficient way. The proactive approach is an ideological shift from the reactive approach normally adopted for events in WSN. This approach has many advantages over the reactive approach as we enlarge substantially the time window in which to react so as to become able to delay or even eliminate the problem that might occur in the network. Its capability to react in time ensures maintaining the QoI that is desired.

The strength of this proactive approach lies in the fact that it is complimentary to most of the other mobility based approaches to enhancing mobility. Because it *predicts* the events rather than detecting the events in real time, it simply changes the context of time in which these existing algorithms will work. The rest of the functionality of the approaches can be kept unmodified or additional optimizations can be made to take the advantage of the predicted state of the network.

To design a predictive approach to support proactive actions we developed a comprehensive framework [110] for predicting network behaviour for a given attribute at the sink. In order to increase efficiency, we designed a localized technique that allows accurate profiling of energy holes (regions with low residual energy compared to the rest of the network) [111]. Both techniques allow us to predict network partitions. Unlike contemporary research focused on 'detecting' network partitions, our work allows one to forecast the accurate shape of future partitions. Relying on such forecasts, mobile nodes can now move proactively (e.g. according to one of the algorithms discussed above) in order to maintain the desired connectivity.

6.5 Cooperation Across Mobile Entities

In this section, we continue considering the emerging mobile Wireless Sensor Network (mWSN) model. However, we assume more than one assist node that cooperate with each other in addition to their cooperation with the static sensor nodes.

6.5.1 Cooperative Path Planning

As we discussed in Section 6.4.1, using mobility in WSN to carry information significantly increases the WSN lifetime. To this end path planning algorithms are needed. In order to provide for fault-tolerance, lower latency, etc, one does not want to be dependent on a single assist node. In this section, we show that path planning for multiple cooperating mobile nodes opens up various improvements for mobility-aided data collection.

In Section 4.1 and [6], we proposed a stepwise decomposition of the problem into different less complex sub-problems, i.e. (1) finding suitable breakpoints with a (2) minimal number of overlappings of transmission areas and (3) planning single shortest optimal path for these breakpoints. The extension to multiple assist nodes only affects sub-problem (3).

As mentioned in Section 4.1, TSP and VRP are the basic problems in this kind of path planning. This also holds true for cooperating mobile entities. An important distinction for instances of TSP is given by the number of salesmen (single TSP, multiple-TSP, k-TSP). Multiple-TSP are usually relevant to the case where the sum of all paths lengths is minimized, but other instances such as the more interesting min-max problem (e.g. [113, 114]), where the length of the longest sub-tour is minimized, were discussed. Other approaches look at multiple-TSPs with multiple depots. TSPs are often seen as special VRPs.

The VRP considers multiple tours, whereby the nodes may be equipped with certain demands and assist nodes are constrained by individual capacities (CVRP = 'capacitated VRP'). Also, VRPs with time windows (VRPTW)

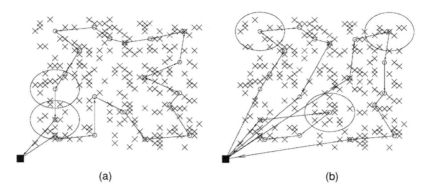

(a)   (b)

Figure 6.4 Approaches for path planning by solving a TSP and CVRP based on a set of breakpoints. (a) Single node path planning. (b) Multiple node path planning.

are very popular. The surveys [71, 115, 116] provide an excellent overview for the interested reader concerning VRP solutions in recent years.

6.5.1.1 Path Planning of Multiple Cooperating Mobile Nodes

Given a fixed set of breakpoints we now discuss extensions from planning a single path to path and trajectory planning for a homogeneous team of cooperating assist nodes.

As an initial approach one might think of solving a TSP or a VRP for multiple assist nodes. Therefore, subsets of waypoints must be allocated to the assist nodes and a TSP has to be solved for each of these subsets. These two tightly coupled sub-problems have to be solved simultaneously such that a general objective function is minimized.

By generalizing the already *NP*-hard single path-planning problem to multiple assist nodes a big jump in the combinatorial complexity of the problem arises. As will be discussed later, additional features may have to be considered in a cooperative multiple assist nodes system.

In Figure 6.4(b) a solution for the same setting as in Figure 6.4(a) is shown but for the case of three assist nodes. Therefore, the problem was modeled as a CVRP, where each of the 21 breakpoints had a demand of 1 and the assist node was constraint by a capacity of 8. Although this problem is much harder to solve than the TSP, the solution was computed by using [117][2] in less than 0.4 *s*. An advantage of using such a CVRP-model is that the user is able to enforce evenly distributed subsets of waypoints to the according sub-tours.

The assist nodes are doing their tours independently and communication among them can only be guaranteed at the sink before the mission starts and after the last vehicle has finished.

6.5.1.2 Important Aspects of Cooperation of Mobile Entities

Under certain conditions real cooperation among mobile entities can be enforced. Namely, communication among vehicles during the mission can increase many of the desired qualities of the system. Especially, in uncertain environments, communication enables the cooperative adaptation of planned behaviour and computed paths to unexpected situations occurring during the mission.

In a setting where assist nodes operate in an area that is considerably larger than the assist node's communication range, it is obvious that a requested radio link during a mission has an influence on path planning. If multiple assist nodes want to communicate, they have to stay close together at a certain time. More precisely, one needs synchronized movement of the assist nodes. Practically, this means that the path planning problem turns into planning of trajectories. Up to now we have neglected the locomotion dynamics of the assist nodes completely and communication as well as synchronization among the assist nodes could only be guaranteed at the beginning and the end of the mission. At the very least, velocities have to be considered if one wants to optimize path planning under the requirement of guaranteed inter-assist nodes-communication. A similar problem with minimal linear dynamics representation was investigated in [82].

---

[2] Running on a standard PC (Intel(R) Pentium(R) M processor 1.86GHz; 1024 MB RAM).

For dealing with the entire problem of trajectory planning for cooperative assist nodes, one in practice has to deal with many non-trivial sub-problems such as collision avoidance, cooperative tasks allocation, formation constraints, or the highly combinatorial character of the resulting problems. For more information on radio-link connectivity in the context of mobility-control, we refer to [118] and the references therein.

6.5.1.3 Planning of Synchronized Movements under Communication Constraints

In the following we discuss the computation of the synchronized movement of multiple assist nodes which guarantees inter-node communication and assist-nodes-to-sink-communication at multiple times during the mission. In addition to sharing operational data among vehicles, this also allows one to transmit collected data to the sink via multi-assist-node-hop-communication and to receive completely new instructions from an operator during the mission.

A suitable approach to the problem of planning an optimal trajectory based on the nodes' topology is using hybrid optimal control theory, that results in solving mixed-integer nonlinear programming (MINLP) in practice [80]. A linear approximation that considers communication ranges and optimality with respect to a specific physical motion dynamics of mobile nodes is shown in [82]. Therefore, an adjusted objective function describes the features that are desired to be minimized such as overlaps or path-length. This solution is well suited to small scale networks but becomes very inefficient for common WSN settings. This is caused mainly by the underlying discrete structure which results in a highly combinatorial character that is tightly coupled to a high dimensional continuous part of the MINLP.

Therefore, we now discuss solving a mixed-integer linear program (MILP) with a minimal dynamic representation for the vehicle under consideration. The model is based on an initially fixed time grid of equally distributed timepoints t_k ($k = 1, \ldots, n_t$). To keep things as simple as possible the linearized first order point-mass model for the i-th vehicle locomotion is modelled as

$$x_{ik+1} = x_{ik} + \Delta_k \cdot v_{xik}, \; y_{ik+1} = y_{ik} + \Delta_k \cdot v_{yik} \tag{6.1}$$

where $x_{i,k} := x_i(t_k)$ (x-coordinate), $v_{x,i,k} := v_{x,i}(t_k)$ (velocity in direction of x), y respectively, $\Delta_k := t_{k+1} - t_k$.
With binary variables b_{ijk} we set up the constraint

$$b_{ijk} = 1 \Rightarrow \left[(x_{i,k} - \xi_j) < \varepsilon \lor (y_{i,k} - \eta_j) < \varepsilon \right] (0 < \varepsilon << 1 const.), \tag{6.2}$$

that can be expressed as linear inequalities by using a Big-M-formulation [119], and together with

$$\sum_{i=1}^{n_v} \sum_{k=1}^{n_t} \sum_{j=1}^{n_b} b_{ijk} = n_b, \forall j : \sum_{i=1}^{n_v} \sum_{k=1}^{n} b_{ijk} = 1 \tag{6.3}$$

they form conditions that effect that each of the breakpoints $\left\{ (\xi_1, \eta_1)^T, \ldots, (\xi_{n_b}, \eta_{n_b})^T \right\}$ is visited once by a vehicle during the mission. Additionally we require that whenever a vehicle stops at a breakpoint, an inter-vehicle multi-hop communication to the sink is possible. By using a radial vehicle communication range with radius R_v ($R_v \gg R$), the expression

$$b^c_{i_1 i_2 k} = 1 \Rightarrow \sqrt{\left(x_{i_1,k} - x_{i_2,k} \right)^2 + \left(y_{i_1,k} - y_{i_2,k} \right)^2} \leq R_v, \tag{6.4}$$

together with a linear approximation for the inequality by using ($l = 1, \ldots, n_l$)

$$\sin\left(\frac{2l}{n_d} \pi \right) \left(x_{i_1} - x_{i_2} \right) + \cos\left(\frac{2l}{n_d} \pi \right) \left(y_{i_1} - y_{i_2} \right) \leq R_v \tag{6.5}$$

and eventually a Big-M-formulation for it, models desired connectivity-links between vehicle i_1 and i_2 as a set of linear inequalities. By some simple constraints on the set of binary variables $b^c_{i_1 i_2 k}$, one can formulate connectivity properties for the entire multi-vehicle system. Formulating constraints has to be done very carefully, e.g. by

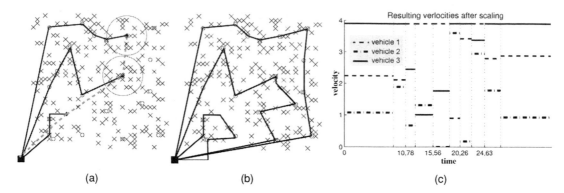

Figure 6.5 Communication with the sink can be guaranteed whenever a vehicle stops to collect data from nodes. (a) Synchronized movement after the 4th visited point. (b) Synchronized movement at the end of the mission. (c) Velocities.

avoiding equivalent global optima, which may occur due to equal vehicles in a symmetric setting. Adding a simple linear constraint may reduce or extend the computational complexity by magnitudes, without restricting any quality of the solution

In the proposed example (Figure 6.4), we minimized the sum of velocities by solving

$$\min_{u_{i,k},v_{i,k}} \left\{ \sum_{i=1}^{n_v} \sum_{k=1}^{n_t} \Delta_k \left((v_{xik}) + (v_{yik}) \right) \right\}$$

subject to the constraints resulting from (6.1)–(6.5).

The number of visited breakpoints per vehicle is at most $n_t - 1$, such that the user is formulate a problem similar to the capacitated CVRP by fixing $n_t \leq n_b$. Furthermore, in a subsequent computation after solving the MILP, the time intervals $[t_k, t_{k+1}]$ can be shortened or stretched according to desired maximal or minimal velocities (Figure 6.5(c)).

Due to the *NP*-hard characteristics of the underlying TSP, the time to compute the guaranteed global optimum explodes with the increasing number of breakpoints and timesteps. For a problem with three mobile entities, the solution of the corresponding MILP[3] (342 variables, 1656 constraints) for a setting with 10 points on a grid of seven timesteps took approximately 5.5 s. For a problem with three mobile entities, the computation for a setting with 21 points on a grid of 10 timesteps (816 variables, 8244 constraints) took approximately 16 500 s. Further improvements in the description of MILP-constraints, the incorporation of heuristics and a careful decomposition into sub-problems are already showing ways to overcome the computational burden in ongoing work.

The proposed approach computes optimized trajectories for the whole system and can be seen as an initial plan for the mission. To be able to react on changes a combination with a distributed approach (e.g. [120, 121]) is target-aiming for many applications. Furthermore synchronization might be an important topic in complex setting an implemented concept like in 0 for a multi-agent setting will increase the cooperative potential significantly.

6.5.2   Data-Based Agreement for Coordination

Another way of implementing cooperation between mobile entities is to exploit a data-based agreement especially in the case of the predefined mobility paths of assist nodes. By data-based agreement we refer to database transactions where mobile entities agree on a set of cooperative tasks that need to be performed by these entities in an atomic way. Atomicity means that all transaction participants agree on a set of tasks which will be performed by them or that no one of them is performing any task. The data about the agreed tasks and their corresponding stakeholders are kept in databases as proof of the obtained agreement. This proof may be of interest to the user, police, insurance companies, etc. We focus in this section on database transactions executed between mobile entities where network

---

[3] Using CPLEX (http://www.ilog.com) running on a PC (Intel(R) Pentium(R) M processor 1.86GHz; 1024 MB RAM).

partitioning (due to either node or link failure/disruption) is a dominant network failure to consider. This failure should be taken into consideration while developing database transaction protocols to use them as a base for data-based agreement coordination. Database transactions require the existence of a coordinator which is responsible for coordinating the execution of the transaction and taking the final decision about the outcome of the transaction that can be either Commit (successful agreement) or Abort (unsuccessful agreement). In case of an unsuccessful agreement the transaction needs to be reinitiated in order to conclude the desired agreement in an atomic way.

We consider that mobile entities communicate with each other if they are in each other's communication range or multihop (where only mobile nodes are relay nodes). For reasons of efficiency, we consider that the mobile nodes have no access to a powerful infrastructure (which would be possible through the sink, but would lead to an unacceptable overhead in the energy-limited static sensor nodes). Such an ad hoc network usually shows frequent and unpredictable network partitioning. Mobile entities in such networks are the only participants in the execution of transactions. We review the design challenges for commit protocols in the ad-hoc environment. Subsequently, we outline existing solutions with their limitations and then provide our solution, presented in [7].

6.5.2.1   Commit Design Challenges

As we assume that mobile entities do not connect to any infrastructure, the coordinator of the transaction is required to be a mobile entity. A mobile entity is not assumed to have stable storage and is therefore not able to play the coordinator's role alone unless specific assumptions of the capabilities of such a mobile entity can be made (however the same capabilities as a fixed entity are in general not realistic). Failures of the single mobile coordinator usually lead to the blocking of all participants. Two extreme cases are possible: only one coordinator is defined by introducing a more powerful mobile entity (with additional assumptions such as stable storage) or every single participant in the transaction is a coordinator. We believe that only a subset of the participants should play the coordinator role, as justified later in this section.

A lifetime concept can be used in ad-hoc networks to control the blocking time of the transaction's participants. Estimating the appropriate lifetime value depends on multiple factors. A key issue is network connectivity, which depends primarily on mobility parameters such as speed of mobile entities, as well as their communication parameters. These variables make estimating lifetime in ad-hoc transaction scenarios a challenge. Applications initiating transactions should be at least able to compute how long they will be able to wait before receiving the results of the initiated transaction. This time can be used as the lifetime of the initiated transaction or can be adapted to the current state of the underlined mobile ad-hoc network. Synchronized clocks are not assumed across mobile entities. Thus the lifetime can elapse at different times for different participant mobiles. This issue should be considered when designing an appropriate mobile ad-hoc transaction solution.

Given frequent network partitioning, it is challenging for ad-hoc scenarios to disseminate the fragments of the transaction to their corresponding mobile entities. For this, partition-aware dissemination protocols can be used such as Hypergossiping [123]. Next, we review the existing commit solutions and sketch our new approach to deal with these challenges.

6.5.2.2   Existing Cluster-Based Transaction Commit

[124] and [125] propose the use of a cluster of coordinators preferably in single-hop distance from each other to avoid blocking of mobile participants in case one coordinator fails. The cluster of coordinators elects a single main coordinator and uses the 3 PC protocol [126] to agree on a consistent decision either to commit or abort the transaction. If the cluster of coordinators is partitioned or the main coordinator fails the authors use a termination protocol based on the Paxos Consensus protocol [127] to elect a new main coordinator. The termination protocol succeeds only if a majority of the coordinators in the cluster of coordinators does not fail and also belongs to the same partition. The assumption on the mobility of the cluster of coordinators made here is not valid in most ad-hoc scenarios. Targeting a more general solution, we relax this assumption and consider a generalized arbitrary mobility model.

6.5.2.3   Proposed Transaction Commit based on Partition Membership

We now present a commit solution assuming that every mobile entity in a partition knows all the members of the partition it belongs to. Later we will relax this assumption and present a corresponding solution. This solution is based partly on the work presented in [128]. Given the partition membership information, the participants in every partition elect a coordinator and send their votes to the elected coordinator which takes a pre-decision on the outcome of the transaction. The pre-decision can be different from the final decision. This temporary decision is communicated to all participants within the partition. If the pre-decision is Abort, then every participant mobile entity that receives this pre-decision can safely abort the transaction. If the pre-decision is Commit, every participant in the partition should wait until all pre-decisions are collected. Alternatively, when two partitions merge, then the pre-decisions are exchanged and if all pre-decisions are collected the outcome of the transaction can be decided safely since these include the votes of all participants in transaction. Now, all the mobile participants must be informed about this outcome which can be achieved through partition-aware communication protocols similar to fragment dissemination. The correctness of the basic solution described above is assured by the partition membership assumption [128]. If this assumption is not valid a participant can be a member of a partition but the coordinator of that partition is not aware of the membership of this participant and subsequently the vote of this participant can be lost, i.e. not included in any pre-decision and consequently not in the final decision.

The assumption that every mobile entity in a partition knows all the members of its partition is crucial for the proposed solution. Some works [129, 130] addressed the problem of group membership in mobile ad-hoc networks; however a general solution for mobile environments remains a challenge. Furthermore, the blocking time of participants is often not considered and in the worst case all participants may be blocked forever if one of the participants disappears forever. As shown in [131], there exists no non-blocking atomic commit protocol if network partitioning may occur for an unpredictable duration. Fortunately, the number of blocked participants can be minimized, as we will discuss later. This approach based on partition membership information is independent of the mobility of nodes in contrast to [124, 125]. However, it is based on the assumption that partition membership is available to all its members. Partitions in mobile ad-hoc scenarios are usually very dynamic as nodes may leave and join partitions arbitrarily. Therefore acquiring global partition membership information becomes very inefficient.

6.5.2.4 Proposed Transaction Commit without Partition Membership Information

We present an approach [7] which (a) does not require partition membership information, and (b) limits the resource blocking time for participants by defining a lifetime for each initiated transaction and electing a coordinator in each partition. The lifetime information is communicated to every participant upon initialization of the transaction but can only be used by partition coordinators. Thus each partition coordinator can abort a transaction if its lifetime expires. Transaction progress of the proposed approach is guaranteed because in each partition a pre-decision is agreed upon as soon as all participants in the partition communicate their votes to an *a priori* elected partition coordinator. Any election algorithm can be used for the selection of the partition coordinator such as election of a random participant or an election based on node IDs, e.g. selecting the participant with the highest ID as a partition coordinator (Nodes 6 and 7 in Figure 6.6(a) are elected as partition coordinators because they have the highest ID in their corresponding partitions). Existing election algorithms for mobile ad-hoc networks such as [132, 133] can also be used for the election of the coordinator. The election can also be optimized according to some factors such as connectivity to other mobile entity participants in the same partition, communication, storage and computation capabilities. The coordinators can also be selected upon initialization of the transaction. In this case the coordinators are not tied to partitions that change dynamically over time; hence the notion of partition coordinator is not appropriate. For this work we assume that sufficient time exists to elect a partition coordinator before the partition composition changes.

As we do not assume the existence of partition membership information, we require that every participant sends its vote to each partition coordinator it encounters (i.e. it is able to communicate with either directly or multi-hop) as long as there is no final decision. In this way even if the original partition coordinator was not aware about this participant's vote, e.g. due to message loss, then the vote information is not lost. The vote information is communicated to other partition coordinators when these are encountered. The pre-decision taken by the partition coordinator can propagate from one partition to another as the partition coordinator moves to another partition. If at least one participant votes to abort the transaction, the partition coordinator decides to abort the transaction and the final decision is propagated to the rest of the participants. If the pre-decision is Commit, the partition coordinator cannot take a decision about the outcome of the transaction locally and proceeds as follows.

The coordinator of each partition maintains a list of all participants from which the partition coordinator received a vote (e.g. in Figure 6.6(a), Node 6 maintains the list {2,4,5,6}). As long as the received votes are Commit votes the pre-decision is Commit. As soon as a participant votes to abort the transaction, the coordinator decides to abort

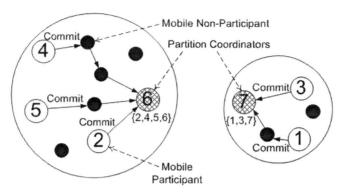

(a) Intra-partition decision (Pre-decisions)

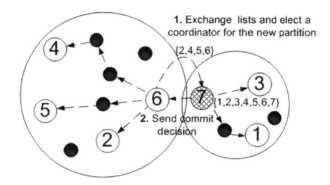

(b) Inter-partition decision (Final decision)

Figure 6.6   Partition-aware ad-hoc commit.

the transaction, as discussed above. The partition coordinator propagates its pre-decision and the list of participants which voted to commit the transaction (if the pre-decision is Commit) to other partitions on partition joins. As shown in Figure 6.6(b), if two partitions join, the corresponding partition coordinators (Nodes 6 and 7 in Figure 6.6(b)) exchange their lists and elect one coordinator among themselves (e.g. using the same strategy as for election of the partition coordinators). In the example of Figure 6.6(b), Node 7 is elected as partition coordinator because it has a higher ID than Node 6. Using this schema for the election of a new partition coordinator if two partitions join, we guarantee that no two or more partition coordinators have the same knowledge about which participants voted to commit the transaction. In the latter case these partition coordinators can take different decisions about the outcome of the transaction which violates the correctness of the proposed solution. Every coordinator can abort the transaction if the lifetime expires before reaching a decision. Once the list of participants that voted for Commit contains all the participants of the transaction, the coordinator (which is the single coordinator in the system by this time) takes the decision to commit the transaction and sends the decision to the participants in its current partition. Only one coordinator remains in the system because all the partition coordinators should meet together in order to exchange their lists and by this time they select one coordinator among them. The lists are merged only if the election succeeds. This guarantees the uniqueness of the taken decision. The final decision is communicated to other participants when they encounter participants which already know this decision. For the dissemination of the decision and for the communication between the participants inside a single partition, either flooding or a routing protocol like AODV [134] are used depending on the ratio of participants to non-participants which exist in that partition.

The proposed approach reduces the resource blocking time of mobile participants because the partition coordinators do not arbitrarily wait long to connect in order to decide about the outcome of the transaction. If the lifetime expires on at least one partition coordinator before reaching a final decision, the transaction is aborted. This is not viable in any existing solution as mobile participants have to meet asynchronously to be able to reach a final decision.

Coordination/cooperation between mobile assist nodes based on data stored on them delivers guarantees especially with respect to identifying responsibilities of different mobile nodes in case of misbehaviour or non achievement of the targets of the mission of these mobile nodes. This data becomes more important if the cooperating nodes belong to different organisations or companies.

6.6 Inter-WSN Cooperation

Obviously, there will be varied WSNs deployed in different or overlapping locations to observe different physical phenomena. Users will benefit from the composition of the varied information delivered by the different WSNs. For example, combining weather information obtained from a city WSN with parking place availability information provided by another WSN would support the decision to drive to that city or not for pleasure. Generally, we observe a trend to integrate sensor data (such as GPS tracking data and webcams) into Geographical Information Systems (GIS), location-based information systems and online services.

In the rest of this section we briefly survey the existing projects to provide for inter-WSN cooperation and present our novel map-based approach which enhances the existing efforts and allows for easy inter-WSN cooperation. We consider more than one WSN that are built according to the WSN or mWSN system models. The inter-WSN cooperation is easily realized through cooperation between the sinks of each elementary WSN (Figure 6.7).

6.6.1   The Sensor-Map Integration Approaches

Many applications such as weather forecasting, precision agriculture and environmental monitoring have generally followed one of two paths. One is based entirely on map based information and the other is based on real-time sensor information. The first approach is based mainly on using maps to represent historical information. The second approach allows one to access real-time data. With the development of networking and communication technologies the trend is to combine both approaches.

There have been many attempts to provide a global platform to integrate different sensors and sensor networks. The SensorMap project [145, 146] allows one to integrate webcams, environmental sensors, transportation traffic sensors, etc. into a global map. Users can then access the map through a Web browser and benefit from the displayed sensor information. A similar project introduced recently by the US homeland security department is SensorPedia [147].

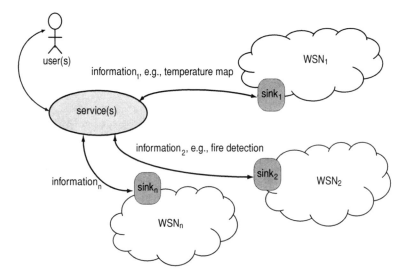

Figure 6.7   Overview on inter-WSN cooperation.

The leading project has been initiated by NASA and is referred to as SensorWeb [149]. SensorWeb is a concept that supports macro scale sensing with web-enabled heterogeneous sensors. Sensor Web Enablement (SWE) is a generic approach designed to allow the integration of maps and real-time sensor data. This has been driven by a working group from the Open Geospatial Consortium (OGC) [148]. A further example is augmenting the navigation map with real-time sensor information from sensors on the road and on vehicles.

6.6.2   Generalized Map-Based Cooperation

Most WSN research is application-oriented and relies on cross-layer approaches. Therefore, the current research does not allow for a widely accepted abstraction. While the design paradigms (WSN as a network, database and event service provider) reduce application dependency and hide low-level communication detail, they still address single sensor nodes (although redundancy of nodes and consequently spatial correlation of sensor readings are inherent in WSN). Subsequently, there is a strong need for a holistic design methodology. Such a methodology should be flexible/abstract and systemize/simplify the design as well as the deployment phases and involve functional as well as nonfunctional attributes. Obviously, the holistic approach should retain the advantages of the existing paradigms. Overall, rather than addressing single nodes, designers should address spatially-correlated and appropriately-grouped nodes. We refer to such groups as regions. A few attempts do exist that address regions instead of single nodes, such as [135, 136].

WSNs, on one hand, are inherently embedded in the real world, with their goal being to detect the spatial and temporal world's physical nature, such as temperature, air pressure and oxygen density. On the other hand, maps present a powerful tool to model the spatial and temporal behaviour of the physical world, being an intuitive aggregated view of it. The map paradigm builds on the region principle and therefore provides excellent modelling primitives for WSNs. Global maps are created for the sake of network monitoring (e.g. residual energy map [137]), event detection (e.g. oxygen map [138, 139]), event boundaries detection [142] and tracking [143], or network protocol optimization [140]. This research highlights the map-based methodology as a powerful and promising abstraction. In [10], we proposed that maps are the natural step towards a holistic Map-based World Model (MWM) and Map-based WSN design.

In [10] we developed the MWM for generalized WSNs. We showed how this model retains the advantages of existing design methodologies while augmenting their efficiency and level of abstraction. We also presented a step-wise design methodology to build an appropriate MWM and the process driving its usage. In particular, we emphasize the following qualities of our MWM: (1) MWM can reflect appropriately both physical and network worlds while being aware of the strong constraints on network and node resources, and (2) the MWM holistic approach provides a natural way to define, detect, query and predict arbitrarily complex real world situations and events. Below, we briefly define our MWM. For details about the generalized architecture and the core primitives for MWM usage and management we refer to [10].

Without loss of generality, we model the world as a stack of user maps (uMAP) presenting the spatial and temporal distributions of the sensed attributes of interest in the physical world (Figure 6.8). We additionally model the spatial and temporal behavior of system properties as a stack of network maps (nMAP) such as residual energy and connectivity maps. MWM is the superposition of all maps of interest. The unified modelling of both physical/network models maximizes the reusability of concepts and techniques.

A *Map* is an aggregated view on the spatial distribution of a chosen attribute at a specific time. From cartography [144], we identify two main classes of maps, i.e. the choropleths (e.g. nMAP 1 in Figure 6.8) and the isomaps (e.g. uMAP 2 in Figure 6.8). The map construction groups spatially correlated sensor nodes with similar attribute's values to regions. In MWM we define a region by its border (a set of spatial points) and an aggregate (e.g. average) of the attribute's values obtained from all sensor nodes located in the region's area. A map is then the collection of all regions of the WSN. We refer to [10] for a brief survey of existing approaches to create maps for WSNs.

The main benefit of MWM is to abstract from single sensor nodes by addressing regions. While losing sampling details, the map abstraction usually sustains acceptable accuracy concerning the spatial distribution of sensor readings. Accordingly, map abstraction presents a natural step to increase the communication efficiency in WSNs. The higher the number of sensor nodes per region, the higher is the benefit. The MWM abstraction level simplifies the design and deployment of WSNs as it can be easily accepted by different parties, ranging from users to network designers, programmers and administrators.

The MWM model presents a widely accepted abstraction level as it converts less comprehensive sensor data into understandable information. Furthermore, MWM can be implemented as middleware which simplifies the

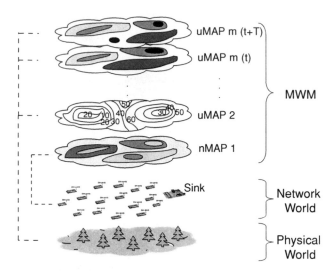

Figure 6.8   Map-based World Model (MWM).

integration of varied applications not only intra-WSN but also inter-WSN. The map-based model is potentially a candidate for developing standards for cooperative WSNs. Such standardization will provide a generic interface simplifying the WSN interoperability through the interconnection of heterogeneous and autonomous WSN systems, which can play a major role in future WSN research (e.g. SensorWEB [149]). This conforms to the trend discussed earlier.

6.7   Conclusions and Future Research Directions

Cooperation between the different WSN entities is a prerequisite for dependable system operation. The traditional approach in the WSN research area is to maximize cooperation between the static nodes and the dedicated sink(s). As this cooperation reveals limitationss due to the nature of fragile sensor nodes and wireless links, the trend is to inject a few powerful reconfigurable nodes in the WSN to catalyze/augment the cooperation. In this chapter, we investigated the key role of node cooperation in providing core functionalities and dependable network operation. Primarily, we studied the core functionalities of sampling the physical phenomena, in-network processing of samples to extract useful information and finally the transport of extracted information to the user. Throughout the chapter, we addressed most of the established system models for WSNs, ranging from the popular static model, to the emerging mobile WSN model.

We have shown how *cooperation between static sensor nodes* allows enabling self-organization in the network and leads to high deployment efficiency and resilience against network perturbations. In addition, this cooperation is a key aspect in maintaining the desired quality of information level, i.e. to provide for efficient QoI provisioning. Considering the example of information transport, we presented our ongoing work that shows how inter-sensor cooperation at local, region and network levels contributes to tune the achievable QoI level through tuning the attainable transport reliability. However, recent efforts to combine all core network functionalities (from sampling to information transport) to define and provision QoI, show that a re-thinking of inter-sensor cooperation design is needed in order to achieve an holistic architecture for QoI provisioning.

Enriching static WSNs with mobility is nowadays simple and extremely useful for sustaining critical function-alities beyond perturbations. Two broad classes of cooperation are possible, i.e. the cooperation of mobile assist nodes with static sensor nodes and among themselves.

We identified two main goals for *cooperation between static and mobile entities*. First, we discussed how mobile assist nodes can contribute to maintain or enhance the different network functionalities. We mainly discussed the repositioning of mobile entities to under-sampled regions in order to improve the spatial sampling and maintain the desired QoI. Furthermore, we presented the existing approaches to utilize mobility so as to physically transport information to the sink. In these approaches cooperation between mobile and static nodes significantly reduces the network load on static sensor nodes and saves valuable resources such as energy and bandwidth. We presented in detail our gMAP approach, where assist nodes collect samples efficiently from all static sensor nodes independent of the considered node deployment. Second, we investigated the key role of mobility in maintaining the dependable operation of WSN. We focused primarily on improving deployment, ensuring network and sensing coverage, extending the lifetime and repairing network partitioning.

Cooperation between mobile assist nodes is investigated less in the literature. In this chapter, we presented our ongoing efforts in cooperative path planning and data-based agreement for coordination. A novel technique has been presented to plan the path of multiple assist nodes in order to collect data from sensor nodes. We also presented a novel commit protocol for transaction-based atomic coordination between mobile assist nodes. This transaction-based cooperation approach allows the permanent and consistent tracing of coordination data in local databases, which is of great importance for future mobile WSN deployments.

Though a lot of effort has been made to drive cooperation between static and mobile nodes, we still observe a lack of systematic investigation of the trade-offs for cooperation among only static, static and mobile and only mobile nodes. In particular, we investigate in ongoing research how to tune information transport timeliness. Our approach consists of finding the appropriate balance between the low-delay multi-hop communication and the high-delay physical transport by mobile nodes. To this end, the network conditions are profiled in order to reposition existing mobile nodes proactively if needed.

A recent case for research is enabling WSN systems for the internet of things. One major step towards this vision is to enable *cooperation between autonomous WSNs* and to allow for sensor webs. We have shown that our map-driven WSN architecture allows for standardization of inter-WSN cooperation and also for the easy integration of WSNs in existing location-aware information systems. Much effort is needed in order to realize the vision of all-map sensing. Especially, real-time integration of sensors into maps and the federation/integration of deployed (wired and wireless) sensor networks open up many challenges for research.

Overall, we have considered only cooperative WSN entities and excluded uncooperative behaviour such as selfishness and deliberate faults. Unfortunately, there are few projects that investigate the impact of security breaches on aspects of cooperation [14, 150]. Hence, we highlight this research gap that should be filled in the near future so as to achieve mature techniques ready for real deployment.

References

[1] W. Wei, V. Srinivasan, and K.C. Chua (2005) 'Using mobile relays to prolong the lifetime of wireless sensor networks,' In *Proceedings of ACM MobiCom.*

[2] R.C. Shah *et al.* (2003) 'Data MULEs: modeling a three-tier architecture for sparse sensor networks,' In *Proceedings of IEEE SNPA.*

[3] Y. Wang *et al.* (2007) 'A survey on analytic studies of delay-tolerant mobile sensor networks,' *Journal of Wireless Communications and Mobile Computing (WCMC)* **7**(10).

[4] W. Zhao *et al.* (2004) 'A message ferrying approach for data delivery in sparse mobile ad hoc networks,' In *Proceedings of ACM MOBIHOC.*

[5] A. Khelil, F.K. Shaikh, A. Ali, and N. Suri (2009) 'gMAP: efficient construction of global maps for mobility-assisted wireless sensor networks,' In *Proceedings of IEEE WONS.*

[6] A. Khelil, F.K. Shaikh, A. Ali, N. Suri, and C. Reinl (2010) 'Delay-tolerant monitoring of mobility-assisted wireless sensor networks,' In *Delay Tolerant Networks: Protocols and Applications* (eds A. Vasilakos, Y. Zhang, T. Spyropoulos), Auerbach Publications, CRC Press, Taylor & Francis Group.

[7] B. Ayari, A. Khelil, and N. Suri (2010) 'ParTAC: a partition-tolerant atomic commit protocol for MANETs,' In *Proceedings of the 11th International Conference on Mobile Data Management (MDM).*

[8] Microsoft. Sensor Map. http://atom.research.microsoft.com/sensewebv3/sensormap/.

[9] D. Guinard, V. Trifa, T. Pham, and O. Liechti (2009) 'Towards physical mashups in the web of things,' In *Proceedings of the IEEE Sixth International Conference on Networked Sensing Systems (INSS).*

[10] A. Khelil *et al.* (2008) 'MWM: a map-based world model for wireless sensor networks,' In *Proceedings of ACM AUTONOMICS.*

[11] T. ElBatt (2004) 'On the scalability of hierarchical cooperation for dense sensor networks,' In *Proceedings of ACM/IEEE IPSN.*

[12] T. Elbatt (2004) 'On the cooperation strategies for dense sensor networks,' In *Proceedings of the IEEE International Conference on Communications (ICC).*

[13] T. Elbatt (2009) 'On the trade-offs of cooperative data compression in wireless sensor networks with spatial correlations,' *IEEE Transactions on Wireless Communication* **8**(5).

[14] Y. Wei and K.J.R. Liu (2008) 'Secure cooperation in autonomous mobile ad-hoc networks under noise and imperfect monitoring: a game-theoretic approach', *IEEE Transactions on Information Forensics and Security* **3**(2).

[15] C. Bisdikian (2007) 'On sensor sampling and quality of information: a starting point,' In *Proceedings of PerCom Workshops.*

[16] V. Srivastava and M. Motani (2005) 'Cross-layer design: a survey and the road ahead,' *IEEE Communications Magazine* **43**(12).

[17] C. Intanagonwiwat *et al.* (2000) 'Directed diffusion: a scalable and robust communication paradigm for sensor networks,' In *Proceedings of ACM MOBICOM.*

[18] S. Madden *et al.* (2005) 'TinyDB: an acquisitional query processing system for sensor networks,' *ACM Trans. on Database Systems* **30**(1).

[19] P. Bonnet *et al.* (2001) 'Towards sensor database systems,' In *Proceedings Of IEEE MDM.*

[20] Y. Yao and J. Gehrke (2003) 'Query processing for sensor networks', In *Proceedings of CIDR.*

[21] S. Madden *et al.* (2003) 'The design of an acquisitional query processor for sensor networks,' In *Proceedings of ACM SIGMOD.*

[22] H. Wu *et al.* (2006) 'Distributed cross-layer scheduling for in-network sensor query processing,' In *Proceedings of IEEE PerCom.*

[23] D. Chu *et al.* (2007) 'The design and implementation of a declarative sensor network system,' In *Proceedings of ACM SenSys.*

[24] P. Costa *et al.* (2005) 'Publish-subscribe on sensor networks: a semi-probabilistic approach,' In *Proceedings of IEEE MASS,.*

[25] E. Souto *et al.* (2005) 'MIRES: a publish/subscribe middleware for sensor networks,' *Personal Ubiquitous Comput.* **10**(1).

[26] M. Sharifi *et al.* (2006) 'A publish-subscribe middleware for real-time wireless sensor networks,' In *Proceedings of The International Conference on Computational Science.*

[27] J.H. Hauer *et al.* (2008) 'A component framework for content-based publish/subscribe in sensor networks,' In *Proceedings of EWSN.*

[28] K. Römer and F. Mattern (2004) 'Event-based systems for detecting real-world states with sensor networks: a critical analysis,' In *Proceedings of DEST at ISSNIP.*

[29] J. Elson and D. Estrin (2004) 'Sensor networks: a bridge to the physical world', Chapter in the book *Wireless Sensor Networks,* Kluwer Academic Publishers.

[30] F. Dressler (2008) 'A study of self-organization mechanisms in ad hoc and sensor networks,' *Computer Communications* **31**(13).

[31] T. He and M. Zafer (2008) 'Adaptive sampling for transient signal detection in the presence of missing samples,' In *Proceedings of the First IEEE Workshop on Quality of Information for Sensor Networks.*

[32] D. Hakkarinen and Q. Han (2008) 'Data quality driven sensor reporting,' In *Proceedings of The First IEEE Workshop on Quality of Information for Sensor Networks.*

[33] S. Zahedi, M.B. Srivastava, and C. Bisdikian (2008) 'A computational framework for quality of information analysis for detection-oriented sensor networks,' In *Proceedings of MILCOM.*

[34] E. Gelenbe and L. Hey (2008) 'Quality of information: an empirical approach,' In *Proceedings of the First IEEE Workshop on Quality of Information for Sensor Networks.*

[35] E. Gelenbe and L. Hey (2008) 'Experimental insights into quality of information,' In *Proceedings of ACITA.*

[36] F. Stann and J. Heidemann (2003) 'RMST: Reliable data transport in sensor networks,' In *Proceedings of SNPA.*

[37] S. Kim *et al.* (2004) 'Reliable transfer on wireless sensor networks,' In *Proceedings of IEEE SECON.*

[38] Y. Sankarasubramaniam *et al.* (2003) 'ESRT: event-to-sink reliable transport in wireless sensor networks,' In *Proceedings of ACM MobiHoc.*

[39] D. Ganesan *et al.* (2001) 'Highly-resilient, energy-efficient multipath routing in wireless sensor networks, SIGMOBILE Mob. Comput. Comm. Rev.* **5**(4).

[40] H. Zhang *et al.* (2005) 'Reliable bursty convergecast in wireless sensor networks,' In *Proceedings of MobiHoc.*

[41] E. Felemban *et al.* (2006) 'MMSPEED: multipath multi-speed protocol for QoS guarantee of reliability and timeliness in wireless sensor networks,' *IEEE Trans. on Mobile Computing* **5**(6).

[42] K. Shaikh, A. Khelil, and N. Suri (2008) 'A comparative study for data transport protocols for WSN,' In *Proceedings of WOWMOM.*

[43] C. Wan, A.T. Campbell, and L. Krishnamurthy (2002) 'PSFQ: A reliable transport protocol for wireless sensor networks,' In *Proceedings of WSNA.*

[44] F.K. Shaikh, A. Khelil, A. Ali, and N. Suri (2010) 'TRCCIT: tunable reliability with congestion control for information transport in wireless sensor networks,' In *Proceedings Of The 5th Annual International Wireless Internet Conference (WICON).*

[45] F.K. Shaikh, A. Khelil, and N. Suri (2009) 'AReIT: adaptable reliable information transport for service availability in wireless sensor networks,' In *Proceedings of the International Conference on Wireless Networks (ICWN).*

[46] C.Y. Wan, S.B. Eisenman, and A.T. Campbell (2003) 'CODA: Congestion detection and avoidance in sensor networks,' In *Proceedings of SenSys.*

[47] J. Paek and R. Govindan (2007) 'RCRT: Rate-controlled reliable transport for wireless sensor networks,' In *Proceedings of SenSys.*

[48] S. Chen and N. Yang (2006) 'Congestion avoidance based on lightweight buffer management in sensor networks,' *IEEE Trans. Parallel Distributed Systems* **17**(9).

[49] R. Kumar *et al.* (2008) 'Mitigating performance degradation in congested sensor networks', *IEEE Trans. on Mobile Computing* **7**(6).

[50] B. Deb *et al.* (2003) 'Information assurance in sensor networks,' In *Proceedings of WSNA*.

[51] N. Tezcan and W. Wang (2007) 'ART: An asymmetric and reliable transport mechanism for wireless sensor networks,' *Int. Journal on Sensor Networks* **2**(4).

[52] B. Marchi *et al.* (2007) 'DTSN: Distributed transport for sensor networks,' In *Proceedings of ISCC*.

[53] Y.G. Iyer *et al.* (2005) 'STCP: A generic transport layer protocol for wireless sensor networks,' In *Proceedings of ICCCN*.

[54] J.-Y. Teo *et al.* (2008) 'Interference-minimized multipath routing with congestion control in wireless sensor network for high-rate streaming,' *IEEE Trans. on Mobile Computing* **7**(9).

[55] C. Hsu *et al.* (2007) 'A multi-path routing protocol with reduced control messages for wireless sensor networks,' In *Proceedings of IIH-MSP*.

[56] Y. Wang, W. Peng, M. Chang, and Y. Tseng (2007) 'Exploring load-balance to dispatch mobile sensors in wireless sensor networks,' In *Proceedings of ICCCN*.

[57] M. Rahimi, R. Pon, W.J. Kaiser, G.S. Sukhatme, D. Estrin, and M. Srivastava (2004) 'Adaptive sampling for environmental robotics,' In *Proceedings of ICRA*.

[58] Z. Butler and D. Rus (2003) 'Event-based motion control for mobile sensor networks,' In *IEEE Pervasive Computing*.

[59] B. Zhang and G.S. Sukhatme (2005) 'Controlling sensor density using mobility,' In *Proceedings of the 2nd IEEE Workshop on Embedded Networked Sensors*.

[60] Y. Wang, H. Dang, and H. Wu (2007) 'A survey on analytic studies of delay-tolerant mobile sensor networks,' *Journal of Wireless Communications and Mobile Computing (WCMC) SI on Disruption Tolerant Networking for Mobile or Sensor Networks* **7**(10).

[61] W. Zhao, Mostafa Ammar, and Ellen Zegura (2004) 'A message ferrying approach for data delivery in sparse mobile ad hoc networks,' In *Proceedings of ACM MobiHoc*.

[62] M.C. Chuah, and P. Yang (2005) 'A message ferrying scheme with differentiated services,' In *Proceedings of MILCOM*.

[63] H. Jun, W. Zhao, M.H. Ammar, E.W. Zegura, and C. Lee (2005) 'Trading latency for energy in wireless ad hoc networks using message ferrying,' In *Proceedings of the Third IEEE International Conference on Pervasive Computing and Communications Workshops (PERCOMW)*.

[64] H. Jun, M. Ammar, and E. Zegura (2005) 'Power management in delay tolerant networks: a framework and knowledge-based mechanisms,' In *Proceedings of IEEE SECON*.

[65] A.A. Somasundara, A. Kansal, D. Jea, D. Estrin, and M.B. Srivastava (2006) 'Controllably Mobile Infrastructure for Low Energy Embedded Networks,' *IEEE Transactions on Mobile Computing* **5**(8).

[66] A.A. Somasundara, A. Ramamoorthy, and M.B. Srivastava (2004) 'Mobile element scheduling for efficient data collection in wireless sensor networks with dynamic deadlines,' In *Proceedings of the IEEE Real Time Systems Symposium (RTSS)*.

[67] A. Kansal, M. Rahimi, W.J. Kaiser, M. Srivastava, G.J. Pottie, and D. Estrin (2004) 'Controlled mobility for sustainable wireless networks,' In *Proceedings of IEEE SECON*.

[68] A. Kansal, A. Somasundara, D. Jea, M. Srivastava, and D. Estrin (2004) 'Intelligent fluid infrastructure for embedded networks,' In *Proceedings of MobiSys*.

[69] R.C. Shah, S. Roy, S. Jain, and W. Brunette (2003) 'Data MULEs: modeling a three-tier architecture for sparse sensor networks,' In *Proceedings of the First International Workshop on Sensor Network Protocols and Applications (SNPA)*.

[70] O. Tekdas, V. Isler, J.H. Lim, and A. Terzis (2009) 'Using mobile robots to harvest data from sensor fields,' In *Proceedings of IEEE Wireless Communications*.

[71] M. Arkin, R. Hassin, and A. Levin (2006) 'Approximations for minimum and min-max vehicle routing problems,' *Journal of Algorithms* **59**(1).

[72] S. Hanoun and S. Nahavandi (2008) 'Dynamic route construction for mobile collectors in wireless sensor networks,' In *Proceedings of Intelligent Robotics and Applications*.

[73] X. Li, A. Nayak, and I. Stojmenovic (2009) 'Exploiting actuator mobility for energy-efficient data collection in delay-tolerant wireless sensor networks,' In *Proceedings of the 5th International Conference on Networking and Services (ICNS)*.

[74] W. Burgard, M. Moors, C. Stachniss, and F.E. Schneider (2005) 'Coordinated multi-robot exploration,' *IEEE Transactions on Robotics* **21**(3).

[75] R. Sugihara and R.K. Gupta (2008) 'Improving the data delivery latency in sensor networks with controlled mobility distributed computing in sensor systems,' In *Proceedings of the 4th IEEE International Conference on Distributed Computing in Sensor Systems*.

[76] M. Bosse, N. Nourani-Vatani, and J. Roberts (2007) 'Coverage algorithms for an under-actuated car-like vehicle in an uncertain environment,' In *Proceedings of ICRA*.

[77] L. Bloebaum and Ulrich Faigle (1997) 'Coverage path planning: the boustrophedon cellular decomposition,' In *Proceedings of the International Conference on Field and Service Robotics*.

[78] M.M. Solomon (1987) 'Algorithms for the vehicle routing and scheduling problems with time window constraints,' *Operations Research* **35**(2).

[79] Elbassioni, A.V. Fishkin, and R. Sitters (2006) 'On approximating the TSP with intersecting neighborhoods,' In *Proceedings of Algorithms and Computation*.

[80] M. Glocker, C. Reinl, and O. von Stryk (2006) 'Optimal task allocation and dynamic trajectory planning for multi-vehicle systems using nonlinear hybrid optimal control,' In *Proceedings of 1st IFAC-Symposium on Multivehicle Systems*.

[81] J.S.B. Mitchell (2007) 'A ptas for tsp with neighborhoods among fat regions in the plane,' In *Proceedings of the 18th ACM-SIAM Symposium on Discrete Algorithms (SODA)*.

[82] C. Reinl and O. von Stryk (2007) 'Optimal control of multi-vehicle systems under communication constraints using mixed-integer linear programming,' In *Proceedings of the First International Conference on Robot Communication and Coordination (RoboComm)*.

[83] B. Yuan, M. Orlowska, and S. Sadiq (2007) 'On the optimal robot routing problem in wireless sensor networks,' *IEEE Transactions on Knowledge and Data Engineering* **19**(9).

[84] A.H. Maja *et al.* (2001) 'An Incremental self-deployment algorithm for mobile sensor Networks,' *Autonomous Robots, SI on Intelligent Embedded Systems* **13**(3).

[85] G. Wang, G. Cao, and T.F. La Porta (2006) 'Movement-assisted sensor deployment,' *IEEE Transactions on Mobile Computing* **5**(6).

[86] S. Poduri and G.S. Sukhatme (2004) 'Constrained coverage for mobile sensor networks,' In *Proceedings of The IEEE International Conference on Robotics and Automation*.

[87] A. Howard, M.J. Mataric, and G.S. Sukhatme (2002) 'Mobile sensor network deployment using potential fields: a distributed, scalable solution to the area coverage problem,' In *Proceedings of DARS*.

[88] B. Zhang and G. Sukhatme (2005) 'Controlling sensor density using mobility,' In *Proceedings of the Second IEEE Workshop on Embedded Networked Sensors*.

[89] W. Wang, V. Srinivasan, and K.C. Chua (2008) 'Coverage in hybrid mobile sensor networks,' *IEEE Trans. Mobile Computing* **7**(11).

[90] S.B. Ayed, M. Hamdi, and N. Boudriga (2008) 'DPRMM: a novel coverage-invariant mobility model for wireless sensor networks,' In *Proceedings of GLOBECOM*.

[91] A. Kansal, W.J. Kaiser, G.J. Pottie, M.B. Srivastava, and G. Sukhatme (2007) 'Reconfiguration methods for mobile sensor networks,' *Transactions on Sensor Networks (TOSN)* **3**(4).

[92] G. Wang, G. Cao, T.L. Porta, and W. Zhang (2005) 'Sensor relocation in mobile sensor networks,' *Proceedings of the IEEE INFOCOM*.

[93] J. Teng, T. Bolbrock, G. Cao, and T.L. Porta (2007) 'Sensor relocation with mobile sensors: design, implementation, and evaluation,' *Proceedings of Fourth IEEE Int'l Conf. Mobile Ad-Hoc and Sensor Systems (MASS)*.

[94] X. Du *et al.* (2007) 'Self-healing sensor networks with distributed decision making', *International Journal of Sensor Networks* **2**(6).

[95] G. Dini, M. Pelagatti, and I.M. Savino (2008) 'An algorithm for reconnecting wireless sensor network partitions,' In *Proceedings of EWSN*.

[96] M. Younis *et al.* (2008) 'A localized self-healing algorithm for networks of moveable sensor nodes,' In *Proceedings of IEEE GLOBCOM*.

[97] Y. Mei, C. Xian, S. Das, Y.C. Hu, and Y.H. Lu (2006) 'Repairing sensor network using mobile robot,' In *Proceedings of the ICDCS Int. Workshop on Wireless Ad hoc and Sensor Networks*.

[98] S. Basagni, A. Carosi, E. Melachrinoudis, C. Petrioli, and Z.M. Wang (2008) 'Controlled sink mobility for prolonging wireless sensor networks lifetime,' *Wireless Networks* **14**(6).

[99] W. Wang, V. Srinivasan, and K.C. Chua (2008) 'Extending the lifetime of wireless sensor networks through mobile relays,' *IEEE/ACM Trans. Netw.* **16**(5).

[100] S. Basagni, A. Carosi, C. Petrioli, and C.A. Phillips (2008) 'Moving multiple sinks through wireless sensor networks for lifetime maximization,' In *Proceedings of IEEE MASS*.

[101] K. P. Shih, H.C. Chen, J.K. Tsai, and C.C. Li (2007) 'PALM: A Partition Avoidance Lazy Movement protocol for mobile sensor networks,' In *Proceedings of WCNC*.

[102] Y. Yang and M. Cardei (2007) 'Movement-assisted sensor redeployment scheme for network lifetime increase,' In *Proceedings of the 10th ACM Symposium on Modeling, Analysis, and Simulation of Wireless and Mobile Systems*.

[103] G. Wang, G. Cao, and T.L. Porta (2004) 'Proxy-based sensor deployment for mobile sensor networks,' In *Proceedings of The IEEE International Conference on Mobile Ad-hoc and Sensor Systems*.

[104] P. Corke, S. Hrabar, R. Peterson, D. Rus, S. Saripalli, and G. Sukhatme (2004) 'Deployment and connectivity repair of a sensor net with a flying robot,' In *Proceedings of the 9th International Symposium on Experimental Robotics*.

[105] G. Dini, M. Pelagatti, and I.M. Savino (2008) 'An algorithm for reconnecting wireless sensor network partitions,' In *Proceedings of EWSN*.

[106] G. Wang, G. Cao, T.L. Porta, and W. Zhang (2005) 'Sensor relocation in mobile sensor networks,' In *Proceedings of INFOCOM*.

[107] Z. Jiang, J. Wu, A. Agah, and B. Lu (2007) 'Topology control for secured coverage in wireless sensor networks,' In *Proceedings of WSNS at MASS*.

[108] A. Sekhar, B.S. Manoj, C. Murthy, and R. Siva (2005) 'Dynamic coverage maintenance algorithms for sensor networks with limited mobility,' In *Proceedings of IEEE PERCOM*.

[109] S. Ganeriwal, A. Kansal, and M. B. Srivastava (2004) 'Self aware actuation for fault repair in sensor networks,' In *Proceedings of the International Conference on Robotics and Automation*.

[110] A. Ali, A. Khelil, F.K. Shaikh, and N. Suri (2009) MPM: Map based Predictive Monitoring for wireless sensor networks,' In *Proceedings of AUTONOMICS*.

[111] P. Szczytowski, A. Khelil, and N. Suri (2010) 'LEHP: Localized Energy Hole Profiling in wireless sensor networks,' In *Proceedings of IEEE ISCC*.

[112] A. Applegate, R.E. Bixby, V. Chvatal, and W.J. Cook (2006) *Concorde tsp-solver*, Technical report, http://www. tsp.gatech.edu/concorde.html.

[113] P.M. Franca, M. Grendreau, G. Laporte, and F.M. Mueller (1995) 'The m-traveling salesman problem with minmax objective,' *Transportation Science*.

[114] V. Bugera (2004) 'Properties of no-depot min-max 2-traveling-salesmen problem,' In S. Butenko, R. Murphey, and P.M. Pardalos, eds, Recent Developments in Cooperative Control and Optimization, *vol. 3 of* Cooperative Systems, Kluwer.

[115] P. Toth and D. Vigo (1987) *The Vehicle Routing Problem*, SIAM..

[116] A. Assad (1988) 'Modelling and implementation issues in vehicle routing,' In B. Golden and A. Assad, eds, *Vehicle Routing: Methods and Studies*, Elesvier.

[117] 'Symphony open-source solver for mixed-integer linear programs,' http://www.coin-or.org/SYMPHONY.

[118] M.M. Zavlanos and G.J. Pappas (2005) 'Controlling connectivity of dynamic graphs,' In *Proceedings of The 44th IEEE Conference on Decision and Control*.

[119] H.P. Williams and S.C. Brailsford (1996) 'Advances in linear and integer programming,' Chapter in *Computational Logic and Integer Programming*, Oxford University Press, Inc., New York.

[120] L.E. Parker (2002) 'Distributed algorithms for multi-robot observation of multiple moving targets,' *Autonomous Robots* **12**(3).

[121] N. Michael, M.M. Zavlanos, V. Kumar, and G.J. Pappas (2008) 'Maintaining connectivity in mobile robot networks,' In *Proceedings of the 11th International Symposium on Experimental Robotics*.

[122] U. Furbach, J. Murray, F. Schmidsberger, and F. Stolzenburg (2007) 'Hybrid multiagent systems with timed synchronization – specification and model checking,' In *Proceedings of the 5th Workshop on Programming Multi-Agent Systems (ProMAS07)*.

[123] A. Khelil, P.J. Marrón, C. Becker, and K. Rothermel (2007) 'Hypergossiping: a generalized broadcast strategy for mobile ad hoc networks,' *Elsevier Journal on Ad Hoc Networks* **5**(5).

[124] J.H. Bose, S. Böttcher, L. Gruenwald, S. Obermeier, H. Schweppe, and T. Steenweg (2005) 'An integrated commit protocol for mobile network databases,' In *Proceedings of IDEAS*.

[125] S. Böttcher, L. Gruenwald, and S. Obermeier (2007) 'A failure tolerating atomic commit protocol for mobile environments,' in *Proceedings of IEEE MDM*.

[126] D. Skeen and M. Stonebraker (1983) 'A formal model of crash recovery in a distributed system,' *IEEE Transactions on Software Engineering* **9**(3).

[127] L. Lamport (1998) 'The part-time parliament,' *ACM Transactions on Computer Systems* **16**(2).

[128] W. Xie (2005) 'Supporting distributed transaction processing over mobile and heterogeneous platforms, dissertation,' Georgia Institute of Technology.

[129] G.C. Roman, Q. Huang, and A. Hazemi (2001) 'Consistent group membership in ad hoc networks,' in *Proceedings of ICSE*.

[130] L. Briesemeister and G. Hommel (2002) 'Localized group membership service for ad hoc networks,' In *Proceedings of IWAHN*.

[131] D. Skeen (1981) 'Nonblocking commit protocols,' In *Proceedings of COMAD*.

[132] N. Malpani, J.L. Welch, and N. Vaidya (2000) 'Leader election algorithms for mobile ad hoc networks,' In *Proceedings of DIALM*.

[133] S.M. Masum, A.A. Ali, and M.T.Y.I. Bhuiyan (2006) Asynchronous leader election in mobile ad hoc networks, In Proc. of AINA, 2006.

[134] E. Perkins and E.M. Royer (1999) 'Ad-hoc on-demand distance vector routing,' In *Proceedings of WMCSA*.

[135] D. Gracanin *et al.* (2004) 'On modeling wireless sensor networks,' In *Proceedings of IPDPS*.

[136] M. Welsh (2004) 'Exposing resource tradeoffs in region-based communication abstractions for sensor networks,' *SIGCOMM Comput. Comm. Rev.* **34**(1).

[137] Y. Zhao *et al.* (2002) 'Residual energy scan for monitoring sensor networks,' In *Proceedings of IEEE WCNC*.

[138] W. Xue *et al.* (2006) 'Contour map matching for event detection in sensor networks,' In *Proceedings of ACM SIGMOD*.

[139] M. Li *et al.* (2007) 'Non-threshold based event detection for 3D environment monitoring in sensor networks,' In *Proceedings of ICDCS*.

[140] M.V. Machado *et al.* (2005) 'Data dissemination in autonomic wireless sensor networks,' *IEEE Journal on Selected Areas in Comm.* **23**(12).

[141] O. Goussevskaia *et al.* (2005) 'Data dissemination based on the energy map,' *IEEE Comm. Magazine* **43**(7).

[142] K. Ren *et al.* (2008) 'Secure and fault-tolerant event boundary detection in wireless sensor networks,' *IEEE Trans. on Wireless Comm.* **7**(1).

[143] X. Zhu *et al.* (2008) 'Light-weight contour tracking in wireless sensor networks,' In *Proceedings of INFOCOM*.

[144] A.H. Robinson *et al.* (1995) *Elements of Cartography*, John Wiley & Sons.

[145] S. Kininmonth, S. Bainbridgea, I. Atkinsonc, E. Gilla, L. Barrald, and R. Vidaude (2004) 'Sensor networking the Great Barrier Reef,' *Spatial Sciences Qld Journal*.

[146] SensorMap, http://atom.research.microsoft.com/sensewebv3/sensormap/.

[147] SensorPedia, http://www.sensorpedia.org.

[148] OGC Sensor Web Enablement Working Group: http://www.opengeospatial.org/projects/groups/sensorweb.

[149] S. Chien *et al.* (2007) 'Lights out autonomous operation of an Earth observing SensorWEB,' In *Proceedings of RCSGSO*.

[150] S. Capkun, J.P. Hubaux, and L. Buttyan (2003) 'Mobility helps security in ad hoc networks,' In *Proceedings of MobiHOC*.

7

An Opportunistic Pervasive Networking Paradigm: Multi-Hop Cognitive Radio Networks[1]

Didem Gözüpek and Fatih Alagöz
Department of Computer Engineering, Bogazici University, Istanbul, Turkey.

7.1 Introduction

The proliferation of wireless applications and services has intensified the demand for radio spectrum. Although licensed spectrum is at a premium, a large portion of spectrum is used sporadically with a high variance of geographical and temporal usage. This ineffective utilization constitutes the very 'raison d'être' of the *dynamic spectrum access (DSA)* concept, which refers to the opportunistic utilization of the spatiotemporally unoccupied portions of the spectrum. The key and jump-start enabler for DSA is cognitive radio, which is the next evolution of adaptive/aware/software-defined radios through a supplemental intelligent layer providing DSA capability.

Cognitive radio devices can operate anywhere and using any portion of the spectrum as long as they guarantee that the licensed owners of the spectrum portion are not disturbed. This ubiquitous and opportunistic nature of cognitive radio devices renders cognitive radio networks a vital paradigm in opportunistic pervasive communications.

A cognitive radio network (CRN) is comprised of primary (PU) and secondary (SU) users. The former is the licensed owner of a frequency band, whereas the latter utilizes the spectrum opportunities during the inactive times of the PUs. A CRN may have a single-hop or multi-hop structure. The emerging IEEE 802.22 standard-based wireless regional area network (WRAN) technology is based on the single-hop CRN concept, in which a centralized cognitive base station (BS) manages the SUs that use the TV bands opportunistically when they are unoccupied by the incumbent TV services. On the other hand, multi-hop CRNs (MHCRN) have no fixed network infrastructure or central controller with an additional requirement that the information needs to be relayed over multiple wireless links. Thus, the SUs in a MHCRN have to coordinate themselves in a distributed manner.

Multi-hop wireless networks gain increasing popularity as multi-hop connections inevitably become necessary in order to maintain a high degree of network connectivity and achieve higher data rates for larger distances. Furthermore, the difficulty of providing infrastructure in certain applications such as emergency situations and battlefields necessitates that the network has an ad hoc structure. Moreover, CRNs promise to obviate radio interoperability problems with the ultimate goal of providing a 'universal wireless device'. For instance, the Joint Tactical Radio System (JTRS) attempts to provide a common architecture to solve the formidable radio interoperability

[1] This work is supported by the Scientific and Technological Research Council of Turkey (TUBITAK) under grant no. 109E256, State Planning Organization of Turkey under grant no. DPT-2007K 120610, and Bogazici University Research Fund (BAP) under grant no. 09A108P.

problems of the US military [1]. Inter-operability problems are also an impediment in joint operations, where each nation typically has its own radio system. Lately, the emphasis on peacekeeping, disaster relief, homeland security and other non-combat military operations has produced additional problems. In these roles, military units have to communicate with humanitarian organizations and public safety agencies, as well as the civilian population. Consequently, the emergency communication networks and military tactical networks of the future are expected to be based on MHCRNs. Moreover, the pivotal role of multi-hop wireless networks in sustaining a high degree of network connectivity and attaining increased data rates coupled with the enhanced spectrum efficiency and radio inter-operability benefits of CRNs make MHCRNs a strong candidate for commercial applications as well.

The primary function of a medium access control (MAC) protocol is to govern the physical layer data transmissions and to provide an access service for error control and recovery at the link layer. The design of the MAC layer for CRNs has additional challenges such as ensuring that the communications of the PUs are not disturbed and collisions among the SUs are avoided so that channel utilization is improved. Unlike single-hop CRNs, which have the emerging IEEE 802.22 standard, the MAC layer of MHCRNs has hitherto been unstandardized. Furthermore, the MAC design of MHCRNs introduces challenges that are non-existent in single-hop CRNs, such as transceiver synchronization, group communication, hidden incumbent node problem, as well as the Clear to Send (CTS) timeout problem in addressing hidden and exposed terminal problems.

The authors in [2] present an overview of the existing work and research challenges with regard to the ad hoc CRNs in general. They outline the research issues concerning spectrum sensing, decision, sharing and mobility in different layers of the protocol stack. On the other hand, the authors in [3] present a survey of the MAC protocols in CRNs in general, where they discuss both infrastructure-based and ad hoc CRNs. In contrast, we focus in this chapter on MAC protocols specific to MHCRNs. In this respect, the area of study in this chapter can be regarded as the intersection of those of [2] and [3]. Hence, we provide a deeper discussion on a narrower and more specific topic.

The rest of this chapter is organized as follows: In Section 7.2, we outline the challenges presented in the MAC design of MHCRNs, accompanied by a comparison of multi-channel networks and MHCRNs. In Section 7.3, we briefly discuss the key MAC protocols for MHCRNs, listing their advantages and disadvantages. We make a comparison of the investigated protocols in Section 7.4, and provide future directions for researchers with regard to open issues that have not been addressed thoroughly. Finally, we conclude the chapter in Section 7.5.

7.2 Overview of Multi-Hop Cognitive Radio Networks MAC Layer

7.2.1 MAC Design Challenges in MHCRNs

7.2.1.1 Common Control Channel (CCC) Problem

In a MHCRN, the SUs need to communicate with each other through control messages for accomplishing tasks such as the negotiation of a common channel available to both parties. For this purpose, a common control channel (CCC) is needed. A separate dedicated control channel would seem to be the proper solution. Although a CCC facilitates numerous spectrum sharing functions such as sender and receiver handshake or sensing information exchange, a dedicated CCC has several drawbacks. Firstly, it is a waste of channel resources. Secondly, a control channel can quickly saturate as the number of SUs increases, which constitutes a big problem, especially for MHCRNs. Thirdly, an adversary can cripple the CCC by flooding it on purpose and can thus severely obstruct the channel negotiation and allocation process, causing Denial of Service (DoS) attacks [4]. These three problems also exist in multi-channel networks. MHCRNs have the additional problem of the possibility of a PU appearance in the CCC. If an incumbent signal is detected in the same band, CCC needs to switch to another band by applying a control channel policy. In order for this control channel hopping pattern to be identical in all the nodes in the MHCRN, a channel selection policy that guarantees this requirement needs to be applied. Furthermore, each SU has to sense the CCC band constantly for a prompt detection of PU appearance. Another approach to obviate the need for a CCC is to choose a channel among the available channels as the control channel. When the PU of that channel returns, a new channel which is available to all users is chosen. Nevertheless, the probability that a certain channel is available to all SUs in a MHCRN is quite low. Furthermore, the available channels may differ in transmission range, operation frequency and bandwidth. Owing to this heterogeneity in the transmission range, the scalability and connectivity of the network is subject to change in accordance with the control channel since a channel with a shorter transmission range may not provide service to all the areas served by a channel with a longer transmission range. Therefore, a better protocol that avoids the use of a CCC, while at the same time taking network heterogeneity into account is essential.

Figure 7.1   Multi-channel hidden terminal problem.

7.2.1.2   Transceiver Synchronization

In order to establish communication and data exchange between two pairs of SUs, both the sender and the receiver node have to tune to the same channel at the same time. Since the nodes do not know which channels are available and which one of the available channels the other node is going to tune to, they need to establish the communication frequency and time period prior to the incipient communication. Transmitter-receiver synchronization challenge is unique to the MAC in opportunistic spectrum access networks and maintaining this without introducing extra control message exchange is a nontrivial task. The problem becomes even more complicated in MHCRNs because of the lack of a centralized controller to govern the transmissions from all nodes.

7.2.1.3   CTS Timeout and Undecodable CTS Problems

The existence of hidden and exposed terminals is a classical problem in MAC design for multi-hop ad hoc networks. In a MHCRN, the hidden terminals are SUs outside the secondary transmitter's range, but inside the secondary receiver's range, exposed terminals are SUs within the secondary transmitter's range but outside the secondary receiver's range. Since hidden terminals can result in collisions and exposed terminals may lead to wasted opportunities, they need to be addressed properly. Approaches such as IEEE 802.11 Request to Send (RTS)/Clear to Send (CTS) might alleviate the problem; nevertheless, this scheme presents two problems in the MHCRN domain. Firstly, the conventional RTS/CTS approach fails when the RTS/CTS packet is not decodable; e.g. when the received signal power is just below what is needed for decoding. In a MHCRN, this might happen when there is a collision due to the PU activity. Secondly, unlike in traditional MAC protocols, the sender in a MHCRN cannot merely set a fixed timeout while expecting a CTS in MHCRNs because the PU activity can inevitably prevent the SU control channel transmissions. Therefore, a more sophisticated mechanism is needed to address this problem in MHCRNs more effectively.

7.2.1.4   Multichannel Hidden Terminal Problem (MCHTP)

The multi-channel hidden terminal problem (MCHTP) was identified initially in multi-channel networks; however, the same problem also exists in MHCRNs. Figure 7.1 illustrates four nodes with their respective Available Channel Lists (ACL). Assume that only the adjacent nodes are in the transmission range. Since channel 1 is available to all nodes, suppose that channel 1 is chosen as the CCC and that nodes C and D are already communicating through channel 3. When node A wants to send a packet to node B using channel 2, it sends an RTS to B on the CCC, which is channel 1 in this case, B suggests channel 2 for data communication by sending a CTS packet. Subsequently, node A sends a confirmation message to node B and to its neighbors indicating that it has reserved channel 2 for data communication. Nevertheless, since C has been communicating using channel 3, it fails to receive the CTS from B. Therefore, it presumes that channel 2 is available and might commence communication with node B using channel 2, and hence yielding a collision. This is called the MCHTP.

7.2.1.5   Hidden Incumbent Node Problem (HINP)

In MHCRNs, when the sender and receiver nodes negotiate to determine an available channel, they select a data channel based on their ACLs in order to avoid causing interference to the PUs. The hidden incumbent node problem (HINP) that arises in this situation was introduced by the authors in [5]. In Figure 7.2, node S is the source node with an ACL of $\{1, 2, 3, 4\}$, which indicates that no signal was detected on that channel, implying the absence of any PU or other SU activity. The circles in the figure represent the transmission range of the nodes in the centre. Besides, PU_i indicates that the PU is operating on channel i. Node S sends its ACL in the RTS packet to

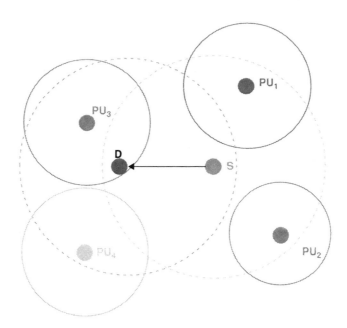

Figure 7.2   Hidden incumbent node problem.

the destination node D using the CCC. The incumbent nodes PU_1 and PU_2 are operating inside the transmission range of the source node S. However, node S cannot detect the signals of these incumbent systems since the radio waves from these signals cannot reach the source S. It is possible that after the RTS and CTS exchange between S and D, D may choose channel 1 as the data channel, which will cause interference directly to the incumbent system working on channel 1. The same also applies for channel 2. Similarly, within the transmission range of the destination node D, channels 3 and 4 are occupied by the incumbent devices. The incumbent signal on channel 3 can be sensed by D; therefore, the ACL of D excludes channel 3. In the case of channel 4, the signal from the incumbent system cannot reach node D or it can be ignored by the destination node as background noise due to weak signal strength. If node D selects channel 4 and includes this in the CTS that it sends to S, then PU_4 will receive harmful interference from the destination node D. This problem is called the Hidden Incumbent Node Problem (HINP).

7.2.1.6   Number of Transceivers

In the single transceiver model, the CR nodes utilize the same transceiver for both control and data channel transmissions. Since the nodes have to switch back periodically to the control channel, this model results in longer transmission delay than the dual transceiver model. Furthermore, a single transceiver model requires exact synchronization between the nodes. Although the dual transceiver model eliminates these drawbacks since one of the transceivers is dedicated to the control channel, this model possesses hardware implementation complexity. A similar situation exists for the sensing task. If a dedicated radio senses the spectrum continually while another radio is involved with the data transmission tasks, more transmission opportunities can be detected and the sensing results have greater accuracy. Furthermore, the multi-channel hidden node problem illustrated in Figure 7.1 can be avoided, since node C does not miss the CTS message transmitted by node B due to its dedicated control channel transceiver. Nevertheless, more radios imply hardware implementation complexity. If a single radio is used for both sensing and accessing, then the data transmission has to be interrupted periodically, which incurs more latency.

7.2.1.7   Coordination of Spectrum Sensing and Accessing Decisions

To achieve optimal performance in a MHCRN, the MAC protocol has to determine a set of channels to sense and a set of channels to access. In a single-hop CRN with a central coordinator, the coordinating node can decide on which nodes will sense and guide a cooperative spectrum sensing process in addition to making the spectrum accessing decisions. Nonetheless, since there is no central coordinator in a MHCRN, the nodes have to make the spectrum sensing and accessing decisions in a distributed manner. Moreover, the spectrum access decisions should take into account not only the availability of a sensed channel, but also the channel fading condition and the nodes' energy constraints. Furthermore, the cognitive nodes have to determine the optimal sensing time and optimal transmission time in a distributed manner.

Distinguishing the PU activity from the transmissions of other SUs is an important problem in CRNs because the PU statistics play an important role in various decision criteria, especially during channel access. That is to say, 'carrier sensing' and 'PU sensing' are two different phenomena in the MHCRN context, and hence, both of them have to be handled separately and effectively. To this end, IEEE 802.22 protocol establishes quiet periods to coordinate the spectrum sensing process. However, unlike the IEEE 802.22 network, MHCRNs lack a central coordinator. Therefore, the coordination of the quiet periods without a central coordinator is a challenging research issue in MHCRNs.

7.2.1.8 Group Communication

The proper operation of higher layer protocols in carrying out mechanisms such as address resolution depends on the existence of a group communication mechanism at the MAC layer. Per contra, finding a channel available to all the nodes in the group to communicate is a daunting task in a MHCRN, especially when taking into account the fact that even communication between a pair of CR nodes requires significant control message exchanges.

7.2.1.9 MAC Layer Authentication

In a single-hop CRN, confidentiality and authentication across the network can be provided by applying cryptographic transforms to the MAC frames. For instance, IEEE 802.22 contains a security sublayer. Unfortunately, such a protocol cannot be implemented in a MHCRN because there is no trusted entity that can act as a server to control the distribution of keying material. Therefore, an adversary node can send spurious control frames to saturate the CCC. If the control frames are exchanged in an unencrypted form, the candidate channel list in the control channel hopping pattern can also be acquired by the adversary. Thus, even if the control channel hops ceaselessly among different frequency bands due to the presence of the incumbent signals, the adversaries still have the capability to saturate the CCC continually [4].

7.2.2 *Comparison of Multi-Channel Networks and MHCRN*

Multi-channel networks and MHCRNs share many common features. In both networks, each user has a set of channels available for communication. When two users want to communicate, they might negotiate via a common control channel (CCC). Furthermore, the CCC and MCHTPs, which are related to a multi-channel network, are common to a MHCRN. Therefore, many MHCRN MAC proposals in the literature are inspired by the work on MAC designs in multi-channel networks [5, 6, 7].

There are two major differences between these two networking environments. Firstly, the number of channels available at each node is fixed in a multi-channel network, while it is variable in a MHCRN. Hence, it is probable that a SU in a MHCRN has no available channels owing to the complete occupancy of the spectrum by the PUs. Secondly, the channels in a multi-channel network generally have equal bandwidths and transmission ranges; however, the environment is heterogeneous in a MHCRN. In this respect, a MHCRN may be considered as an amalgamation of multi-hop and multi-channel networks together with the additional challenge of varying spectrum availability.

7.3 Proposed Mac Layer Protocols

In this section, we briefly describe a wide range of MAC protocols designed for MHCRNs by stating the essential behavior of the protocols wherever possible. Moreover, we also present the advantages and disadvantages of the protocols.

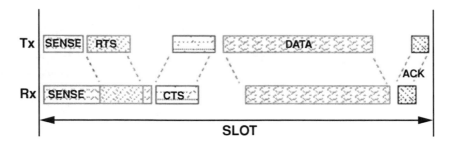

Figure 7.3   DC-MAC operation phases [8].

7.3.1   POMDP Framework for Decentralized Cognitive MAC (DC-MAC)

The authors in [8] formulate a Partially Observable Markov Decision Process (POMDP) framework for modeling the channel sensing and accessing decisions in a MHCRN. They model the PU activity as a two state (occupied/idle) Markov chain. Since in general the network state cannot be observed completely due to partial spectrum monitoring or sensing error, the model is a POMDP. At the beginning of each time slot, each user firstly determines a set of channels to sense. On the basis of the sensing observations, each node then selects a subset of the available channels to access. At the end of the time-slot, in response to these sequential sensing and accessing actions, each user obtains a reward, which is equal to the number of bits delivered. The objective is to maximize the total expected reward accumulated over a certain number of slots, where the spectrum occupancy statistics remain unaltered. The constraint is to bound the probability of collision below the permissible maximum value. Depending on the received reward, each user updates its belief vector, which represents its knowledge about the network state. Figure 7.3 illustrates the sequence of operations in this protocol.

In order to ensure transceiver synchronization, both the transmitter and the receiver nodes update their belief vector using only the common information. This way, they tune to the same channel in the next time-slot. Moreover, this framework addresses the CCC problem since it does not require a dedicated communication channel. Additionally, the proposed framework deals with the conventional hidden and exposed terminal problems using an RTS/CTS exchange mechanism.

Advantages: Spectrum sensing and accessing decisions are coordinated in a distributed manner without additional control message exchanges through a CCC. When the sender and receiver nodes are neighbours to the same PU, they are guaranteed to access the same channel without explicit synchronization. However, when this is not the case, they still need a handshaking procedure because a PU that is idle for one party may be occupied for the other one. The authors implement this handshaking procedure through RTS/CTS exchanges. Furthermore, the proposed protocol can be implemented using a single transceiver at each SU.

Disadvantages: The transmission pairs need RTS/CTS messages not only to handle the conventional hidden and exposed terminal problems, but also to negotiate the communication channel when they do not have a common neighboring PU. Since RTS/CTS message exchanges require a common communication channel, the authors do not completely solve the CCC problem, although they partially address it. Furthermore, the authors also do not address the CTS timeout problem mentioned in Section 7.2, which is specific to MHCRNs.

7.3.2   DCR-MAC

The authors in [5] focus primarily on dealing with the HINP discussed in Section 7.2. They state that it is necessary to exchange the incumbent system information with one-hop neighbors from the source and destination nodes. To this end, they propose a reactive and collision-free reporting mechanism. The neighbor nodes of the source overhear the RTS message. If they detect an incumbent signal at the pth listed sub-channel in the ACL of the RTS, they transmit a short pulse at the pth reporting slot of the reporting phase in the CCC, which is right after the RTS phase. For a certain reporting slot, if the source node detects a pulse signal, then it indicates that there exists an incumbent system around the neighbor of the source that uses the corresponding sub-channel in spite of the fact that the source did not sense it. Therefore, the source node updates the ACL and transmits an RTSu (RTSupdated) signal. Similarly, the neighbor nodes around the destination overhear the CTS and report the incumbent system information to the destination node through short pulses. The RTS/CTS exchange also serves the purposes of maintaining transceiver synchronization and resolving the conventional hidden and exposed terminal problems. The authors consider both single and dual transceiver models. The dual transceiver helps in combating the MCHTP. The single transceiver model that the authors consider adapts the MMAC ad hoc traffic indication message (ATIM) mechanism, proposed originally by the authors in [9] for multichannel networks. Besides, the authors assume a CCC in the ISM band.

Advantages: The authors address the HINP, which had previously received little attention. Since the proposed sensing information exchange mechanism between the neighbor nodes and the source/destination nodes is reactive, it does not require periodic MAC control message transmissions, which would occur in a proactive protocol.

Disadvantages: Because of the additional reporting slots in the RTS/CTS exchange, the proposed protocol incurs a greater access delay than a conventional RTS/CTS message exchange. Furthermore, the usage of a CCC makes the protocol prone to DoS attacks, and the reporting slots in the CCC exacerbate the CCC saturation problem. It is also noteworthy to mention that the proposed protocol does not entirely solve the HINP because the one-hop neighbors also may not be in the transmission range of the PU. It may even be the case that no SU is in the transmission range of the PU. Therefore, a more sophisticated transmission power control mechanism needs to be employed by the SUs in order to completely solve the HINP.

7.3.3   Cross-Layer Based Opportunistic MAC (O-MAC)

The authors in [10] propose PHY and MAC layer integrated spectrum sensing policies for MHCRNs. Each SU is equipped with two transceivers: one transceiver for spectrum sensing and data transmission and one transceiver for control channel transmissions. Firstly, the authors make a Markovian analysis of a simple random sensing policy, where each SU randomly selects one of the licensed channels to sense. The authors prove that when the number of SUs is large enough, the SUs can sense all of the licensed channels even using the simple random sensing policy. Nonetheless, this policy is inadequate when the number of SUs is smaller than or close to the number of licensed channels. To amend this weakness, the authors then propose a negotiation based sensing policy and analyze it through an $M/G^y/1$ queuing model. The basic idea is to let the SUs know which channels are already sensed by their neighbouring SUs and then select different channels to sense in the next time-slot. At the very beginning, the SUs randomly select a licensed channel to sense and report the channel state by sending beacons in the reporting phase of the control channel. During the negotiation phase, the SUs encapsulate the channel sensing information into the RTS/CTS packets. The neighboring nodes that overhear these packets learn about whether they have sensed the same channel. If there are neighboring SUs that sense the same channels as the sender in a particular time slot, each of them will sense another different licensed channel in the subsequent time slot, which is randomly picked up from the rest of the channels that have not been sensed. If the number of SUs is larger than or equal to the number of licensed channels, the negotiation based sensing policy eventually reaches the desired state where all the licensed channels are sensed by all the SUs.

Advantages: The rigorous throughput and delay analysis provides insights into under which circumstances a simple random sensing policy is enough and when a more sophisticated negotiation based sensing policy is needed.

Disadvantages: In addition to the inclusion of the channel sensing information into the RTS/CTS packets, the usage of the control channel for the reporting and negotiation phases aggravates the CCC saturation problem. Furthermore, the authors assume that the licensed channel availability information is consistent among all SUs; i.e. all SUs utilize the licensed channels used by the same set of PUs. This assumption may hold only for small scale MHCRNs. This entire analysis is invalid for situations where a licensed channel is occupied by a PU and hence unavailable for the SUs in some part of the MHCRN, but it is available in another part of the MHCRN.

7.3.4   HC-MAC

The authors in [11] propose a cognitive MAC protocol that determines the optimal spectrum sensing decision for a single secondary transmission pair with single radios that cannot sense and transmit simultaneously. If more channels are sensed in a certain time period, more channels may be available for transmission. Nevertheless, sensing consumes time and if the additionally sensed channels are unavailable it is a waste of time since this time could be used for data transmissions instead. The contention, channel sensing and data transmission phases are sequential, as illustrated in Figure 7.4. The authors formulate this spectrum sensing decision problem as an optimal stopping problem, which can be solved by backward induction. Three types of control messages are used.

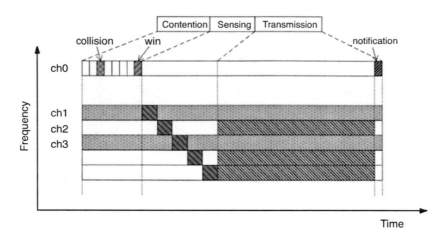

Figure 7.4   HC-MAC operation phases [11].

C-RTS/C-CTS messages are used for contention and spectrum reservation. Any SU hearing either of these two messages defers its operation and waits for the notification message. S-RTS/S-CTS messages are used to exchange the channel availability information between the sender and receiver in each sensing slot. T-RTS/T-CTS messages are used to notify the neighbouring nodes the completion of the transmission.

Advantages: The influence of sensing overhead for the multi-channel opportunity is considered. The approach requires little hardware complexity, since the hardware constraints such as a single radio for both spectrum sensing and data transmission, as well as partial spectrum sensing ability are taken into account.

Disadvantages: Additional control message exchanges in the decentralized version of the approach have a detrimental impact on the CCC. Furthermore, HC-MAC does not take the impact of SU spectrum usage into account. A secondary pair A-B that wins the contention period senses the spectrum with the neighboring nodes silenced. Nevertheless, the two hop away nodes that do not receive the C-RTS/C-CTS messages can still perform their operations. If these two hop away nodes operate on the same channels as the ones sensed by the pair A-B, then the sensing results of these A-B nodes will inaccurately indicate that there is a PU transmission on these channels. This situation is referred to as the sensing exposed terminal problem and it is not handled effectively by HC-MAC.

7.3.5   C-MAC

The major component of the Cognitive MAC (C-MAC) protocol proposed by the authors in [6] is the rendezvous channel (RC). In the C-MAC protocol, each channel has a superframe structure and one channel is identified as the RC, which is decided dynamically and in a distributed manner. The RC has functionalities such as synchronization among the available channels, discovery of the neighbor nodes, channel load balancing, group communication support (multicast and broadcast), and mitigation of the conventional hidden terminal problem as well as the MCHTP. The superframe structure also includes a slotted beaconing period, through which the nodes exchange information during the channel negotiation.

Advantages: One of the major merits of the RC is the support for group communication, which is often neglected in other existing work in the literature. Although not stated explicitly by the authors, the proposed distributed beaconing approach can help alleviate the HINP because the nodes acquire the information about their neighbors' neighbors such as occupied beacon slot and transmission schedules through this approach.

Disadvantages: C-MAC overcomes the MCHTP by having one transceiver tuned perpetually to a pre-determined RC. In other words, only when at least two transceivers are available, can C-MAC mitigate this problem. With a single transceiver, SUs have to switch back periodically to the RC both for control messages and re-synchronization. Therefore, when C-MAC is employed with a single transceiver, the nodes can miss the control frames informing them about the data transmissions among their neighboring nodes, which can possibly lead to a MCHTP.

7.3.6   DOSS-MAC

The Dynamic Open Spectrum Sharing MAC (DOSS-MAC) protocol proposed by the authors in [13] is based on the busy tone concept. Each node has three transceivers; i.e. for the busy tone channel, control channel and data channel. When a node is receiving data on a particular frequency band, it also transmits a busy tone signal in another narrowband frequency, which is found by mapping the wide-band data channel via utilizing a transformation function. A node that has data to transmit observes these busy tone bands and hence it is apprised of all the data reception activities in its neighborhood. This way, hidden and exposed terminal problems are eliminated.

The authors also state that multicasting and broadcasting are addressed by having the transmitter send a multicast/broadcast request packet over the CCC and all the pertinent nodes adjust their receivers accordingly. No busy tone is used in these message exchanges. However, if the spectrum declared by the transmitting node is not available at some recipients of the broadcasting request packet, then this message exchange process fails to address this problem. Determining a frequency band that is available for all the recipients of the multicast/ broadcast packet is a nontrivial task that has not been handled by the DOSS-MAC protocol.

Advantages: Since the receiver node sends the busy tone, DOSS-MAC successfully overcomes the problem that arises when the RTS/CTS packet is not decodable while trying to handle the conventional hidden node problem.

Disadvantages: The major drawback of this protocol is the need for three dedicated transceivers, which yields a costly hardware implementation. Besides, if a PU appears on the busy tone band after a data receiving SU turns on its busy tone signal or if a PU already occupies that frequency band at the time that the data receiving SU implements the spectrum mapping, then the data receiving SUs cannot send their busy tone signals in the corresponding band. The SUs have the additional burden of being obliged to monitor the busy tone bands continuously in order to ensure that the busy tone signals are turned off in a timely manner as soon as a PU appears in that band. Moreover, when a PU is operating on the busy tone band, the other SUs that observe this band may decide erroneously that an SU is receiving data in the corresponding wide-band data channel and refrain from using the data channel. Therefore, these situations constitute a waste of opportunities and additional complexity in the SUs.

The authors also state that multicasting and broadcasting are addressed by having the transmitter send a multicast/broadcast request packet over the CCC and all the pertinent nodes adjust their receivers accordingly. No busy tone is used in these message exchanges. However, if the spectrum declared by the transmitting node is not available at some recipients of the broadcasting request packet, then this message exchange process fails to address this problem. Determining a frequency band that is available for all the recipients of the multicast/broadcast packet is a nontrivial task that has not been handled by the DOSS-MAC protocol.

7.3.7   SCA-MAC

The authors in [12] propose a channel allocation strategy for ad hoc cognitive radio networks. Their proposed method predicts the successful rate of a channel by first calculating the probability of that particular spectrum band being idle and then the probability that a packet with a specific length will fit the spectrum hole during the idle PU period. Subsequently, they bundle several continuous idle channels to expedite the data transmission. This way, the authors ensure that the interference to the PUs is limited by a predetermined acceptable rate. The cognitive nodes use Control-channel-Request-To-Send (CRTS) and Control-channel-Clear-To-Send (CCTS) messages to coordinate the access to the channel through the control channel. Access to this control channel is implemented by a CSMA/CA mechanism. The control packets carry the information of packet length and channel aggregation, whose expected successful rate meets the interference limit. After the exchange of the control packets, the sender and receiver nodes tune their transceivers to the agreed channels. Figure 7.5 depicts the operation phases of the SCA-MAC protocol.

Advantages: The ability of the protocol to guarantee that the interference imposed to the PUs is bounded is an essential feature for CRNs, since it ensures that there is no noticeable deteriorating impact on the QoS of the PUs. Furthermore, incorporating the channel statistics into the decision criteria for channel access enables quality assessments of the channels. This way, the SU acquires the ability to wait intelligently for a busy channel with a high successful rate to become idle again when the currently existing idle channels have low successful rate.

Disadvantages: The authors assume that a CCC exists. When the CRTS/CCTS packets fail due to collision among the other SUs when they are trying to be transmitted over the CCC, the SCA-MAC protocol simply restarts the negotiation process. Nevertheless, this renegotiation exacerbates the CCC saturation problem.

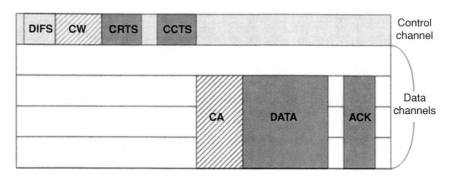

Figure 7.5 SCA-MAC operation phases [12].

7.4 Open Issues

Table 7.1 provides a comparison of the MAC protocols investigated in Section 7.3, which address various aspects of the MHCRN specific research challenges discussed in Section 7.2. Each protocol is marked as either (+), indicating that the corresponding research issue is addressed or (−), indicating that the issue is not addressed.

MAC Layer Authentication: None of these studies solves completely all the vital aspects related to MHCRN operation. Transceiver synchronization is addressed by all the investigated protocols; however, none of the work addresses the MAC layer authentication problem, which is a challenging issue in the absence of a centralized controller.

Quiet Period Coordination in Spectrum Sensing and Accessing: The nodes in a MHCRN have to distinguish between the transmissions of PUs and the transmissions of other SUs. To this end, other SUs may be forced to be silent when an SU senses the channel. Coordination of these quiet periods in a distributed manner is a possible research issue. The only work that claims to have addressed this problem is C-MAC [6], where the authors mention that they use the RC for this purpose. Nevertheless, they do not discuss about how this coordination would be implemented using the RC and without the presence of a central entity.

CTS Timeout Problem: Conventional hidden and exposed terminal problems are addressed by most of the work through IEEE 802.11 such as RTS/CTS mechanisms; nevertheless, none of these studies considers the cognitive radio specific fact that no fixed timeout can be applied while expecting the CTS message because of the PU activity. A possible research issue might be to incorporate the predicted channel usage pattern into the calculation of the CTS timeout value. In other words, a method that changes the CTS timeout value dynamically according to the predicted channel usage pattern of that frequency might be considered as a possible research issue. For instance, the work in [12] might serve as a basis and be modified to be incorporated into the CTS timeout value calculation.

Tradeoff Between MCHTP and the Number of Transceivers: MCHTP can be entirely obviated with the usage of a dedicated CCC transceiver. Nevertheless, an additional dedicated transceiver implies hardware implementation complexity. Thus, these two problems are inter-related and there is usually a tradeoff in their design and implementation. Most of the investigated protocols address either the CCC problem or the MCHTP, but not both of them [5, 8, 13]. The only protocol that addresses both problems is C-MAC [6]. Although C-MAC avoids the usage of a CCC while combating the MCHTP, this capability comes at the hardware expense of having a dedicated transceiver tuned to a predetermined rendezvous channel. Since this extra transceiver does not exist in the actual protocol but mentioned by the authors as a possible extension to alleviate the MCHTP, we marked the number of required transceivers for C-MAC as one in Table 7.1. Combating the MCHTP without a dedicated control channel transceiver is a promising research issue.

HINP: HINP is specific to MHCRNs; therefore, there are no other previously proposed protocols in other realms, such as multi-channel networks or IEEE 802.11 MAC, which can be adapted to the MHCRN framework. HINP is also not solved completely by the investigated protocols, although it is partially addressed by the authors in [5] and [6].

Group Communication: Group communication has also received little attention. Determining a communication channel for a group of SUs without introducing extra control message overhead is an open and challenging research issue. C-MAC [6] seems to be the only work that addresses group communication, considering that the multicasting/broadcasting capability of DOSS-MAC [13] entails some drawbacks, as discussed in Section 7.3.

Table 7.1   Comparison of MAC protocols for multi-hop cognitive radio networks

Issues +: Addressed −: Not addressed	DDC-MAC [8]	DCR-MAC [5]	O-MAC [10]	HC-MAC [11]	C-MAC [6]	DOSS-MAC [13]	SCA-MAC [12]
CCC Problem	+	−	−	−	+	−	−
Transceiver Synchronization	+	+	+	+	+	+	+
Conventional Hidden/Exposed Terminal Problem	+	+	−	+	+	+	+
CTS Timeout Problem	−	−	−	−	−	−	−
Undecodable CTS Problem	−	−	−	−	−	+	−
Multichannel Hidden Terminal Problem (MCHTP)	−	+	−	−	+	+	−
Hidden Incumbent Node Problem (HINP)	−	+	−	−	+	−	−
Number of Transceivers	1	Both 1 and 2	2	1	1	3	1
Coordination of Spectrum Sensing and Accessing Decisions	+	−	+	+	+	−	−
Group Communication	−	−	−	−	+	+	−
MAC Layer Authentication	−	−	−	−	−	−	−

7.5 Conclusions

To put it in a nutshell, MAC design for MHCRNs carries with it the challenges of multi-channel networks and multi-hop networks in addition to the complications that stem from the varying spectrum availability of cognitive radio networks. The major challenging research issues are the common control channel (CCC) problem, transceiver synchronization, conventional and MCHTPs, the hidden incumbent node problem (HINP), the number of transceivers, coordination of spectrum sensing and accessing decisions, group communication and MAC layer authentication. There is as yet no study that addresses all of these issues concurrently and effectively. Moreover, there is currently no existing standard for the MAC design of MHCRNs. Therefore, protocols that handle all of these research challenges are imperative and crucial in actualizing the opportunistic pervasive networking paradigm of multi-hop cognitive radio networks.

References

[1] L. Pucker (2003) 'Applicability of the JTRS software communications architecture in advanced MILSATCOM terminals,' *Military Communications Conference (MILCOM)*.

[2] I. Akyildiz, W. Lee, and K. Chowdhury (2009) 'CRAHNs: Cognitive radio ad hoc networks,' *Ad Hoc Networks*.

[3] C. Cormio and K. Chowdhury (2009) 'A survey on MAC protocols for cognitive radio networks,' *Ad Hoc Networks*.

[4] K. Bian and J. Park (2006) 'MAC-layer misbehaviors in multi-hop cognitive radio networks,' *US-Korea Conference on Science, Technology, and Entrepreneurship (UKC)*.

[5] S. Yoo, H. Nan, and T. Hyon (2008) 'DCR-MAC: distributed cognitive radio MAC protocol for wireless ad hoc networks,' *Wiley Journal on Wireless Communications and Mobile Computing*.

[6] C. Cordeiro and K. Challapali (2007) 'C-MAC: A cognitive MAC protocol for multi-channel wireless networks,' *IEEE Symposium on New Frontiers in Dynamic Spectrum Access Networks (DYSPAN)*.

[7] A. Mishra (2006) 'A multi-channel MAC for opportunistic spectrum sharing in cognitive networks,' *Military Communications Conference (MILCOM)*.

[8] Q. Zhao, L. Tong, A. Swami, and Y. Chen (2008) 'Decentralized cognitive MAC for opportunistic spectrum access in ad hoc networks: a POMDP framework,' *IEEE Journal on Selected Areas in Communications (JSAC)* **25**(3): 589–99.

[9] J. So and N. Vaidya (2004) 'Multi-channel MAC for ad hoc networks: handling multi-channel hidden terminals using a single transceiver,' *5th ACM International Symposium on Mobile Ad hoc Networking and Computing*, pp. 222–33.

[10] H. Su and X. Zhang (2008) 'Cross-layer based opportunistic MAC Protocols for qos provisionings over cognitive radio wireless networks,' *IEEE Journal on Selected Areas in Communications (JSAC)* **26**(1): 118–29.

[11] J. Jia, Q. Zhang, and X. Shen (2008) 'HC-MAC: A hardware-constrained cognitive MAC for efficient spectrum management,' *IEEE Journal on Selected Areas in Communications (JSAC)* **26**(1): 106–17.

[12] A. Hsu, D. Wei, and C. Kuo (2007) 'A cognitive MAC protocol using statistical channel allocation for wireless ad-hoc networks,' *IEEE Wireless Communications and Networking Conference (WCNC)*.

[13] L. Ma, X. Han, and C. Shen (2005) 'Dynamic open spectrum sharing MAC protocol for wireless ad hoc networks,' *IEEE Symposium on New Frontiers in Dynamic Spectrum Access Networks (DYSPAN)*.

8

Wearable Computing and Sensor Systems for Healthcare

Franca Delmastro and Marco Conti
Institute of Informatics and Telematics (IIT) at the National Research Council of Italy (CNR), Pisa, Italy.

8.1 Introduction

In recent years wearable computing and sensors technology has gone hand in hand with medical healthcare representing an innovative and leading solution designed to improve the quality of life of patients and prevent critical situations both inside and outside medical facilities. However, more work is needed in this research field to respond appropriately to both medical and technological challenges in deploying efficient and reliable e-health systems. First of all, we have to identify the main issues in current healthcare systems in order to design and adapt the technology in this direction. Following the international reports on healthcare [1], national health systems need essentially a reduction of costs for maintaining high quality of treatment and guaranteeing high quality of life for patients. They should also provide easy access to care for as many people as possible, anywhere and anytime, addressing in particular the increase in the aging population and the care of chronic diseases, which represent one of the major causes of death worldwide. With this aim in mind, healthcare professionals are trying to improve the efficiency of the patient care, focusing on the *'continuum of care'* [2]. This requires continuous medical assistance to the patient, from the beginning of his/her hospitalization to discharge and consequent rehabilitation at home, increasing the demand for portable and versatile medical devices that can support both the patient and the doctor in continuous monitoring, thus becoming the medium of communication between these two entities. The American Institute of Medicine [2] summarized six aims for healthcare improvement identified by: safety, effectiveness, patient-centered, timeliness, fairness and efficiency across the different nations. To address all of these features the experts claimed that it is necessary to define a mobile information infrastructure tailored to the individual's requirements that can take advantage of the advances in telemedicine systems and information processing techniques. Following these guidelines, wearable and ubiquitous sensors, together with personal mobile devices, represent the new frontier moving towards a novel definition of e-health systems that we can define as *pervasive healthcare systems*. With this definition we can aggregate both patient-centered and hospital-centered systems. The former are mainly dedicated to remote and continuous monitoring of patients outside the medical facility and during daily activities, while the latter are designed to improve the medical workflow inside the facility. The former, defined also as *Personal Health Systems*, represent the most challenging solutions in terms of pervasive technologies and communications, with reference especially to wearable computing. These systems will provide the collection, processing, storage and transmission of medical information, maintaining the fundamental requirements of user acceptability and comfort. In fact, the main functionality of wearable sensors for healthcare (i.e. vital signal monitoring, storage, communication with external devices, low power consumption) must be defined in accordance

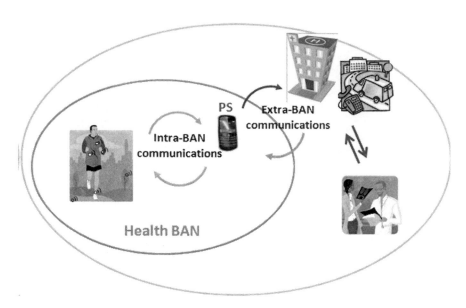

Figure 8.1 Personal Health System architecture.

with the patient's basic needs (e.g. wearability, unobtrusiveness, no skin irritation, easily maintainability) in order to guarantee the correct and constant usage of the system. In this chapter we survey the technological aspects of the design and development of wearable sensor networks for healthcare, analysing and comparing the different solutions proposed in the literature and trying to envisage some future trends in this research area. Specifically, we introduce the concept of the Health Body Area Network (Health BAN) in Section 8.2, providing a general overview of the system architecture of Personal Healthcare Systems, as illustrated in Figure 8.1. Then, focusing on wearable technologies, we describe the medical and technological requirements of health sensors in Section 8.3, analyzing the correspondence between most common diseases and the physiological and non-physiological parameters that are useful in monitoring applications. Further to this, we refer to three main categories of wearable sensors designed for three different purposes: vital signals monitoring (Section 8.4), activity recognition (Section 8.5) and emotion recognition (Section 8.6). In these sections, existing wearable computing solutions are presented, highlighting the main advantages, drawbacks and possible enhancements for the design and development of the next generation. In addition, the analysis and correlation of the parameters derived from these different categories allows the system to have a complete clinical profile of the patient; however, some technological issues have to be addressed in order to guarantee an efficient and reliable system. Specifically, wireless communications in Health BAN represents one of the main aspects to be investigated with particular attention given to power consumption, reliability, data rate and delays. In Section 8.7 we describe the consolidated wireless standards for sensor networks (i.e. Bluetooth and ZigBee) and their performance evaluations in healthcare environments, followed by a description of the novel specification of Bluetooth aimed at low energy consumption as the most promising technology for Health BANs. We conclude the chapter with a brief discussion on users' acceptance of wearable technology as further inputs for new design and development.

8.2 The Health Body Area Network

Initially, sensor networks were designed for collecting sensing information from stationary nodes, spread in the environment, transmitting data at a relatively low data rate to achieve best effort data collection at a central base station. Medical monitoring and related applications, essentially working with critical information, have more stringent requirements (e.g. reliable communications, high data rate, small dimensions and the possibility of sending data to multiple receivers), thus introducing the concept of *Health BAN* [3]. It consists mainly of a network of communicating devices (sensors, actuators and other mobile devices) generally worn on or implanted in the body providing mobile health services to the user. To guarantee efficient and reliable services, the Health BAN needs to communicate with a remote medical information system where care professionals can monitor the patient's status constantly and provide the correct care plan or related actions. To realize this aim, an additional component of the Health BAN is selected to collect and transmit sensed data to the medical server. Generally, this role is undertaken by a mobile device with a discrete storage and processing capacity, able to support multiple wireless protocols and inter-device communications. It has been defined in the literature in several ways: Mobile Base Unit (MBU) [3], Personal Mobile Hub (PS) [4], Personal Server [5] and others, depending on its technical features, from the basic collection of sensed data [3, 4] to the more sophisticated elaboration and correlation of several values to provide at least a preliminary feedback to the patient [5]. In this chapter we generally refer to it as Personal Server (PS) since this definition includes the most complete set of functionalities associated with its role. Therefore, these three elements (i.e., Health BAN, PS and medical server) represent the three tiers of the architecture of a pervasive Personal Healthcare System (see Figure 8.1), dedicated mainly to patient monitoring.

Another important factor in the system is represented by communication protocols between the components. Specifically, we refer to the communication protocols among health sensors and the PS as Intra-BAN communications, and those between the PS and the medical server as Extra-BAN communications. Since the main target of this work is to describe the main impacts of wearable and ubiquitous technology on e-Health, we focus on the description and evaluation of health sensors, specifying their medical, technological and communication requirements (especially in terms of Intra-BAN communications), with detailed examples of developed technologies for specific application scenarios.

8.3   Medical and Technological Requirements of Health Sensors

With regard to the general technological features of a sensor node, we can describe its architecture as a set of components: the sensing element that collects an analog signal, an analog-digital converter, a processor, a wireless transceiver, a flash memory, an antenna and the battery. Indeed, depending on the way they are empowered (local or shared power supply), they are divided into two main categories: *self-supporting* and *front-end supported* sensors. Self-supporting sensors have their own power supply and represent independent building blocks of the BAN, guaranteeing a high configurability but, at the same time, characterized generally by independent internal clocks and sample frequency, thus requiring synchronization mechanisms. By contrast, front-end supported sensors share a common power supply and generally also share data acquisition procedures, operating on the same front-end clock and providing multiplexed samples as a single data block; thus they do not require synchronization procedures. Health sensors contain both of these categories, and different systems exploit them depending on their requirements and final objectives. In general, the main features of a sensor node and its performance depend strictly on the specific application scenarios it is designed for. In case of healthcare systems, there are so many differences among the physiological signals to be monitored and the possible system configurations that there does not exist a single scenario involving all possible diseases and medical characteristics; thus every system defines a set of customized features for the sensors involved. However, before analyzing the existing solutions and their detailed characteristics, we identify a set of medical and technological requirements that are shared among most health sensors:

- *Wearable*: in order to guarantee user acceptability and the correct use of this technology in healthcare applications, sensors must be characterized by very small dimensions, light weight and possibly be integrated in the textile fabric.
- *Reliable wireless communications*: initial solutions for pervasive healthcare systems proposed wearable sensors connected through wires integrated into the textile fabric, but they displayed some drawbacks related to interference created by wires that act as antennae in the woven, fixed positions of the sensors, and possible structural damage caused during patient activity [6, 7]. Thus, to solve these problems and improve ability to wear the sensors, wireless communications are needed. However, at the same time, they should guarantee reliable transmission of data, avoiding persistent packet loss (due to network congestion or node mobility) and consequent loss of important vital signals.
- *Efficient power consumption*: pervasive healthcare systems need power saving policies designed for increasing the life time of the Health BAN reducing the dimensions of battery packs. This is also a general feature for sensor networks, but its implementation depends strictly on the data delivery model and communication standards used for the specific application scenario, since it has been proven that most of the energy is consumed during transmissions [8].
- *Multiple receivers*: patient data can be sent to a central server through the PS as a central aggregation point and then forwarded to the interested care givers or, in some (emergency) cases, it could be necessary to transmit critical

data directly to the interested users such doctors, nurses and family care-givers. Thus, multicast transmission should be also considered as a general feature of Health BANs.

- *Mobility*: data communication should address the mobility issue both for the patient and the care-givers, establishing ad hoc communications or exploiting the network infrastructure, if it exists.
- *Security/Privacy*: patient information is highly sensitive both in terms of security and privacy, since malicious use could even cause the death of a patient (e.g. in the case of data manipulation in the re-programming phase of implantable devices) and violate the professional confidentiality of the doctor. Therefore a general framework including authentication, authorization and appropriate security schemes must be defined for Health BANs.
- *Adaptability*: health sensors should allow custom calibration and tuning of the sensing procedure depending on the patient's status at a specific period of time or during a particular activity.
- *Interoperability*: sensors should be easily integrated and able to interoperate among them and with the PS. The definition of standard interfacing protocols for the configuration of the sensing platform and their possible interactions would favour vendor competition resulting in more affordable systems for pervasive healthcare.

These requirements address the main technical and medical issues in designing Health BANs and related systems. However, the existing solutions do not address all of these aspects, especially adaptability and interoperability.

In the following we present a set of health sensors developed within the framework of projects and experimental activities in the e-health research field in order to give an overview of the state-of-the-art and discuss possible enhancements, related mainly to the integration of several sensors in different healthcare solutions in order to improve the accuracy and reliability of measurements. To better understand the objectives and implementations of current Health BANs, it is important to have a view of the relationship between the diseases that can benefit from continuous monitoring of the patient's status, and the specific signals that should be measured. In this way, we can have an overview of the sensors that can be used for specific medical conditions. Specifically, in Table 8.1 we identify the physiological and non-physiological parameters that are generally associated with the most common diseases.

Most physiological parameters listed can be measured directly by wearable sensors (e.g. ECG, blood pressure) or can be derived from the analysis and correlation of different signals (e.g. heart rate derives from ECG). However, in specific cases, the evaluation of particular biochemical parameters is necessary in order to obtain a complete diagnosis and monitoring of the disease. For example, cardiac and tumour markers, for ischemic heart disease and cancer respectively, can be measured through the use of implantable biosensors. Biosensors are used to transform biological actions or reactions into signals that can be processed to improve the accuracy of specific physiological measurements. For example, an implanted 'excitable-tissue' biosensor can be used as a real-time, integrated bioprocessor to analyze the complex inputs regulating a dynamic physiological variable such as the heart rate [9]. The study and development of biosensors are addressed by bioengineering researchers working in harness with medical specialists owing to the intrusive features of these devices. However, the analysis and correlation of biosensor output, together with physiological and additional parameters that characterize the history and clinical profile of a

Table 8.1 Relationships among diseases and signals to be monitored

Disease	Physiological parameters
Hypertension	Blood pressure
Ischemic Heart disease	Heart rate, Electrocardiogram (ECG), cardiac markers, cardiac stress testing, coronary angiogram
Heart failure, cardiac arrhythmias	Heart rate, blood pressure, ECG, fluid balance, body weight
Cardiovascular diseases	Heart rate, blood pressure, life style, ECG
Post-operative monitoring	Heart rate, blood pressure, ECG, oxygen saturation, body temp.
Cancer (breast, prostate, Lung, Colon)	Weight loss, tumor markers, blood detection (urine, feces...)
Asthma/ Chronic Obstructive Pulmonary Disease (COPD)	Respiration rate, oxygen saturation
Diabetes and obesity	Dietary and activity parameters, blood glucose value
Neurological diseases	Emotional parameters, ECG, EEG, heart rate, Blood Volume Pulse (BVP), skin temperature, Galvanic Skin Response, Pupil Diameter

patient, represent the focal point of pervasive healthcare systems, and they can lead to improving the accuracy of single measurements and the diagnosis of related diseases.

Looking at Table 8.1 we may notice that there are two main categories of diseases: chronic diseases [10] (e.g. cardiovascular diseases, heart failure, hypertension, cancer, diabetes) that mainly need the monitoring of vital signals such as heart rate, ECG and blood pressure, in addition to life style information, and neurological (e.g. Parkinson, Alzheimer) and neuro-psychological (e.g. obesity and dietary) diseases, which are related mainly to physical activity, motion-analysis and emotion recognition.

This distinction is reflected in the solutions proposed in the literature. In fact, some propose general platforms that can be used to monitor patients affected by different cardiovascular diseases. They focus mainly on sensing the most common physiological signals that are also identified as 'vital signals' (i.e. heart rate, ECG, blood pressure). Instead, specific solutions are proposed to address neurological and neuro-psychological diseases, analysing activity and/or emotional signals to prevent critical episodes and make the patient able to react accordingly. The details of these solutions and related wearable sensors are described in the following sections.

8.4   Wearable Sensors for Vital Signals Monitoring

Vital signals monitoring represents the basic feature of all pervasive healthcare systems, and the evolution of wearable technology has improved its accuracy and reliability significantly. In Table 8.4 (at the end of the chapter) we summarize the most important solutions of Health BANs designed mainly for physiological monitoring, analyzing the main differences in sensors development and communication strategies. The Georgia Tech Wearable Motherboard (GTWM) [7], also known as 'Smart Shirt' (see Figure 8.2), represents one of the first applications of wearable sensors for healthcare. Designed originally to improve medical assistance in military scenarios (such as detecting the penetration of a projectile and monitoring the vital signals of soldiers on the battlefield), it introduced the concept of sensors array integrated in the garment. It exploits commercial off-the-shelf sensors such as ECG and pulse oximeter[1] for vital signals monitoring, a microphone for voice recording, and it integrates plastic optical fibres in the fabric and a low power laser for penetration sensing [7]. In this case, sensors are connected to the Smart Shirt Controller that acts as PS to collect and transmit data to the central medical server exploiting Bluetooth or 802.11b wireless communications. This solution presents some problems related mainly to the integration of sensors in the textile fabric. First of all, the ECG sensor was developed using conventional electrodes that can suffer from noise during movement, causing the corruption of the signal. In addition, the physical structure of the shirt requires fixed positions for the sensors and the wires used to connect them can generate interference. The idea of sensors integrated in the textile (also known as textile sensors) was then evolved in other projects: Smart Vest [11], MagIC [12], WEALTHY [13] and MyHeart [14]. The first follows the Wearable Motherboard model, increasing the number of integrated sensors. It is able to monitor ECG, PPG (photoplethysmogram), heart rate, blood pressure, body temperature and Galvanic Skin Response (GSR) continuously, all integrated in specific locations on a shirt [11]. The measurement and analysis of PPG waveform, correlated with the ECG signal, allows the system to implement a non-invasive method of monitoring blood pressure, without using the conventional cuff method [15]. For this purpose, the PPG sensor is realized mainly through a pulse oximeter placed on the patient's finger/ear lobe which is connected to the shirt. The authors also developed a customized ECG sensor in the form of two belts of silicon rubber with pure silver fillings designed to improve the accuracy of measurement during the patient's movements with respect to traditional electrodes. Finally, the GSR is measured by passing a small current through a pair of electrodes placed on the skin and measuring the conductivity level. The shirt has a wired connection to the 'Wearable Data Acquisition Hardware' (the correspondent PS) that transmits sensed data wirelessly to the central server, and all sensors are powered by a rechargeable battery. The MagIC system integrates only sensors for ECG and respiration monitoring in the vest, and provides a portable electronic board designed for data collection and motion detection. More specifically, the ECG sensor consists of two woven electrodes made by conductive fibers; the elastic properties of the garment guarantee direct contact with the patient's body (thorax) without requiring gel or other medium. The respiratory frequency is then measured through a textile transducer analyzing the assessment of the changes in the thorax's volume. Finally, the electronic board, in charge of collecting sensed data, is also equipped with a two-axis accelerometer to detect the subject's movements. In fact, to evaluate

---

[1] Pulse oximetry is generally used to assess heart rate and blood oxygen saturation (SpO2) reliably. It consists of monitoring the pattern of light absorption by hemoglobin. The level of SpO2 is measured detecting the amount of absorbed light at two different wavelengths, while the heart rate is determined observing the pattern over time since blood vessels contract and expand with the pulse [11].

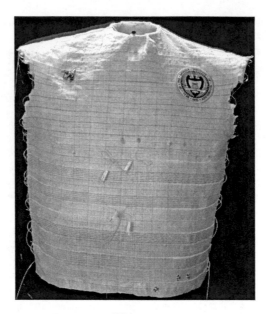

Figure 8.2 Georgia Tech Wearable Motherboard™. Reproduced by permission of © Georgia Institute of Technology.

the physiological parameters correctly, especially in the case of patients affected by cardiac diseases, it is important to correlate the information with the activity and movements of the patient. The board has a fixed position on the shirt and is connected to the sensors through the conductive fibers of the garment. It communicates with the medical server through a wireless connection. Experimental evaluations demonstrated a close similarity of the ECG signal measured by the MagIC system and a traditional ECG recorder, also during physical exercise, with an accuracy of about 95% [16]. Following the same model, the WEALTHY and MyHeart projects developed a shirt equipped with textile sensors. The former proposes an innovative method for sensor development based on the standard textile industrial processes, while the latter focuses on the enhancement of signal quality and management of a huge quantity of sensed data, which is one of the main challenges for smart clothing. To resolve this issue the MyHeart project proposes two embedded signal processing techniques designed to extract relevant data from physiological sensed data before being transmitted outside the Health BAN (see [14] for detail).

In all of these examples we may note that, even though sensors are developed to be as integrated as possible in a vest that is comfortable for the patient, there are drawbacks that represent a limitation for both user acceptance and continuous and long-term monitoring of physiological signals (especially in daily life). They are related mainly to wires integrated into the fabric, the fixed position of sensors, the presence of a centralized processing unit to elaborate, digitize and transmit all data and the specific structure of the vest [6]. Recent advances in integration and miniaturization of sensors (see Micro-ElectroMechanical Systems (MEMS) and Nanotechnology) allow for the definition of a new generation of wireless sensor networks suitable for several application scenarios, especially in healthcare. Following this line, several alternative solutions to textile sensors for health monitoring have been proposed in the literature. For example, the CodeBlue project [17] proposes a set of wearable health sensors based on the Mote technology. They developed a pulse oximeter integrating an available OEM module for heart rate and SpO2 calculations (BCI Medical Board [18]) whose board is able to relay data on a serial line that can interface with the mote platform. They also developed Mica2/MicaZ and Telos mote platforms providing continuous ECG monitoring by measuring the differential across a single pair of electrodes (Mote-EKG). Finally, they developed a motion-analysis sensor board containing a 3-axis accelerometer, single-axis gyroscope and an EMG (electromyographer[2]). The triaxal accelerometer is used to measure the orientation and movement of a body segment; a gyroscope measures the angular velocity and, combined with an accelerometer, is used to improve

---

[2] Electromyography (EMG) detects the electromechanical properties of muscle fibres. It requires correct positioning and excellent contact with the skin, in addition to complex signal processing that make the devices bulky and expensive.

accuracy in limb position measurement. The EMG sensor consists of two surface electrodes designed to capture the electrical field generated by depolarized zones in muscle fibres during contraction. Since the root mean square of EMG data is nearly proportional to the force exerted by the monitored muscle, EMG analysis is especially useful in activity recognition. Code Blue motion board was designed originally for monitoring stroke patients' rehabilitation and the efficacy of the care plane for patients affected by Parkinson's disease, but it can be applied in several other scenarios. Finally, it also proposes a RF-based location system (MoteTrack [17]) to locate patients using only the low-power radios already incorporated in the previous sensors (performance analysis shows an 80th percentile location error).

In order to coordinate and support these different sensor platforms, CodeBlue also proposes a middleware framework implemented in TinyOS [19]. It implements a publish/subscribe routing framework to allow multiple sensors to relay data to all interested receivers, a discovery protocol to make the Personal Server able to discover available sensors and a query interface to select sensed data to be downloaded on the mobile device through the use of filters, or specifying the source sensor through a physical address. Thus, CodeBlue represents one of the most complete pervasive healthcare solutions, from the design and development of customized wearable sensors to the definition of a general software architecture involving all the components of the Health BAN. However, some issues arise from an initial evaluation of the system: lack of reliable communications due mainly to patients' mobility, the need for efficient techniques for sharing bandwidth among sensors (e.g. prioritize critical message with respect to standard physiological values) and the lack of security (this is a critical aspect that affects all pervasive healthcare systems).

An extension of CodeBlue software and hardware has been proposed in AID-N (Advanced Health and Disaster Aid Network) [5] as an electronic triage system for medical support in case of disaster. In this case, the three-tier architecture differs mainly from the classical pervasive healthcare system in the second tier, where the concept of PS for data gathering and transmission is designed for the medical personnel and not for the patient. AID-N exploits CodeBlue sensor technology to develop an ETag sensor board able to embed mote-based pulse oximeter and mote-EKG creating low-power and low-data rate ETag devices. In fact, the sensor board communicates with the mobile personal server of care-givers through the IEEE 802.15.4 standard with a maximum data rate of 250 kbps (referring to MicaZ and TmoteSky motes), while the personal server communicates with the remote medical server using the IEEE 802.11 standard. CodeBlue software then allows the care-givers to control multiple ETags simultaneously, creating a mesh network between patients and care-givers in mass casualty events, able to support hundreds of patients. The miTag device in [20] further enhances the electronic triage system of AID-N, introducing additional tags with temperature sensor and GPS to track patients at every stage of disaster recovery (i.e. disaster scene, ambulance, hospital). Regarding the organization of the Health BAN, the system elects one of the miTags on the patient's body to operate as the hub for data aggregation before transmission to the personal server of the care-giver. In addition, miTag introduces the concept of the dynamic health monitoring platform, making the sensor hardware and software able to adjust their configuration dynamically to suit the current scenario. For example, miTag can increase the sensing frequency when the patient's status deteriorates, select the appropriate sensors for the current patient conditions, and enter the sleep-mode appropriately so as to save energy.

Another example of a remote patient monitoring platform has been proposed in the MobiHealth project [3] and the subsequent Awareness project [21]. The former proposes a general monitoring platform based on the classical Health BAN architecture, i.e. self-supporting and front-end supported sensors, SpO2 and ECG respectively, communicating to the PS that transmits data to the central server. Here, intra-BAN communications exploits Bluetooth technology, while extra-BAN communications use UMTS/GPRS standards. Awareness evolves MobiHealth BAN introducing EMG, respiration and temperature sensors in addition to various motion sensors (step counter and triaxal accelerometer), since it is designed essentially for neurological diseases (e.g. epilepsy, spasticity and chronic pains). Here, the main objective is to predict a critical condition, and alert the patient, just before it happens to make him/her able to react accordingly. For this reason the system also provides vibration and auditory signals as biofeedback to the patient.

In contrast with the previous solutions, based on the use of separate wearable sensors for the different physiological signals to be monitored, other health monitoring systems tried to reduce the number of separate devices, integrating all of the necessary sensors in a single wearable device. The Scalable Medical Alert and Response Technology (SMART) system [22] proposes a Waist pack containing SpO2 and ECG sensors, a sensor box with two AA batteries to power the sensors, and a PDA (equipped with a location tag) to collect sensed data. Depending on the chosen SpO2 sensor (located on the finger), its communication with the sensor box and the PDA can be wired through a serial line, or Bluetooth, even though this last solution results in unreliable communications and high power consumption. Regarding the ECG sensor, the sensor box contains a Cricket Mote processor to analyze data and transmit them to the PDA used as PS. SMART is designed for monitoring patients in the waiting areas of emergency rooms, to provide a constant monitoring of unattended patients, at least regarding the most significant vital signals. This is an important application scenario, considering that patients spend several hours in emergency rooms before being visited, based on a preliminary diagnosis. Thus, the monitoring platform can help care professionals prevent critical situations inside the facility.

As a further evolution of integrated wearable sensors for healthcare, the AMON system [23] proposes a health monitoring system integrating basic physiological and activity sensors, communication and processing modules in a single wrist-worn device. AMON monitors SpO2, skin temperature, blood pressure, ECG and the level of physical activity through an accelerator (it mainly detects walking or running pace). The device is also equipped with a cellular engine connected to the GSM network to transmit data to the central server. Experimental evaluation of this wearable device highlights the difficulties in signal acquisition (especially for ECG) with respect to previous distributed systems that place sensors in specific and favourite positions on the body. However, the study demonstrates the feasibility of the integration of multiple health sensors in a single device, improving user acceptance and comfort, especially in case of daily monitoring.

In the next section we present specific examples of wearable sensors and systems for activity and emotion recognition. Regarding environmental sensors useful in health monitoring, their features depend mainly on the specific application scenarios and their possible correlations with the related diseases, thus involving many possible combinations and correlations that we cannot address in this chapter due to lack of space. However, we can say that the use of basic environmental sensors (like temperature, light and humidity) is made easier by their integration in commercial sensor boards, such as the Telos platform used in ActiS sensor nodes [25, 26], or by their deployment in the home environment of the patient (as presented in ALARM-NET system for assisted living and residential monitoring [27]).

8.5   Wearable Sensors for Activity Recognition

As explained previously, the correct diagnosis and monitoring of many diseases can depend strictly on the relationship and correlation of physiological signals and specific user's movements or activities (see Table 8.1). In addition, for specific neurological diseases, emotional parameters can further enhance the description of the patient's health status. Wearable technology can support both activity and emotion recognition establishing accurate relationships among different physiological signals, or developing novel dedicated sensors.

Several of the systems presented in the previous section take into account activity recognition, with particular attention paid to neurological scenarios such as stroke rehabilitation, Parkinson's and epilepsy. However, these solutions exploit inertial sensors (e.g. accelerators, gyroscopes) to evaluate the level of activity of the patient (i.e. walking, running, sitting), but they are not able to detect the exact movement or the specific activity. Where this feature is available we can talk about activity recognition and motion characterization systems [28]. Detailed information about user motion can improve the accuracy of activity recognition providing additional indicators (e.g. user fatigue). Some of the work in this direction has been proposed in the literature, designing and developing innovative wearable sensors. However, the main innovation is represented by the correlation of data derived from motion analysis with the patient's health profile.

As with vital signals monitoring, sensors and systems for activity and motion recognition depend on the application scenario they are designed for. For example, [28] explores the possibility of using wearable force sensors placed on the muscle surface to obtain information about locomotion problems and user fatigue (this could be especially useful in the rehabilitation of patients affected by disorder of the central nervous system, e.g. stroke). Wearable force sensors can be realized as ultra thin foils, or even in textiles, using capacitance change between two conductive layers. Sensed data can be used both as additional source of information, to improve the accuracy of measurements derived from standard techniques (generally EMG and MMG[3]), and as a novel type of information, providing behavioral and physiological indicators. Specifically, the work in [28] demonstrates that locomotion modes can be derived from the relationship between signals from front-leg and back-leg muscles, and that long-term muscle inflation can be used as a muscle fatigue indicator. Thus, it shows that the information derived from this kind of sensor goes beyond that derived from inertial motion sensors, enriching the context derived from classical activity recognition procedures. As a more specific application, [29] proposes the use of force sensors

---

[3] Mechanomyography (MMG) detects the mechanical oscillation over a contracting muscle by attaching electrodes on the skin overlying the selected muscle.

and fabric stretch sensors (attached to the lower arm) to capture muscle contractions during specific hand and arm movements. Most of the work on wearable sensing for activity recognition is based only on arm motion analysis, since movement of the lower arm can be measured easily and unobtrusively through inertial sensors, while hand action recognition is generally implemented using gloves or other obtrusive instruments. In [29] the main idea is to exploit the analysis of lower arm muscle contraction and circumference variations to detect hand motion. The proposed sensors provide an alternative method to implement motion and activity recognition with respect to the more obtrusive EMG technique that requires several electrodes and complex signal processing schemes. It could be especially useful in cases of rehabilitation to support patients in executing physical exercise correctly, or monitoring their daily activities. However, this work represents only a first step towards specific wearable sensors for muscle and motion analysis and it must be validated in larger studies, considering also that muscle properties depend greatly on usage, age and sex, thus providing different results for different subjects.

As a more general solution for activity recognition in terms of daily activities and life style [30] proposes a wrist-worn device that is able to monitor and log the patient activities continuously using a model based on the 'user's rhythms'. The system exploits environmental sensors such as light and temperature in addition to inertial sensors (accelerometer and tilt switch, used alternatively to save energy) able to detect the level of activity and posture information. The correlation of sensed data improves the accuracy of activity recognition corresponding to specific circadian frequencies (heart rate). The basic rhythm model corresponds to a single day sampled in a 5-minute time slot. In order to initialize the model and support subsequent sensed data, the user is requested to register the high-level information of planned and usual activities following the same time scale, to make the system able to correlate the user's information and sensed data. Specifically, the system stores the start, stop and duration of an activity providing three probability distributions for each activity in relation to heart rate and circadian rhythm. Activities include eating (breakfast, lunch, dinner), driving, sleeping, taking shower, sauna and so on. The rhythm model is trained and evaluated for long periods (by weeks to months) and past activities are used to improve accuracy in recognition of current activities. Preliminary results show that accuracy in specific activity recognition depends mainly on the type of activity and its duration within the one-day period of time (e.g. sleeping is easily recognized since it has a long duration). However, the rhythm model improves the recognition of specific activities with respect to the exclusive use of sensors' data in the case of usual and recurrent activities, while it has no effect on recognizing activities that do not belong to the user's habits. This solution is not designed specifically for the healthcare environment, but it can be integrated in pervasive healthcare systems, especially in the case of remote monitoring applications, enriching the system with information about the life style of the patient. This could improve the monitoring system, adding coaching and counselling features to classical medical feedback, such as alert and updates of the care plan, especially in the case of chronic diseases where a correct life style can improve both the quality and duration of the patient's life.

Nowadays, one of the main diseases highly influenced by life style is obesity. Estimations presented in [31] account for over one billion overweight and 400 million obese patients worldwide, and this trend is increasing. This is mainly due to wrong dietary habits that can involve other important diseases such as diabetes mellitus, cardiovascular diseases and several types of cancer. Thus, dietary monitoring can be seen as another possible application for activity recognition in pervasive healthcare. Original solutions designed for this purpose are based on food intake questionnaires, specifying an estimation of the calories as a manual acquisition method, shopping receipts scanning [32] as well as products' bar codes or patient's voice log [33]. These methods require a constant interaction with the patient and they are generally prone to errors due to imprecision and missing detail. To overcome these limitations [31] proposes an innovative solution based on the use of wearable sensors. More specifically, it focuses on three main activities:

- arm and trunk movements associated with food intake actions. This information is obtained by inertial sensors (accelerator, gyroscopes);
- chewing of foods, recording the sound of food breakdown with an ear microphone[4];
- swallowing activity obtained by EMG electrodes and stethoscope microphone integrated in a sensor-collar.

Using the sensed data the system is able to derive pattern models for dietary activities and improve the event recognition procedures. The approach to detect and classify the activities is divided into three steps: signals segmentation to define search bounds, event detection based on a feature similarity search algorithm and event

---

[4] The ear canal has been proven to provide the best SNR between chewing and user speaking considered as a noise [28].

Table 8.2   Performance evaluation of dietary activity recognition [32]

	Movement recognition	Chewing recognition	Swallowing recognition
Recall	80%	93%	68%
Precision	64%	52%	20%
Recognition rate	75–82%	85–87%	64%

fusion. Event fusion procedures can combine events of different types (*competitive fusion*) or different modalities of the same type of activity (*supportive fusion*). It is important to note that by using multiple detectors for each event, a competitive fusion method is used to select the final event, while supportive fusion is used to combine the detection of different modalities related to the same activity, to reinforce the final selection. To evaluate the performance of the activity recognition procedure based on these methods, the authors define two indexes: *Recall* as the ratio between the number of recognized events (i.e. events returned and recognized correctly by the system) and the number of relevant events (i.e. events annotated manually as they actually occurred) and *Precision* as the ratio between the number of recognized events and the number of retrieved events (i.e. events returned by the event recognition procedure). For each activity a set of event categories have been defined; regarding movement recognition, four categories were defined for each intake session (i.e. eating meat with fork and knife, fetching a glass and drinking from it, eating a soup with a spoon and eating slices of bread with one hand); regarding chewing recognition, food is divided by consistence (i.e., dry, wet, soft); swallowing recognition does not require specific categories even though its frequency depends on food category. Experimental evaluations show that good results can be obtained in specific dietary activity recognition exploiting different event fusion methods, but further studies are necessary in this field both to identify the categories inside each activity and to correlate the information derived from each activity recognition in order to improve the overall accuracy (in addition to basic information such as meal schedule, intake timing and food quality). Table 8.2 shows a summary of the performance results presented in [32]. Note that we present only the best results among those obtained with the different event fusion methods.

8.6   Sensors and Signals for Emotion Recognition

Emotion recognition has a great importance in healthcare systems dedicated to neurological and neuro-psychological diseases. Since the relationship among physiological parameters, activities and the emotional status of a person is completely subjective and dependent on the health conditions of the patient, it is especially hard to define a set of objective parameters to be monitored and their relationship with specific emotional concepts. In addition, based on medical experience, these patients generally reject the use of pervasive technology, thus few solutions in this field have been proposed so far, maintaining classical solutions based on questionnaires related to patients' activities and feelings. The starting point is to identify possible relationships among physiological signals and emotional status. A summary derived from most of the work presented in the literature is shown in Table 8.3. A more invasive solution is proposed in [37], based mainly on EEG (electroencephalogram) signal analysis, since it has been demonstrated that it generally contains emotional markers. The system is based on a commercially available EEG wearable sensor consisting of an EEG cap, an amplifier and an analog-digital converter, and it is able to distinguish among five different classes of emotions on both valence and arousal dimensions exploiting the IAPS method.

The combination of these parameters is generally used to determine a set of features and inputs for specific learning classifiers able to learn and derive the 'best' result in terms of emotion recognition (in this specific application scenario). Neural networks, Support Vector Machine (SVM), Naïve Bayes and Decision Tree [34] are some examples of learning classifiers used in the systems presented in the following. The classification of the emotional status of a patient is generally defined in terms of *arousal* and *valence* values, whose reference values are obtained through some classical methods. One of the most famous methods is known as the IAPS photo set [35]. It consists of a set of 800 photos classified by a large number of users evaluating how strong the content is (*arousal*) and how positive or negative the content is considered to be (*valence*). In [36] this method is used to initialize the system recording physiological values of the patients looking at pictures with different levels of valence and arousal, to train a neural network classifier, and then test the system on other patients. This work focuses mainly on monitoring EMG, skin temperature and conductivity, Blood Volume Pulse (BVP), ECG and respiration rate. Results presented in [36] show that the estimation of valence values from physiological signals is

Table 8.3 Possible relationships among physiological signals and emotional status

Physiological signals and sensors	Emotional status
EMG as muscle tension measure	High muscle tension generally occurs under stress, but the reference value for this parameter depends greatly on the muscle where it is measured.
Skin Conductivity (GSR)	It increases if the skin is sweaty. It can help differentiating between conflict/no-conflict situations and anger/fear. It is influenced by external parameters (e.g. environmental temperature).
Skin temperature	It decreases when the muscles are tense under strain. It depends also on external parameters.
PPG and ECG to measure heart rate, blood pressure and Blood Volume Pulse (BVP)	Low heart rate variability can indicate a state of relaxation, while high variability can indicate a stress situation.
Respiration rate	Fast and deep breathing → excitement (anger, fear, joy) Rapid shallow breathing → tense anticipation (panic, fear, concentration) Slow and deep breathing → states of withdrawal (depression, calm happiness)
Pupil diameter	Pupil size variation is related to cognitive information processing that, in turn, relates to emotional states (e.g. frustration, stress). It can also be used as indication of affective processing.
EEG	Recognizing emotional markers in EEG signal.

much more difficult than arousal (89.7% with respect to 63.8% considering an error range of 10%), even though the distance between the two parameters decreases greatly considering a light increase in the error range (96.6% with respect to 89.9% with 20% error range).

A more invasive solution is proposed in [37], based mainly on EEG (electroencephalogram) signal analysis, since it has been demonstrated that it generally contains emotional markers. The system is based on a commercially available EEG wearable sensor consisting of an EEG cap, an amplifier and an analog-digital converter, and it is able to distinguish among five different classes of emotions on both valence and arousal dimensions exploiting the IAPS method.

Evaluating the percentage of samples in which the emotion is recognized correctly as classification rate, the system shows a 90% classification rate for both valence and arousal in a case where the same data is used for both training and testing the system, while it falls down to 30% using different data, either using different classifiers or grouping some classes of emotions in a single super-class, simply identifying notions of positive, negative and neutral emotions. Therefore, further work is needed in evaluating the relationship between EEG signals and emotion to improve the accuracy of recognition, but this study has also to take into account user acceptance in wearing this obtrusive equipment or find a more comfortable and unobtrusive solution.

Emotion recognition is an important feature also addressed by the Human Computer Interaction research field, in order to adapt dynamically the user interface and the system's reaction to the affective state of the user. The results obtained in this field exploit wearable sensor technology to define relationships between physiological values and emotional status, and they can be also applied directly in pervasive healthcare systems. For example, [38] proposes a system architecture for the definition of multi-modal affective user interfaces based on emotion recognition. It collects physiological signals – skin temperature, galvanic skin response (GSR), heart rate – in addition to facial expression, vocal intonation and language, to define a database of emotion concepts that maintains a mapping between these parameters and related emotional status. A SenseWear armband is used to collect physiological values while users observe short segments of movies, following the model presented in [39] and selecting a specific set of emotions (e.g., sadness, amusement, fear, anger, surprise). Three different pattern recognition algorithms have been used to evaluate the performance of the system in terms of accuracy of recognition, which has been measured in the range [71%, 83%], comparing system results with emotions reported by users. However, it has been observed that not all people are able to identify accurately single emotions with respect to general feelings (i.e. positive and negative), thus influencing the objectivity of the system evaluation.

Another physiological signal that can be used in emotion recognition is Pupil Diameter (PD). Medical studies showed that pupil size variation is related to cognitive information processing that greatly influences emotional status [40]. In addition, in the Human Computer Interaction field it has been shown that it can be seen as an indicator of affective processing [41]. On this basis, [34] proposes an automatic stress detection system that exploits PD observation in addition to more common physiological parameters (GSR, BVP, skin temperature). To measure PD, an infra-red eye tracking sensor is used to collect eye movements and point of gaze information (see [42] as an example of an eye-tracking commercial device). Considering the combination of all the parameters, the accuracy achieved by the system is in the range [78.65%, 90.1%] using three different learning classifiers. The recognition rate drops dramatically to [53.65%, 58.85%] excluding PD measurement, while it increases to [82.81%, 90.1%] excluding skin temperature.

These results demonstrate the importance of PD as a physiological parameter to be considered in stress and emotion recognition, and considering the new unobtrusive sensor developed for this purpose, additional investigation in this direction could be especially useful for emotion recognition systems. All of the previous systems are based on data acquired from a single subject, thus obtaining a user-dependent system, and requiring from the user a long training procedure before the system is able to recognize his/her emotions. As an alternative solution [43] proposes a user-independent system for emotion recognition based on physiological signal databases obtained by tens to hundreds of subjects following a multi-modal approach (audio, visual and cognitive stimuli) instead of the classical IAPS method. The system exploits skin temperature variation, GSR, ECG and facial EMG as physiological parameters. Applying detection, feature extraction and pattern classification algorithms to those data, the system should be able to improve the classification rate. However, the results show comparable rates with systems based on a single subject. Since it is not feasible to obtain a rigorous definition of emotional status and their relationship with physiological parameters, the possibility of having a general database containing emotions classifications, derived from different subjects and with different shades in terms of valence and arousal, is an important objective for emotion recognition systems, and further investigation in this direction could be especially useful.

Emotion recognition is thus an important part of the multi-parametric monitoring necessary for pervasive healthcare systems, especially in the case of patients affected by neurological and neuro-psychological diseases. These patients are generally characterized by repeated phases of mania or depression, and the best therapy is the ability to recognize the transition phases between normal, manic and depressed conditions in order to make the patient able to react accordingly, even from the psychological point of view. To allow this, a continuous monitoring system is necessary to support patients in their daily life, and several parameters must be involved. Preliminary work focused on the early diagnosis of bipolar disorder is presented in [44], studying the feasibility of a system involving parameters related to insomnia and sleep disorders (heart and respiration rates monitoring in addition to capacitive pressure sensors to monitor sleep motion), activity and emotion recognition, and environmental parameters (to enrich the description of the surrounding context). A first tentative step to evaluate and recognize emotional state in this scenario is based on the analysis of the verbal activities and social contacts or conversations of the patient through automatic speech character identification. In [45], the authors present preliminary results focusing on a set of basic emotional states, but the correlation of this data with sleep analysis and environmental conditions is still the subject of ongoing work. Currently the definition of pervasive solutions for monitoring both mental and stress-related disorders stirs up the research community. This is demonstrated also by the funding of new European projects such as MONARCA [46], focused on bipolar disorder, and INTERStress [47] based on the use of pervasive technologies and virtual reality to support patients affected by stress in their daily activities. Other projects (e.g. Chronius [48] and METABO [49]), dedicated chiefly to chronic diseases, are currently studying the involvement of environmental conditions in health monitoring as further input to multiparametric monitoring. However, this fundamental feature of pervasive healthcare systems requires not only signal analysis and processing, but also reliable communication platforms and interoperability of wireless sensors. Thus, to give a complete overview of wearable computing for healthcare, in the next section we analyse the characteristics of current communication protocols for short-range wireless sensor networks and their main issues for health monitoring.

8.7 Intra-BAN Communications in Pervasive Healthcare Systems: Standards and Protocols

Wireless communications in pervasive healthcare systems represent the conjunction between the generation of sensitive and personal data and its elaboration in order to provide a feedback both to patients and doctors. However, the wireless standards currently available for sensor and personal area networks have several limitations when used in healthcare environments due mainly to the strict application requirements. Specifically, one of the main requirements of Health BANs is low power consumption and radio communications have the greatest impact on it. Thus, low-power communication protocols are necessary to design an efficient and reliable Health BAN. Original solutions aimed at reducing power consumption by exploiting low data rates. In fact, most of them are based on Bluetooth and ZigBee (IEEE 802.15.4) standards to define Intra-BAN communication protocols, while the Extra-BAN communications, involving the rest of the system (i.e. the Personal Server and the medical server), are based mainly on infrastructured networks since those devices have no strict requirements for power consumption. However, in recent years other standards have emerged that are promising for of Intra-BAN communications, such as Bluetooth Low Energy, an evolution of classical Bluetooth standard that greatly reduces power consumption. In this section we provide an overview of all of these standards highlighting their features and their use in e-health systems.

8.7.1   IEEE 802.15.4 and ZigBee

IEEE 802.15.4 represents one of the reference standards for sensor networks communications. In fact, it was designed originally to define physical and MAC layers for very low-power and low-duty network connections in order to allow the deployment of long-lived systems with low data rate requirements. These features fit well the requirements of the Health BAN but the technical detail must be analyzed so as to better understand its impact in this field.

IEEE 802.15.4 can operate in three different frequency ranges: 868 MHz, 902–928 MHz and 2.4–2.4835 GHz. The first band has a single communication channel with a data rate of 20 Kbps, the second has 10 channels with a 40 Kbps data rate each, and the third is divided in 16 channels, each with a 250 Kbps data rate. A device can assume two types of role: Full Function Device (FFD) and Reduced Function Device (RFD). The former can talk to all the others and can operate both as the network coordinator and a simple device. The latter can only talk to an FDD to send small amount of data. Thus, the network must contain at least one FFD device. It is expected that RFD devices will spend most of their operational life in a sleep state to save their batteries, and only wake up periodically to listen the channel in order to determine whether a message is pending.

These roles create two possible network topologies: star topology and peer-to-peer. On the one hand, star topology is the best choice for low-latency communications between a more powerful device and its peripherals. On the other hand, in peer-to-peer topology the network is organized as a multi-hop ad hoc network in which each device communicates directly with the others within its transmission range. Generally, the latter is used to cover large areas in which a single device has not enough power to communicate with all the others. Instead, in the case of a Health BAN, the star topology has the best advantages in terms of higher data rate and the presence of an external coordinator of the network (generally the personal server) that could also be used to access an external power supply.

In this configuration, there are two communication modes: *beacon* and *non-beacon*. In the former, the network coordinator controls the communication directly by transmitting regular beacons for synchronization and control messages such as the start and end of a *superframe*.[5] In this way, the coordinator can communicate with the nodes whenever it is necessary, but the nodes must wake up to receive beacons at regular time intervals. Instead, in non-beacon mode, a node can send data directly to the coordinator using the CSMA/CA technique if required (to avoid power-consuming collisions in case of simultaneous transmissions), but it must wake up and poll the coordinator to receive data from it. In this case the receiver node does not have to wake up periodically to receive the beacons, but the coordinator cannot decide autonomously when to communicate with the nodes. Therefore, the choice between the two communication modes is generally given to upper-layer protocols, depending also on the final application features and requirements.

In fact, on top of 802.15.4 physical and MAC layers, the ZigBee Alliance has defined additional layers dedicated mainly to routing, security and application features. In this chapter we focus mainly on the performance results of this technology in Health BANs, both with simulative and experimental analysis. Specifically, [50] and [51] present the performance results of ZigBee and Bluetooth technologies in order to highlight the advantages and drawbacks of the initial competitors for Health BAN communications (the results related to Bluetooth will be analyzed in the next section).

---

[5] A superframe is defined as the interval of time between two beacons sending in which nodes are able to transmit (see [63] for details).

The work in [50] focused mainly on battery lifetime evaluation in the case of implanted sensors communicating with an external device as network coordinator (star topology) using ZigBee. In this case, battery lifetime is expected to be on the order of tens of years, to avoid multiple and intrusive operations on the patient. However, analytical model and simulation results showed that, by maximizing the beacon period in order to reduce the number of times the receivers must wake up, the battery is able to reach 15 years' lifetime only under very tight data rate restrictions (on the order of 10 bps for 2.4 GHz band[6]). Therefore, further optimizations of the MAC protocol based on the beaconing approach have been proposed in the literature (S-MAC, T-MAC and B-MAC are some examples summarized in [52] but they mainly present low network throughput and non-negligible delays). Instead, the non-beacon mode results in more efficient performance exploiting larger packets and achieving higher upload and download rates (approximately 20 bps). However, these results are far from the requirements of medical applications that involve a huge quantity of data to be transmitted in almost a continuous way. Even the simulation results presented in [51] revealed the scalability problems of this technology due mainly to bandwidth limitations. In this case, considering an ECG application on a sensor network configured in a star topology, in which the wearable device generates 4 Kbps of data and requires that the additional latency introduced by packing samples and transmission is less than 500 ms, the efficiency of the network drops when more than three sensors are used. To overcome these limitations it is necessary to operate the network protocols in correlation with the requirements of medical applications in terms of data rate and delay. Therefore, it could be useful to move signal processing and preliminary elaboration of sensed data directly to the sensor board. For example, a preliminary analysis could be related to the classification of sensed data based on the frequency of their transmission, e.g. periodic and sporadic, required by the specific application or the amount of data that really needs to be transmitted. For example, [53] proposed an Adaptive Dynamic Channel Allocation (ADCA) algorithm on ZigBee based on the transmission frequency differentiation applied to data derived from an ECG monitoring application. In this physiological signals are classified using two threshold values indicating three health status levels for the patient: good, fair and critical. If a number of consecutive values of a vital signal exceed a specific threshold, the patient status is upgraded/downgraded, indicating also a possible alert to the medical server. Thus, periodical transmissions can be used to monitor the status of the patient and, in case there is a variation in the referenced health status, the application can require an immediate transmission to the central server, impacting thus on the bandwidth requirement of the application. The proposed algorithm tries to reserve bandwidth on the wireless medium in case of emergency transmissions, requesting bandwidth dynamically from neighboring nodes and migrating the transmission on the available bandwidth. In this case nodes share channels and are responsible for their dynamic allocation, guaranteeing a low blocking probability for emergency data transmissions, maintaining low power consumption and reducing critical delays. Naturally, complex processing analysis is required to correlate different physiological signals, that also have external parameters that can influence the health status of a patient. Thus, we have to evaluate carefully the correct trade-off between communications and processing in the development of a Health BAN.

8.7.2   Bluetooth

Bluetooth technology was designed originally for cable replacement and short-range ad hoc connectivity. It operates in the 2.4 GHz ISM frequency band exploiting 79 RF channels of 1 MHz width each, thus defining a maximum transmission rate of 1 Mbps. The building block of a Bluetooth Personal Area Network is represented by the Piconet, i.e. a set of up to eight devices sharing the same physical channel. One of these devices assumes the role of Master (in charge of establishing and managing the communication), and all the others play the role of Slave. These devices are synchronized on the same clock and adopt the same frequency hopping scheme based on a Time Division Multiplexing technique that divides the channel in 625 μsec slots. Transmissions occur in packets that occupy an odd number of slots (up to five) and are transmitted on different hop frequencies with a maximum hop frequency rate of 1600 hops/sec. The communication protocol is divided mainly to in two phases: the discovery phase in which the master device discovers up to seven active slaves in its transmission range and exchanges data necessary for synchronization, and the data exchange. Therefore, data can be transferred between the master and one slave; then the master switches from one slave to another in a round-robin fashion. Two types of communication link are defined: Asynchronous Connection-Less (ACL), an asymmetric point-to-point

---

[6] These results are obtained by considering a crystal tolerance of better than 25 ppm defined as the initial deviation of the crystal or oscillator frequency as compared to the absolute at 25C.

link between the master and the active slaves using retransmissions to guarantee data integrity; Synchronous Connection-Oriented (SCO), a symmetric point-to-point connection between the master and a specific slave at a regular time interval. The latter is designed mainly for supporting real-time traffic (e.g. voice), while the former is dedicated to data communications as necessary for healthcare applications. Evaluating end-to-end delays, packet loss at the receiving node, and efficiency in terms of the ratio between the successful received data packets and the number of data packets generated by the application layer for that receiver, the simulation results in [51] show that Bluetooth technology presents mainly scalability issues related to the limited number of sensors for the collecting device (max seven slaves for each Piconet) and packet loss problems in case of multiple piconets for the same receiver (master) due to interferences. Even though power consumption during transmission is quite low, the need for Bluetooth devices to be active continuously for device discovery or join new piconets implies higher power requirements. Nevertheless, Bluetooth is one of the most frequently used technologies in Health BAN, probably because it is integrated natively in a high number of mobile devices, while it is not easy to find a smartphone equipped with a ZigBee wireless card.[7] In the last few years, Bluetooth SIG has worked on a new standard aimed at reducing the power consumption of Bluetooth devices, as well as meeting medical application requirements. This standard is called Bluetooth Low Energy and it mainly redesigns the communication protocol while maintaining its basic features.

8.7.3   Bluetooth Low Energy

Bluetooth Low Energy (LE) wireless technology represents the main feature of the latest Bluetooth Core specification (v.4.0) released in December 2009 [54]. It extends the applicability of Bluetooth into low power and low cost applications introducing important features that heated up the competition with ZigBee in the development of efficient and reliable Health BANs. It inherits from the standard Bluetooth specification the operating spectrum (2.4 GHz) and the basic structure of the communication protocol, but it implements a completely new lightweight Link Layer that provides ultra-low power idle mode operation, simple and fast device discovery and reliable and secure point-to-multipoint data transfers. One of the main disadvantages of classical Bluetooth is the time required by devices to discover and synchronize themselves on the same channel, due to the frequency hopping procedure and, consequently, the energy consumed during its execution. In Bluetooth-LE the same procedure is maintained but the number of channels is reduced from 32 to only three. These channels are used for advertising the presence of slaves available to communicate with the master, so that the master node only has to scan three channels to open a connection and start to exchange messages. This procedure takes few msec (less than 3 msec), thus allowing the devices to further save power by staying asleep most of the time and waking up quickly in case of events. The reduction of channels in frequency hopping is also due to a larger modulation index implemented in Bluetooth-LE, increasing the frequency spectrum, and the definition of low-energy consumption filters that require the separation of channels of 2 MHz instead of 1 MHz. In addition, the hopping sequence is simplified in a wrap around sequence instead of pseudo-random, and a functionality similar to the Adaptive Frequency Hopping is maintained in order to detect channels used by other devices and technologies (e.g. WiFi).

Regarding data transfer, Bluetooth-LE inherits 1 Mbps data rate from classical Bluetooth and, in order to provide an ultra-low power transmission, it utilizes short data packets with a dynamic length (8 octects minimum up to 27 octects maximum). This feature, in addition to the fast establishment of a connection, makes this technology especially effective in situations of burst data transfers. In fact, the specification declares that a device can wake up, connect, send some application data and then disconnect again within 3 msec. In this way, the lowest amount of energy is used, maintaining the fastest transmission of event-based data.

Regarding the hardware characteristics of these devices, there exist two types of implementation: single-mode and dual-mode. Single-mode is pure low-energy implementation providing a dedicated controller in highly integrated and compact devices. By contrast, in dual-mode implementation, the LE functionality is integrated into an existing classic Bluetooth controller, allowing the coexistence of these two technologies in the same device, even using the same radio and antenna, resulting in a substantial enhancement of current chips by the new low energy stack with a minimal cost increase. The dual-mode is designed to enable fast adoption of LE functionality in classic Bluetooth applications, such as mobile phones and PDA, through a low cost modification, removing the need to add another radio. Several major mobile phone vendors have indicated that they will adopt dual mode devices in

---

[7] Mini- and Micro-SD ZigBee cards have recently been launched on the market by a Taiwanese company [64], but the firmware and the related software support is still limited to only a few operating systems.

their upcoming phones [55]. Owing to the double nature of Bluetooth-LE devices, they can generate two types of network topology: star and star-bus. The star topology is considered to be mainly that among single-mode devices only, in which one device assumes the role of Hub and the others the role of Nodes (comparable to the Bluetooth piconet concept, although here there is no a theoretical limit on the number of active slaves). By contrast, the star-bus topology can include both single- and dual-mode devices and even classical Bluetooth devices. Generally, dual-mode devices act as Hubs and the single-mode as Nodes, establishing Bluetooth-LE connections between a Hub and a Node, while the backbone connection between different Hubs is a classic Bluetooth connection. For this reason, in this topology the role of Hub can also be assumed by a classical Bluetooth device connected to dual-mode devices. An example of this topology can be represented by a mobile phone equipped with a dual-mode device that maintains LE connections to single-mode wearable sensors and, at the same time, a classic Bluetooth connection to other standard devices.

This technology fits well with the features and functionalities of pervasive healthcare systems in the case of low data rate requirements. However, where it is necessary to obtain higher data rates we have to take into consideration other technologies such as Ultra Wide Band and 60 GHz millimeter-wave [56], but they are generally used for hospital-centered systems and not for Health BAN communications.

8.7.4 Integrated and Additional Solutions for Health BAN Communications

As shown in Table 8.4, current e-health solutions exploit mainly ZigBee and Bluetooth technologies, but a further enhancement of these systems can be achieved by developing integrated solutions where multiple wireless interfaces on a single mobile device are exploited to communicate with different networks at the same time. For this purpose, the Simple Sensor Interface (SSI) protocol has been proposed within the framework of the MIMOSA FP6 European project [57] as an application protocol that allows a mobile device to communicate and read data from wireless sensors independently of their type, location or network protocol. However, the approach to multiple wireless connections in a Health BAN still has many open issues, related mainly to interference caused by the coexistence of different technologies such as WLAN (mainly for communication with a remote medical server) and ZigBee (as observed in [58]), or ZigBee and Bluetooth [51], in addition to attenuations caused by the human body. Moving around this problem, other work has investigated some alternative communication solutions, such as the body-coupled wireless communication protocol proposed in [59] to improve the trade-off between network performances and power consumption by exploiting the human body as a communication channel. In this case, a method based on the electromagnetic signalling between the polarized contacts of a transmitter and a receiver located on the body is used, allowing a high data rate independent of external influences. Preliminary results show that this method can achieve the same throughput as ZigBee with a high reduction in transmission power. However, the obtrusiveness of this method can impact greatly on its practical applicability in Health BANs, due to its negative influence on user acceptance.

However, in addition to the interference caused by different wireless technologies, a further limitation of the current solutions is represented by the lack of interoperability among different devices. This is due mainly to the proprietary nature of communication protocols and data formatting used by different vendors, so that devices developed by different vendors cannot interact through the same application, even though they support the same wireless technology. To overcome this limitation the IEEE 11073 Personal Health Device group [60] is currently working to develop guidelines and requirements for wireless technologies and applications in healthcare environments, and several industrial partners participating in the Continua Health Alliance [61] are working to develop a unique profile for pervasive healthcare devices.

8.8 Conclusions

Wearable sensors in the healthcare environment obviously improve the quality of measured data allowing patients to move and behave in a manner close to their normal routines while being monitored constantly. Several advances in this technology have been achieved in recent years, making health monitoring one of the hottest research topics in pervasive and ubiquitous computing. However, it is very important to understand the impact of the technology on its final users, evaluating their experience both in 'wearing the technology' and 'wearing it during their daily life'. In fact, a patient's acceptance greatly influences the correct and continuous use of the technology, playing a major role in the effectiveness and success of the system. Thus, patients' experiences with current technologies must be taken into account for the design and development of new wearable technologies, in order to make

Table 8.4 Health sensors and platforms for vital signals monitoring

Project	Application scenario	Physiological signals	Wearable Sensors	Networking communications
Georgia Tech Wearable Motherboard (GTWM) [7]	General vital signals monitoring platform with textile sensors.	Heart rate, respiration rate, body temperature, SpO2, voice	EKG, pulse oximeter, voice recorder (commercial off-the-shelf sensors integrated in the garment).	Intra-BAN: wired connections of sensors to the Smart Shirt controller (PS) integrated in the garment (Intra-BAN) Extra-BAN: Bluetooth or 802.11b wireless communication from the controller to the central server (Extra-BAN).
Smart vest [11]	General vital signals monitoring platform with textile sensors.	ECG, PPG, heart rate, blood pressure, body temperature, Galvanic Skin Response.	Development of customized textile sensors (ECG, pulse oximeter, electrodes for GSR measurement) in a shirt.	Intra-BAN: wired connections of sensors to the Wearable Data Acquisition Hardware (PS). Extra-BAN: wireless communication from the PS to the central server.
MagIC [12]	Textile sensors dedicated to patients affected by cardio-respiratory diseases.	ECG and respiration rate. Motion sensors integrated in the electronic board.	textile ECG sensor and a textile transducer for respiration acitivity monitoring. A portable electronic board is equipped with 2-axis accelerometer for motion detection. signal preprocessing module and wireless data transmission to a remote monitoring station.	Intra-BAN: wired connections of sensors to the electronic board (PS). Extra-BAN: wireless communication from the PS to the central server.
WEALTHY [13]	General vital signals monitoring platform with textile sensors.	ECG, EMG, motion detection and temperature.	Strain fabric sensors based on piezoresistive yarns, and fabric electrodes realized with metal based yarns.	Intra-BAN: wired connections of sensors with the WEALTHY box realized with the same conductive yarn used for electrodes. Extra-BAN: wireless communication from the WEALTHY box to the central server.
MyHeart [14]	General vital signals monitoring platform with textile sensors.	ECG and motion detection.	Textile sensors. The main innovation of this system consists in two embedded signal processing techniques to reduce the quantity of data to be transmitted to the medical server.	Intra-BAN: Bluetooth communication between the sensors and a mobile phone. Extra-BAN: GSM communication between the mobile phone and the central server.

(Continued)

Table 8.4 (Continued)

Project	Application scenario	Physiological signals	Wearable Sensors	Networking communications
Codeblue [17]	General physiological monitoring and motion-analysis dedicated to stroke rehabilitation and Parkinson's disease.	Heart rate, blood oxygen saturation (SpO2), motion analysis	Mote-based pulse oximeter, mote-EKG, motion-analysis sensor board (3-axis accelerometer, single-axis gyroscope, EMG), RF-based localization system (Mote-Track).	Intra-BAN: wireless communication based on ZigBee standard between sensors and PS. Extra-BAN: wireless communications between the PS and the central server.
AID-N [5]	Physiological monitoring for medical support in case of disaster or mass-casualty events.	Heart rate, blood oxygen saturation (SpO2).	Use CodeBlue mote-based pulse oximeter and mote-EKG to develop ETag devices.	Intra-BAN: ZigBee standard between ETags and the care-giver personal server. Extra-BAN: 802.11 between the personal server and the central medical server.
miTag (medical information Tag) [20]	Patient tracking and monitoring in disaster recovery steps (disaster – ambulance – hospital)	GPS, pulse, blood pressure and SpO2, temperature, ECG.	miTag devices as extension of ETags with GPS and temperature sensor. Dynamic configuration and adaptation of the sensors to the specific scenario.	One of miTag devices on the patient's body is elected as data aggregation hub before transmitting to the personal server.
MobiHealth [3]	General physiological monitoring.	Heart rate, blood oxygen saturation (SpO2).	Pulse oximeter and ECG.	Intra-BAN: Bluetooth communications. Extra-BAN: UMTS/GPRS.
Awareness [21]	Epilepsy and neurological applications.	Extension of MobiHealth adding respiration rate, temperature, and motion-analysis.	Pulse oximeter, ECG, EMG, temperature, motion sensors (step counter, 3D accelerometer).	Intra-BAN: Bluetooth communications. Extra-BAN: UMTS/GPRS. Auditory and Vibration biofeedback to the patient.
SMART [22]	Monitoring in waiting areas of emergency rooms.	SpO2, ECG, location.	Waist pack containing the sensor box (Finger sensor for SpO2 connected to the box, ECG) and a PDA	Intra-BAN: wired or Bluetooth communications. Extra-BAN: wireless communications.
AMON [23]	High-risk cardiac/respiratory patients	Heart rate, blood pressure, SpO2, skin temp; Activity recognition	ECG, pulse oximeter, temperature, 2-axis accelerator, all integrated in a wrist-worn single wearable device.	GSM communications from the device to the telemedicine center (TMC).
ALARM-NET [27]	Assisted-living home monitoring.	Pulse, SpO2, ECG, accelerometer, environmental parameters.	ECG, pulse oximeter, accelerometer. Environmental sensors: temperature, dust and light.	Intra- and Extra-BAN wireless communications.

them able to reduce their impact on daily life, further improving the quality of sensed data in terms of accuracy and reliability. There is little work that addresses this issue in the literature, and in particular [62] proposes a model of user acceptance based on a pilot study that compares patients' responses to long-term ambulatory Holter arrhythmia procedures with those derived from the use of a wireless ECG sensor for the same purpose. The acceptance model is based on two questionnaires focused on five 'dimensions': hygienic aspects, physical activity, skin reactions, anxiety and equipment. These parameters, collected by asking questions, are then correlated with patient characteristics, like gender and age, and information related to their physical and mental status, in addition to an index called *Pretrial Expectation* in order to evaluate the user's perceptions before wearing the sensors. All of these components form the *Sensor Acceptance Index (SAI)* [62]. The results refer to patient experiences using ECG wireless sensor continuously for three days compared with other remote-care systems. Some important aspects arise from this study. First of all, some patients wish to hide the wireless sensor from the eyes of other people, revealing embarrassment about wearing it in public. Until patients overcome this condition, they will not be able to accept using the technology in their daily activities. Patients also show a need for constant feedback from the system or care professionals; thus the system must be able to provide constant support to patients, both in terms of alerts and suggestions (even as a psychological support). In addition, the strict interaction between the healthcare system and the medical server must be clearly visible to the patients so as to improve their trust in wearing the system. Finally, patients display overall a good degree of confidence in the wearable sensor owing to ease of use and improvement in aspects of hygiene and comfort, with respect to the classical Holter device. Another important aspect in designing novel wearable sensors is the possibility for patients to carry out daily activities such as participating in physical sports, sauna etc, without damaging the system.

All of these aspects influence both the hardware and software design of wearable sensors for healthcare, from wearable characteristics to the ability to correlate physiological, neuro-psychological and environmental parameters that greatly influence the health status of the patients. Low-power, reliable and secure wireless communications, both inside and outside the Health BAN, are also necessary to achieve these objectives. Current technologies and communication protocols focus mainly on optimizing power consumption, moving away from the requirements of medical applications that usually involve a huge quantity of data to be transmitted in almost a continuous way. To better support these requirements we can focus on different research areas. Regarding communication protocols, both novel technologies (e.g. Bluetooth-LE) and the possibility of using multiple wireless interfaces should be investigated further to obtain higher data rates while maintaining low consumption, and to improve the interoperability of wearable sensors with the rest of the system. Signal processing and elaboration of sensed data represent other important research fields in wearable computing. These techniques can be implemented with different features on the sensor board, on the PS, and finally on the central server depending on the hardware and software characteristics of the devices involved. These techniques also enable the correlation of different physiological and non-physiological signals exploiting complex processing analysis, data fusion algorithms and expert systems to make the system able to react correctly to situations that are critical for patient health. In addition, they can be defined so as to improve the adaptability of the system allowing, for example, custom calibration and tuning of the sensing procedure depending on the patient's status within a specific period of time or during a particular activity, or to adapt the power saving policies to the current context (related to both the patient and the surrounding environment).

Therefore, further work on wearable computing is necessary to design and develop a complete and effective Pervasive Healthcare system, and the current results represent a good starting point from which to address the emerging issues.

References

[1] 'The world health report,' http://www.who.int/whr/en/.

[2] S. Park and S. Jayaraman, 'Enhancing the quality of life through wearable technology,' *IEEE Engineering in Medicine and Biology Magazine*, **22**: 41–8.

[3] A. Van Halteren, R. Bults, K. Wac, D. Konstantas, I. Widya, N. Dokovsky, G. Koprinkov, V. Jones, and R. Herzog (2004) 'Mobile patient monitoring: The mobihealth system,' *The Journal on Information Technology in Healthcare* **2**: 365–73.

[4] D. Husemann, C. Narayanaswami, and M. Nidd, (2004) 'Personal mobile hub,' *ISWC 2004, Eighth International Symposium on Wearable Computers*.

[5] T. Gao, T. Massey, L. Selavo, D. Crawford, B. Chen, K. Lorincz, V. Shnayder, L. Hauenstein, F. Dabiri, J. Jeng *et al.* (2007) 'The advanced health and disaster aid network: A light-weight wireless medical system for triage,' *IEEE Transactions on Biomedical Circuits and Systems*, vol. **1**, pp. 203–16.

[6] P.G. P.S. Pandian K.P. Safeer and V.C. Padaki (2008) 'Wireless Sensor Network for Wearable Physiological Monitoring,' *Journal of Networks* **3**: 21.

[7] C. Gopalsamy, S. Park, R. Rajamanickam, and S. Jayaraman (1999) 'The Wearable Motherboard: The first generation of adaptive and responsive textile structures (ARTS) for medical applications,' *Virtual Reality* **4**: 152–68.

[8] E. Jovanov, A. Milenkovic, C. Otto, and P.C. De Groen (2005) 'A wireless body area network of intelligent motion sensors for computer assisted physical rehabilitation,' *Journal of NeuroEngineering and Rehabilitation* **2**: 6.

[9] D.J. Christini, J. Walden, and J.M. Edelberg (2001) 'Direct biologically based biosensing of dynamic physiological function,' *American Journal of Physiology- Heart and Circulatory Physiology* **280**: 2006–10.

[10] 'Preventing Chronic Diseases: a vital investment, the World Health Organization Report,' http://www.who.int/chp/chronic_disease_report/media/

[11] K.P. P.S. Pandian K. Mohanavelu (2008) 'Smart Vest: Wearable multi-parameter remote physiological monitoring system,' *Medical Engineering and Physics* **30**: 466–77.

[12] M.D. Rienzo, F. Rizzo, G. Parati, G. Brambilla, M. Ferratini, and P. Castiglioni (2005) 'MagIC system: a new textile-based wearable device for biological signal monitoring. Applicability in daily life and clinical setting,' *27th Annual International Conference of the Engineering in Medicine and Biology Society, IEEE EMBS 2005*, pp. 7167–9.

[13] R. Paradiso, G. Loriga, N. Taccini, A. Gemignani, and B. Ghelarducci, 'WEALTHY, a wearable health-care system: new frontier on Etextile,' *Journal of Telecommunications and Information Technology* **4**: 105–13.

[14] J. Luprano, J. Sola, S. Dasen, J.M. Koller, and O. Chetelat (2006) 'Combination of body sensor networks and on-body signal processing algorithms: the practical case of MyHeart project,' *International Workshop on Wearable and Implantable Body Sensor Networks, BSN 2006*.

[15] C.W. Crenner (1998) 'Introduction of the blood pressure cuff into US medical practice: technology and skilled practice,' *Annals of Internal Medicine* **128**: 488–93.

[16] M. Di Rienzo, F. Rizzo, P. Meriggi, B. Bordoni, G. Brambilla, M. Ferratini, and P. Castiglioni (2006) 'Applications of a textile-based wearable system in clinics, exercise and under gravitational stress,' *3rd IEEE/EMBS International Summer School on Medical Devices and Biosensors*, pp. 8–10.

[17] V. Shnayder, B. Chen, K. Lorincz, T.R. Fulford-Jones, and M. Welsh (2005) 'Sensor networks for medical care,' *Proceedings of the 3rd international conference on Embedded networked sensor systems – SenSys'05*, p. 314.

[18] 'Smiths Medical PM, Inc., BCI Digital Micro Power Oximeter Board,' http://www.smiths-medical.com/Userfiles/oem/OEM-WW3711sell%20sheet.

[19] 'TinyOS Operating System,' http://www.tinyos.net/.

[20] T. Gao, C. Pesto, L. Selavo, Y. Chen, J. Ko, J.H. Lim, A. Terzis, A. Watt, J. Jeng, B. Chen *et al.* (2008) 'Wireless medical sensor networks in emergency response: Implementation and pilot results,' *Proceedings of the 2008 IEEE Int. Conf. Technologies for Homeland Security*.

[21] V.M. Jones, T.M. Tonis, R.G. Bults, B.J. van Beijnum, I.A. Widya, M.M. Vollenbroek-Hutten, and H.J. Hermens (2008) 'Biosignal and context monitoring: Distributed multimedia applications of body area networks in healthcare'.

[22] D.W. Curtis, E.J. Pino, J.M. Bailey, E.I. Shih, J. Waterman, S.A. Vinterbo, T.O. Stair, J.V. Guttag, R.A. Greenes, and L. Ohno-Machado (2008) 'SMART – an integrated wireless system for monitoring unattended patients,' *Journal of the American Medical Informatics Association* **15**: 44–53.

[23] U. Anliker, J.A. Ward, P. Lukowicz, G. Troster, F. Dolveck, M. Baer, F. Keita, E.B. Schenker, F. Catarsi, L. Coluccini *et al.* (2004) 'AMON: a wearable multiparameter medical monitoring and alert system,' *IEEE Transactions on Information Technology in Biomedicine*, vol. **8**, pp. 415–27.

[24] D. Curtis, E. Shih, J. Waterman, J. Guttag, J. Bailey, T. Stair, R.A. Greenes, and L. Ohno-Machado (2008) 'Physiological signal monitoring in the waiting areas of an emergency room,' *Proceedings of the ICST 3rd international conference on Body area networks table of contents*.

[25] C. Otto, A. Milenkovic, C. Sanders, and E. Jovanov (2006) 'System architecture of a wireless body area sensor network for ubiquitous health monitoring,' *Journal of Mobile Multimedia* **1**: 307–26.

[26] A. Milenković, C. Otto, and E. Jovanov (2006) 'Wireless sensor networks for personal health monitoring: Issues and an implementation,' *Computer Communications* **29**: 2521–33.

[27] A. Wood, G. Virone, T. Doan, Q. Cao, L. Selavo, Y. Wu, L. Fang, Z. He, S. Lin, and J. Stankovic (2006) 'ALARM-NET: Wireless sensor networks for assisted-living and residential monitoring'.

[28] P. Lukowicz, F. Hanser, C. Szubski, and W. Schobersberger (2006) 'Detecting and interpreting muscle activity with wearable force sensors,' *Lecture Notes in Computer Science*, vol. **3968**, p. 101.

[29] O. Amft, H. Junker, P. Lukowicz, G. Troster, and C. Schuster (2006) 'Sensing muscle activities with body-worn sensors,' *International Workshop on Wearable and Implantable Body Sensor Networks – BSN 2006*.

[30] K. Van Laerhoven, D. Kilian, and B. Schiele (2008) 'Using rhythm awareness in long-term activity recognition,' *Proceedings of the 12th IEEE International Symposium on Wearable Computers – ISWC 2008*.

[31] O. Amft and G. Troster (2008) 'Recognition of dietary activity events using on-body sensors,' *Artificial Intelligence in Medicine*, **42**: 121–136.

[32] J. Mankoff, G. Hsieh, H.C. Hung, S. Lee, and E. Nitao (2002) 'Using low-cost sensing to support nutritional awareness,' *Lecture notes in computer science*, pp. 371–8.

[33] K.A. Siek, K.H. Connelly, Y. Rogers, P. Rohwer, D. Lambert, and J.L. Welch (2006) 'When do we eat? An evaluation of food items input into an electronic food monitoring application,' *PHC*, pp. 1–10.

[34] A. Barreto, J. Zhai, and M. Adjouadi, (2007) 'Non-intrusive Physiological Monitoring for Automated Stress Detection in Human-Computer Interaction,' *Lecture Notes in Computer Science*, vol. **4796**, pp. 29–38.

[35] P.J. Lang, M.M. Bradley, and B.N. Cuthbert (1999) *International Affective Picture System (IAPS): Technical manual and affective ratings*, Gainesville, FL: 1999.

[36] A. Haag, S. Goronzy, P. Schaich, and J. Williams (2004) 'Emotion recognition using bio-sensors: First steps towards an automatic system,' *Lecture Notes in Computer Science*, vol. **3068**, pp. 36–48.

[37] R. Horlings, D. Datcu, and L.J. Rothkrantz (2008) 'Emotion recognition using brain activity,' *Proceedings of the 9th International Conference on Computer Systems and Technologies and Workshop for PhD Students in Computing*.

[38] F. Nasoz, K. Alvarez, C.L. Lisetti, and N. Finkelstein (2004) 'Emotion recognition from physiological signals using wireless sensors for presence technologies,' *Cognition, Technology & Work*, **6**: 4–14.

[39] J.J. Gross and R.W. Levenson (1995) 'Emotion elicitation using films,' *Cognition & Emotion*, **9**: 87–108.

[40] W.W. Grings and M.E. Dawson (1978) *Emotions and Bodily Responses: A psychophysiological approach*, Academic Press.

[41] T. Partala and V. Surakka (2003) 'Pupil size variation as an indication of affective processing,' *International Journal of Human-Computer Studies* **59**: 185–98.

[42] 'ASL Mobile Eye,' http://asleyetracking.com/site/Products/MobileEye/tabid/70/Default.asp.

[43] K.H. Kim, S.W. Bang, and S.R. Kim (2004) 'Emotion recognition system using short-term monitoring of physiological signals,' *Medical and Biological Engineering and Computing* **42**: 419–27.

[44] D. Tacconi, O. Mayora, P. Lukowicz, B. Arnrich, G. Troster, and C. Haring (2007) 'On the feasibility of using activity recognition and context aware interaction to support early diagnosis of bipolar disorder,' *In Proceedings of UbiComp 2007*.

[45] D. Tacconi, O. Mayora, P. Lukowicz, B. Arnrich, C. Setz, G. Troster, and C. Haring (2008) 'Activity and emotion recognition to support early diagnosis of psychiatric diseases,' *Second International Conference on Pervasive Computing Technologies for Healthcare – PervasiveHealth 2008*, pp. 100–2.

[46] 'European FP7 MONARCA project,' http://www.monarca-project.eu.

[47] 'European FP7 INTERStress project,' http://interstress.eu/.

[48] 'European FP7 Chronius Project,' http://www.chronious.eu/.

[49] 'European FP7 Metabo Project,' http://www.metabo-eu.org/.

[50] N.F. Timmons and W.G. Scanlon (2004) 'Analysis of the performance of IEEE 802.15. 4 for medical sensor body area networking,' *First Annual IEEE Communications Society Conference on Sensor and Ad Hoc Communications and Networks – IEEE SECON 2004*, pp. 16–24.

[51] N. Chevrollier and N. Golmie (2005) 'On the use of wireless network technologies in healthcare environments,' *Proceedings of the fifth IEEE workshop on Applications and Services in Wireless Networks – ASWN2005*, pp. 147–152.

[52] P. Baronti, P. Pillai, V.W. Chook, S. Chessa, A. Gotta, and Y.F. Hu (2007) 'Wireless sensor networks: A survey on the state of the art and the 802.15. 4 and ZigBee standards,' *Computer Communications* **30**: 1655–95.

[53] S. Davgtas, G. Pekhteryev, Z. Sahinovglu, H. Cam, and N. Challa (2008) 'Real-time and secure wireless health monitoring,' *International Journal of Telemedicine and Applications*.

[54] 'Bluetooth Core Specification v.4.0,' http://www.bluetooth.com.

[55] 'Nordic Semiconductor,' http://www.nordicsemi.com.

[56] C. Park and T.S. Rappaport (2007) 'Short-range wireless communications for next-generation networks: UWB, 60 GHz millimeter-wave WPAN, and ZigBee,' *IEEE Wireless Communications*, vol. **14**, pp. 70–78.

[57] I. Jantunen, H. Laine, P. Huuskonen, D. Trossen, and V. Ermolov (2008) 'Smart sensor architecture for mobile-terminal-centric ambient intelligence,' *Sensors & Actuators A: Physical* **142**: 352–60.

[58] N. Golmie, D. Cypher, and O. Rebala (2005) 'Performance analysis of low rate wireless technologies for medical applications,' *Computer Communications* **28**: 1266–1275.

[59] A.T. Barth, M.A. Hanson, H.C. Powell Jr., D. Unluer, S.G. W., and J. Lach (2008) 'Body-coupled communication for body sensor networks,' *BodyNets '08: Proceedings of the ICST 3rd international conference on Body area networks, ICST*, pp. 1–4.

[60] L. Schmitt, T. Falck, F. Wartena, and D. Simons (2007) 'Novel ISO/IEEE 11073 standards for personal telehealth systems interoperability,' *Proceedings of Joint Workshop on HCMDSS-MDPnP*.

[61] 'Continua Health Alliance,' http://www.continuaalliance.org/.

[62] R. Fensli, P.E. Pedersen, T. Gundersen, and O. Hejlesen (2008) 'Sensor acceptance model-measuring patient acceptance of wearable sensors.,' *Methods of Information in Medicine* **47**: 89.

[63] 'IEEE 802.15.4 (2006) Wireless Medium Access Control (MAC) and Physical Layer (PHY) Specifications for Low-Rate Wireless Personal Area Networks (WPANs),' http://www.ieee802.org/15/pub/TG4.html.

[64] 'SPECTEC products,' http://www.spectec.com.tw/zigbee.htm.

9

Standards and Implementation of Pervasive Computing Applications[1]

Daniel Cascado,[1] Jose Luis Sevillano,[1] Luis Fernández-Luque,[2]
Karl Johan Grøttum,[2] L. Kristian Vognild,[2] and T. M. Burkow[3]
[1]*Robotics and Computer Technology Laboratory, University of Seville, Seville, Spain.*
[2]*Northern Research Institute – Norut, Tromsø, Norway.*
[3]*Norwegian Centre for Integrated Care and Telemedicine, UNN, Tromsø, Norway.*

9.1 Introduction

Pervasive Computing is a new paradigm based on the idea of providing access to applications everywhere, anytime, by means of any device, and through natural interactions so that users may not even be aware that they are using computational devices. Related concepts include Ubiquitous Computing or Ubicomp, Ambient Intelligence [1], Everyware [2], etc. As a new paradigm, it requires fundamental changes in many different aspects of computer science and engineering. Many papers in the literature have addressed these challenges and have identified the main characteristics of ubiquitous and pervasive computing [3–5]. This chapter does not provide a detailed analysis of all these challenges and characteristics; instead, we describe the main standards and technologies that are currently available with their advantages and disadvantages from the point of view of Pervasive Applications. To do this, first of all we gather these characteristics into the following more general issues: Mobility, Context-awareness and Heterogeneity. Of course, these issues are interrelated, and there are also characteristics that cannot be easily placed in just one of these 'fundamental' categories. Where this is the case, we simply consider them as part of these fundamental issues. Note that the aim of using only three categories is to avoid a detailed list of features and characteristics. Instead, we prefer a brief discussion of the main features and challenges of pervasive applications. In this way we facilitate the description of the different standards and technologies in subsequent sections, where the identified issues will be used as a way of comparing and selecting the better options depending on the application.

9.1.1 Pervasiveness and Mobility in Computing and Communications

Pervasive applications are meant to be integrated into everyday objects and activities. This means that a user may want to access services and applications at anytime and from any place or (mobile) device. These mobile devices tend to have limited resources: bandwidth, computational and battery power, screen size, etc. Mobility also imposes important requirements on communications technologies and standards in order to achieve seamless

[1] This work was partially supported by Project TIN2006–15617-C03–03. The work related to MyHealthService was supported in part by the Norwegian Research Council (Tromsø Telemedicine Laboratory, project 174934). The Project N4C is funded by the European Commission's FP7 ICT programme's FIRE initiative, Grant No. 223994.

access: efficient roaming, tolerance to intermittent failures, ability to change network addresses when using different access points, etc. But it's not just a problem of physical mobility but also of 'logical' mobility, that is, mobility with respect to code and data [3]. For instance, it's not only that connections should be maintained, with users 'on the move', but also that applications should keep working or data should still be valid. Sometimes a copy of relevant data has to be kept in different locations including mobile devices such as the user interface, so the problem of data coherence and synchronization arises.

Another important issue is so-called pervasive dependability [6]. Instead of specific solutions for safety-critical applications, this concept focuses on the dependability of every-day systems. Therefore, solutions should be low cost and targeted at relatively simple devices with limited resources. Users should be allowed to configure or select the most appropriate service as there will be no system administrator and it's difficult to automate the selection of the most suitable solution in such dynamic environments in common, daily activities. Finally, common 'failures' such as broken links (e.g. due to mobility) should be considered normal and the system should provide ways of adapting to these frequent situations.

Analogously, security mechanisms must be adapted to the specific requirements of pervasive systems (mobile devices, with limited resources, etc.). Particularly, the user's expectation of an effortless, spontaneous access, together with the heterogeneity of devices, standards and connections used, raise many difficulties. Proposed solutions include the use of trust management systems, biometric-based security systems, etc. [7].

We could also have included privacy in this category, a concept related to security concerns. However, privacy is particularly relevant in pervasive systems due to the context-awareness features, so it will be discussed in the following paragraph.

9.1.2   Context Awareness

Context awareness is defined as the use of information to characterize the situation of an entity: person, place, object, etc. [8]. The situation may have several 'dimensions', the most important being *location*. Location awareness ranges from people or vehicle tracking to offering specific services depending on the location (*location-based services*), which requires less accurate location information. This location information may be provided by a specific system (GPS, sensor network) or directly by the underlying wireless network, for instance measuring the power of the incoming radiofrequency signal. Note that the software infrastructure will have to elaborate this kind of measure and extract higher-level information in order to make it available to the application level. This context management is sometimes considered a separate issue.

Another important dimension is Personal awareness which includes dynamic adaptation to user needs, abilities or preferences. The most appropriate service is not always the nearest or the highest quality. Rather, it varies according to the user's preferences. More importantly, user criteria change with their context [9].

A related issue, sometimes listed separately, includes all the features related to 'user interaction' or HCI in the context of pervasive systems: transparent and natural user interaction, spontaneous and occasional use, etc. Other less frequent dimensions include temporal awareness (time scheduling of events), device awareness (processing power, battery), etc.

We include in this discussion on context awareness the problem of privacy. The reason is that with personal awareness, personal information is expressed in digital form and this information could be communicated either directly or by inference. Furthermore, location information used for context-aware applications can be used to infer identity or personal information [10]. Depending on the case, this identification may be an added value, for instance offering adapted services to people with disabilities. But often this is not the case. A key concept is the user's expectation of privacy [11], which may vary depending on the user location. For instance, in public places the user may have lower expectations of privacy than in a private place. In order to meet these expectations, possible solutions may include deleting the data after a certain period of time, guaranteeing anonymous recording, etc.

9.1.3   Heterogeneity

Given the huge number of different types of devices, networks, systems and environments available, it's no surprise that most references include heterogeneity as one of the most important issues in pervasive systems. Pervasive computing environments are far more dynamic and heterogeneous than other environments [12]. The common solution is to use a unifying middleware layer placed on top of the devices, services and resources. This layer abstracts device-dependent features and provides a homogeneous interface to the upper layers of the system, minimizing software complexity. However, as we will discuss later, many different incompatible standards exist. This fact, together with the need for higher-level abstractions of discovery (e.g. users searching for useful devices with a particular context), make an integration framework necessary.

The rest of this chapter is organized as follows. First, we briefly discuss available standards and technologies, focusing on two key issues: wireless connections for the lower layers and middleware for the higher layers. Taking into account our previous discussion, we consider the behaviour of these technologies, with their advantages and disadvantages, from the point of view of pervasive applications. Finally, we consider two very different pervasive applications: Hiker's PDA with Hiker's Applications, developed in the Networking for Communications Challenged Communities: Architecture, Test Beds and Innovative Alliances project (http://www.n4c.eu); and MyHealthService, an ongoing research project in the Tromsø Telemedicine Laboratory (http://www.telemed.no/ttl). These two applications are very illustrative in terms of the use of state of the art technologies and standards, and serve as examples of current pervasive computing applications.

9.2   Wireless Technologies and Standards

The communication system is a fundamental element of every pervasive or mobile system. It is the ability to transmit data and code, as well as location and context information, which allows pervasiveness and mobility. For this reason, being able to select the right choice of network technologies is a prerequisite when designing pervasive systems.

However, in most cases the design of the communication subsystem is not as simple as selecting one candidate from the available standards. It should be remembered that, barring a few cases, there will be many different kinds of devices connected through a heterogeneous set of networks. Each device will have one or more links to communicate with the other elements of the system. Some devices will serve as gateways (allowing communications between two or more sub-networks) and other devices will be end nodes that only use one network connection to transmit or receive its data or code. End nodes will be connected through gateways to the rest of the system in a structure similar to that shown in Figure 9.1.

It is possible that all the devices of the system will use the same network to share their data but, generally, the system will consist of a hierarchy of networks connecting its nodes. The most powerful nodes and networks will be placed in the upper levels, while, on the contrary, in the lower levels the nodes with less computing power and lower data rate networks will be used. Even in pervasive systems, not all network levels have to be wireless. In order to provide robustness to the system, some wired networks may exist. Wireless devices may use one of the available wireless networks to connect with other devices in the same wireless network, or they may connect with a *backbone* network to communicate with devices connected in other networks. These backbone networks may be wired in order to provide enough bandwidth or security (Figure 9.2).

Usually, gateways will be fixed (not mobile), plugged-in devices, with relatively high computing power. End nodes will have many different characteristics regarding computing power, location, energy supply and consumption. In the case of mobile nodes, many questions must be taken into account: what will be the mobility range of these nodes? Will the migration between networks be allowed in the system (e.g. between networks in L3)?

At first sight, it is obvious that power consumption, range, mobility and data rate are issues to keep in mind when designing these systems. However, many other issues related to pervasive systems must be taken into account at the design stage. One of them is the resistance to frequent packet and connection loss: what happens when one mobile

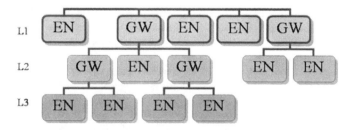

Figure 9.1   System structure, with gateways (GW) and end nodes (EN).

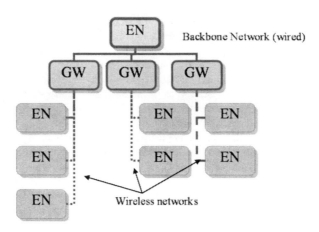

Figure 9.2   Use of wired/wireless networks in pervasive systems.

node looses its connection as a consequence of its movement? In this case we need to guarantee the reconnection of this node without loss of its data or identity (e.g. network address). Related to this, problems arise when our devices operate in a noisy environment or in coexistence with other wireless technologies.

Coexistence between wireless technologies of the same system is another important issue that must be studied carefully. Most times, this issue is solved choosing systems that operate in different frequency bands but, there are some carrier sense mechanisms that allow the coexistence between network technologies operating in the same frequency bands. In other cases, computation of the impact on the network throughput due to interference may be enough to consider whether the coexistence between systems is possible. In this regard, a lot of papers in the literature may shed some light on the subject [13–18].

Context awareness capability is also a desirable feature in a wireless network. Many services in a pervasive application are based on the context information received from the environment. Usually, this information can be gathered by external devices (like for instance a GPS) but the wireless network may need to supply this information itself without having to resort to more external devices. The following section gives a brief description of the most usual wireless networks, taking into account all the ideas mentioned above.

9.2.1   A Simple Classification of Wireless Networks

We have made a simple classification of the different wireless network technologies on the basis of radio coverage, although they could be classified in other ways, as proposed in [19, 20].

- **Global cell:** characterized by radio coverage of hundreds of kilometres, they are suitable for communication with remote areas where there is low traffic density or between countries. Satellites are used for this type of network. Thereby, there is no possibility of movement in their nodes, neither possibility of migration between satellites without connection losses. Data rates are very high in these networks but the latency time (time elapsed between a request and a reception of data) are in the order of dozens of seconds. Great power consumption.
- **Macro-cellular:** characterized by radio coverage of dozens of kilometres, these networks are used to communicate between different cities, with low or medium traffic density. These networks are employed by cellular telephone systems. Mobile nodes can migrate between fixed stations and can take speeds up to 250 Km/h. Data rates are medium (up to 2 Mbps with mobiles of 2.5 or 3G). Location information can be obtained in these networks by triangulation. Great power consumption.
- **Micro-cellular:** characterized by radio coverage of kilometres, these networks are used to replace RDSI/ADSL or wired telephone connections, with high or medium traffic density. Mobile nodes can have speeds up to 50 Km/h in the second case.
- **Pico-cell:** characterized by a small radio coverage (less than 100 meters), they are used to establish high speed data links between devices in the same building. They are the substitutes of local area network, with high data rates and limited movement speed (up to 10 Km/h). Capability of roaming mobile nodes exists, depending of

Table 9.1 Classification of wireless networks

	Global	Macro-cell	Micro-cell	Pico-cell	Personal	WSN
Radio coverage	Hundreds of Km.	Dozens of Km.	< 10 Km.	< 100 m.	< 10 m.	< 100 m.
Mob. speed	Fixed	< 250 Km/h.	< 50 Km/h.	<10 Km/h	<10 Km/h.	<10 Km/h
Data Rate	Several Mbps	< 2 Mbps	< 250 Mbps	<100 Mbps	<10 Mbps	< 256Kbps
Application	Satellite	GSM, GPRS, UMTS	DECT, HIPERACC, 802.16	802.11, HIPERLAN, MMAC, HomeRF	Bluetooth, UWB	802.15.4, ANT, SP100, Zigbee, Wibree, Zwave, INSTEON, nanoNET

the system chosen. Although these networks can be implemented in battery supplied nodes, their high data rates of operation make them unsuitable for devices with small batteries or where a long operation time is required.

- **Personal area network:** dedicated to connecting devices over a small coverage area (less than 10 meters) with a medium data rate link (up to 10 Mbps). These networks have been designed for implementation in low power computing devices thereby allowing better operation time compared to Pico-cell networks. Given their nature, the mobility speed of nodes is limited to 10 km/h.
- **Wireless sensor network (WSN):** designed to connect low cost devices (e.g. physical magnitude sensors) spread over an extended area. Data rates (up to 250 kbps) and radio coverage (up to 10 m, although there are WSN with coverage up to 100 m) are very small in these networks. Nodes are very resistant to disconnection and movement speed is limited (up to 10 km/h), depending on the frequency of operation. Given their nature, these networks allow very long operation time with ultra low power consumption, and have experienced rapid growth in recent years, being the focus of current research (like energy conservation, auto organization, time scheduling, synchronization, multimedia transmission, routing, etc.) [21].

Summarizing the classification given above, Table 9.1 shows their most important features and their most representative application examples.

From Table 9.2 we can assert that no problems of coexistence arise in bands where only one wireless technology is operating at a time and in the same area of influence. For example, this is true in the 1880–1900 MHz band, where only DECT operates: there will be no problems if one DECT network exists alone or if two or more DECT networks operate at the same time but in different areas, far enough away to avoid mutual interference (because the

Table 9.2 Frequency spectrums of wireless technologies

FREQ (GHz)	USED BY	FREQ (Ghz)	USED BY
0.880–0.915	GSM, 802.15.4, Z-wave	**2.170–2.2**	Satellite
0.925–0.960	GSM	**2.4–2.4835 (ISM)**	Bluetooth, 802.15.4, 802.11b/g, HomeRF, Wibree, ANT, SP100, NanoNET, UWB
		3.1	UWB
1.710–1.785	GSM	**5.15–5.3**	HIPERLAN, UWB
		5.725–5.875 (ISM)	802.11a/n, MMAC, UWB
1.805–1.880	GSM		UWB
1.880–1.9	DECT	**10–17.7**	HIPERLAN, HIPERACCESS, MMAC, UWB
1.9–1.980	UMTS	**59–62**	MMAC, UWB
1.980–2.010	Satellite		
2.010–2.025	UMTS		
2.110–2.170	UMTS		

signals of the outer network are too weak to interfere with the 'local' network). In the remaining cases, spectrum management is necessary to avoid interference. There are several types of management:

- **Manual:** users configure their systems to avoid interfering with others. Configuration can be made in terms of strength or frequency. For example: if more than two DECT networks must coexist, their users must configure their networks to work on different channels of that frequency band. Another option is to adjust their transmit power to make the signals weak enough so as not to interfere with the other network. Something similar happens with 802.11.
- **Centralized assignation of frequencies:** there are entities that manage the use of frequencies and assign them to their nodes to avoid interferences. This is the case of GSM and UMTS where fixed stations assign different operation channels to their mobile nodes to avoid interference problems.
- **Distributed assignation of frequencies:** each network senses the medium and gathers information about which channels (frequencies) are crowded. It then selects an un-crowded channel in which to operate. This is the case of SP100 or Bluetooth 1.2.

In this way, the coexistence problem in pervasive systems is a problem of selecting the appropriate network systems for coexistence. For example, it is not very suitable to choose 802.11 and Bluetooth 1.1 to coexist in the same system because they cannot avoid each other, and will cause interference and performance problems. 802.11g and Bluetooth v1.2 could be a good choice because the latter has an adaptive frequency hopping that allows it to avoid the frequencies used by 802.11. Consistently, network technologies to be used in a pervasive environment have to find a way of coexisting, whether this is by employing different bands or adapting their channel utilisation so as not to interfere with other networks in the system. However, we should always bear in mind that other external network technologies may exist and we must consider the degradation of our system where they are present. In this case, taking an estimated calculation of this degradation may help to avoid further problems [13].

9.2.1.1 Global Cell Systems

Nowadays, satellites are used to supply global coverage of service. While a great variety of satellites exist, supplying very different services and coverage, two types of service can be distinguished:

- **Geo-stationary satellite service:** allows the undefined and permanent use of one satellite, reachable in one specific geographic area.
- **Non geo-stationary satellite service:** connections to non-stationary satellites assume that a receptor will be available in the satellite's coverage area sooner or later. For this reason, a receptor must be used to deal with automatic connections and disconnections.

In both services, high data rates (several Mbps) and very good signal qualities can be obtained. However, the main drawback with these systems is the signal delay generated by the large distance between devices, which is dependent on the type of orbit the satellite is placed in:

- **Geo-stationary (35863 Km):** allows data stream with data rates up to 50 Mbps and delays of 0.25 s.
- **Medium (5000 to 12 000 Km) or low (less than 2000 Km):** used to send voice, global positioning systems and short data messages with data rates up to a few Mbps with delays of 20 ms to 50 ms.

These systems cannot be employed when real time or low power is required. This technology presents a weak resistance to interference, given the weak signal received and its problems for fine-tuning the antenna. Coexistence is not a problem here. Therefore, satellites must be used when global access is required in our system at the cost of very expensive devices and long response times.

9.2.1.2 Macro-Cellular Systems

Technologies like GSM [22], IS-54 (also known as TIA-EIA-95[1]), IS-95 (TIA-EIA-95), and IS-136 (TIA-EIA-136) emerged with the sole purpose of transmitting voice, so data rates were low (9600 bps in GSM). Further

---

[1] http://www.tiaonline.org.

improvements in GSM gave data rates up to 22 Kbps, maintaining the same infrastructures as a traditional GSM network.

A further development of GSM (called 2.5G) is GPRS[2] (General Packet Radio System). In GPRS, data rates of 171 Kbps can be obtained maintaining a permanent data connection (therefore allowing Internet connections through mobile phones) at the same time, avoiding the temporal cost of call establishment.

Other developments in 2.5G are HSCSD[3] (High Speed Circuit Switched Data) and EDGE (Extended Data Rates for GSM Evolution). HSCSD is a development of ETSI (European Telecommunication Standards Institute) born in 1997 oriented to improve the data transfer rates of GSM up to 57 Kbps and the behavior of terminal devices in bad signal conditions (error correction methods improved) where, like in GSM, a permanent data connection is not allowed. EDGE[4] is an improvement of GPRS born in 2003 that achieves data rates up to 348 Kbps in permanent data connections, extending the performance of GPRS up to 3G.

UMTS[5] is the most representative example of 3G and was launched in 2000 by IMT (International Mobile Telecommunications) to achieve data rates from 144 Kbps up to 2 Mbps (with reduced mobility speeds) depending of the type of connection. UMTS was designed as a global communication system allowing a permanent data connection using the best communications technology available (wireless personal area networks, 2.5G networks or satellite links were considered).

GSM and GPRS systems use TDMA (Time Division Multiple Access) making use of frequency and time prefixed by base stations. UMTS uses TDMA and CDMA (Code division multiple access) to improve coexistence with older mobile phone technologies.

Summarizing, only 2.5G or 3G technologies are suitable for pervasive and mobility applications. In all of them we can obtain permanent data connections with a good level of coexistence with other wireless systems and a good resistance to bad signal conditions. Location information is possible through triangulation of signals but with very poor accuracy (error margin of hundreds of meters). User mobility is reduced with these technologies (the higher the data rate, the lower the speed allowed).

These technologies can be implemented in a wide range of devices (from mobile phones to laptop computers) and can provide us with greater flexibility for implementing them in gateways or terminal nodes in our system.

9.2.1.3   Micro-Cellular Systems

Two kinds of systems exist in this section: one dedicated to replacing wired LAN networks and the other to replacing wired residential telephony networks. DECT[6] and PACS [23] are two examples of the second kind and they transmit voice digitally achieving data rates of 288 up to 552 Kbps. These systems allow roaming of terminal devices between stations and include security features such as user authentication and data encryption. No location information has been developed for these systems which can be completed with wired data networks to extend their operational range and functionality.

BRAN HIPERACCESS [24] (ETSI) and WiMAX[7] are examples of replacement RDSI/ADSL technologies; used for instance in rural areas where the distance to central nodes prevents the use of wires. For this reason, nodes of these technologies can be used as distributors of data links for small areas (perhaps using wired connections). Both of them allow high speed multipoint or point-to-point connections for sending data through fixed nodes.

9.2.1.4   Pico-Cellular Systems

Designed to extend LAN connectivity to laptop computers, these systems became an alternative to LAN networks with the increasing use of laptop computers. Data rates in these systems are similar to wired LAN but at the cost of reduced user mobility and medium-high power consumption. These power requirements restrict its implementation to powerful battery devices such as PDAs, laptops and similar devices (some mobile phones can implement it but with a short battery duration during operation).

---

[2] http://www.etsi.org/WebSite/technologies/gprs.aspx.

[3] http://www.etsi.org/WebSite/Technologies/hscsd.aspx.

[4] http://www.etsi.org/WebSite/technologies/edge.aspx.

[5] http://www.umts-forum.org/.

[6] http://www.etsi.org/website/technologies/dect.aspx.

[7] http://www.ieee802.org/16/.

IEEE 802.11[8] is one of the most significant technologies. Created by IEEE during the period 1990–97, this standard specifies several PHY layers and one MAC layer. The objective of this is to have the possibility of adapting the system to signal conditions, giving data rates from 1 Mbps up to 100 Mbps (802.11n). Topologies allowed are distributed (Ad-Hoc) or centralized (infrastructure mode) with access points as master nodes of the network. In this latter mode, roaming of mobile devices is possible when signal strength is weak enough. Location services are a recent addition to this standard. Security and synchronous data transmissions modes are possible, so some kind of soft real time applications are possible. Coexistence with other systems that operate in the same band is possible by choosing the right transmission channel.

HIPERLAN [25] was designed by ETSI between 1992 and 1996. The first version was thought to work with data rates up to 19Mbps in centralized or distributed mode, making it possible to include power saving modes, roaming and multi-hop packet transmission in synchronous and asynchronous fashion. The second version was designed to use a centralized architecture, to run over ATM networks and give support to UMTS. However, backward compatibility with the first version (distributed mode) was maintained. Coexistence in this system is achieved through automatic frequency selection but no location information is possible. Power consumption requirements make it possible to implement this technology in laptop devices.

MMAC[9] was developed in Japan (2002) designed to replace LAN networks. Its specification contains several PHY layers in 5, 25, 40 and 60 GHz bands. The main objective of this technology is to transport multimedia information, making very high data rates a primary target (20 to 156 Mbps can be achieved). ATM wireless mode is allowed with synchronous and asynchronous transmission modes. Centralized or Ad-Hoc topologies are allowed and user mobility is reduced to 5–10 Km/h.

9.2.1.5   Personal Area Network Systems

So far, all systems considered use a carrier to transmit their data through the air. In this way, transmissions only occupy a narrow band within the spectrum of available frequencies. These systems are called narrow band transmission systems. Therefore, a multitude of systems can be transmitting at the same time if everyone uses a different frequency. Transmissions can be made without carriers, only transmitting Gaussian pulses within specific slots of time. This way, pulses are used to codify the transmission of one symbol in time (e.g. 1 or 0); frequency depends on the number of transmitted pulses, with the maximum frequency being the inverse of the time-slot employed.

UWB[10] uses this transmission method. At present, Task group 802.15.3 is working on this kind of system in order to achieve better signal error resistance, low cost transmitters and receptors, low consumption and very high data rates (20 to 350 Mbps). The only drawback with these systems is the low power transmission and therefore the very small coverage area (less than 10 m, making it suitable for personal area networks). Difficulties over the implementation of low cost chips, problems with performance and the dissolution of 802.15.3a WorkGroup have limited the use of UWB in consumer products and have led several vendors of UWB chips to cease production during 2008.

Bluetooth[11] is the most representative example of WPAN networks. The success of this technology is due to the fact that it allows implementation in the same processor with very low additional cost and power consumption.

Bluetooth specification 1.0 consisted of one PHY, MAC, LM, upper convergence layers and a set of profiles that specified the minimum set of protocols to be used (RFCOMM, File transfer, Internet access point, LAN access point, synchronization, phone, and headset). It supports TCP/IP and WAP connections, and security features such as user authentication and data encryption. Data rates were 721 Kbps for data and 64 Kbps for voice. Network topology was centralized having one master and up to seven slaves (each of them could be a master of another network). Synchronous and asynchronous transmission modes were available (guaranteeing a fixed data rate and thus, soft real time applications). Coexistence with other ISM networks was facilitated by the use of FHSS (frequency hopping spread spectrum).

Further evolutions of Bluetooth specification 1.0 include network coexistence, better data rates, and security improvements, as can be seen in the following list:

- Bluetooth v1.2: adaptive frequency hopping to avoid using busy frequencies, and extended synchronous connections with packet retransmissions.

---

[8] http://grouper.ieee.org/groups/802.11.

[9] http://www.arib.or.jp/mmac/e/index.htm.

[10] http://www.uwbforum.org.

[11] http://www.bluetooth.com.

- Bluetooth v2.0: compatibility with v1.2, enhanced data rate up to 3 Mbps.
- Bluetooth v2.1: security improvements.
- Bluetooth v3.0: enhanced data rates up to 24 Mbps, based on an 802.11 link (AMP technology). Video streaming available.

The target devices were handsets, PDAs, mobile telephones and laptops, so it can be implemented in medium or high battery capacity devices. In further specifications more PHY layers were added to increase data rates (up to 20 Mbps) or decrease power consumption (using UWB technology as transmission system). Furthermore, coexistence has been improved by an automatic frequency selection mechanism that eliminates busy bands of the FHSS scheme.

9.2.1.6 Wireless Sensor Networks

802.15.4[12] is the most representative example of Wireless Sensor Networks (WSN). It is a standard that covers several PHY layers and one MAC layer, aiming to give low rate network service to devices with low capacity – and therefore low cost – batteries. Expected battery operation time is four to six months for button batteries. Data rates were defined to 250 Kbps at maximum, but lower data rates are possible by choosing the appropriate PHY layer or changing the MAC parameters. Communications security is ensured by AES encryption and coexistence is solved by channel selection and features such as quality of service or noise carrier sensing help to avoid busy channels. Location information is possible through triangulation of RSSI (as implemented in Chipcon CC2431 chips [26, 27]). Network topologies allowed are centralized (star topology with a network master) or Ad-Hoc (peer to peer communications without master). In star topology, beacon enabled communications make it possible to reserve transmission slots, guaranteeing data rates and making soft real time applications possible.

This standard is subject to continual innovation and enhancement. See [28] for other recent strategies to implement real time or [29] for a survey on multimedia transmissions. 802.15.4 is the base of Zigbee[13] that tries to give a complete solution (with more layers and profiles) to low-rate, low-power personal area networks. A profile is a set of protocols and definitions (such as type of messages, IDs, etc.) that must be implemented in case of adopting a specific profile in order to achieve interoperability between devices. One example of this is the recently approved Zigbee Health Care Profile,[14] which offers an open standard for health monitoring and management devices, offering a wide variety of health-care oriented services and protocols.

Another innovation in wireless sensor networks is the implementation of protocol IPv6 over this kind of network. In this sense, 6LoWPAN[15] is an IETF initiative that tries to implement IP networks on 802.15.x devices. This goal is achieved by placing an adaptation layer between the MAC layer in the 802.15.4 protocol stack and the network layer (IPv6 layer) to deal with the limited sizes of 802.15.4 PHY packets (Figure 9.3). This way, the adaptation

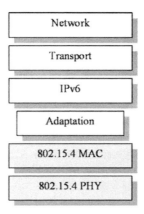

Figure 9.3 Protocol stack of 6LoWPAN.

---

[12] http://www.ieee802.org/15/pub/TG4.html.

[13] http://www.zigbee.org.

[14] http://www.zigbee.org/Markets/ZigBeeHealthCare/Overview.aspx.

[15] http://www.6lowpan.org/.

layer is responsible for header compression, fragmentation, reassembly and routing. More details can be obtained in [30, 31].

The implementation of IP over sensor networks makes it possible for small devices to become part of the Internet. Thus, an Internet consisting not only of medium and large form factor systems but also millions of small computers comes to mind: the Internet of Things [32]. This concept will lead to a huge growth in the number of devices connected to the Internet (some analysts talk about large market opportunities) and the exhaustion of the old range of IP addresses (32 bits). IPv6 is a necessary foe in this case, because it provides a wider range of IP addresses but it also causes problems for small devices due to larger header sizes and computation power consumption. In the near future, smart devices will be incorporated in clothing, appliances, furniture, cars and in just about everything we use or see. All the data from these devices could be sent through Internet where a multitude of applications could make use of them (for our benefit). Security and other challenges will arise but one thing is sure: wireless sensor networks will be one of the main actors in this new global revolution.

Wibree[16] is a Nokia proposal for WSN using Bluetooth PHY layer and simplifying upper layers to guarantee very low power consumption. Wibree allows total compatibility with traditional Bluetooth devices at similar data rates (than 1.0 specification) but at very low power consumption (similar to 802.15.4). The main advantage of this technology is the compatibility with all the already existing Bluetooth devices. Enhancements in upper protocol layers (simplifying and making them more efficient) reduce power consumption by a factor of 20. Enhancements in reliability are achieved through an adaptive frequency hopping (that avoids busy channels). There will be two kinds of devices: single mode (only capable of using low power specification) and dual mode (capable of using both specifications). Nowadays, this technology is being joined to Bluetooth specification for ultra low power consumption.

ANT[17] is a WSN based on Dynastream[18] wireless technology. It provides lower power consumption than 802.15.4 (3 years of battery life) and data rates of 1Mbps. The key point of this technology is the protocol efficiency and ultra low power sleep modes. It supports star and mesh topologies (like 802.15.4) and also user authentication and synchronous time slots to soft real time applications. Coexistence with other networks is possible through adaptive channel techniques.

ISA SP100[19] is an emerging standard for WSN oriented to industrial control applications. In version 11.a, it uses the 802.15.4 2006 PHY and MAC layer and defines upper layers of OSI model up to the application level. The objective is to provide a complete solution (with security, reliability and predictability) for industrial environments where a wide variety of control protocols are used. This way, security is improved with respect to 802.15.4 and coexistence is enforced with techniques such as spectrum organization, intelligent channel selection and frequency hopping. Traffic predictability is guaranteed for latencies of 100 ms and up.

Z-wave[20] is a proprietary technology for remote control applications, such as remote home control and management, energy conservation, home safety and security and home entertainment (control of video players, televisions, and so on). A Z-wave network is a mesh network composed by controllers and controlled devices that operates in the 868 MHz band (selected in order to allow maximum range and maximum battery-life), with a range of 100 meters for each device in open air. This technology is supported by more than 160 manufacturers that agreed to build devices according to the Z-wave standard, such as light control devices, HVAC and security control, as well as home theatres, automated window treatments, pool and spa controls, garage and access controls and more.

INSTEON[21] is a technology for home management and control. Its most remarkable feature is the dual band, which allows the devices to be connected by powerline (with an enhanced version of the X10 standard) or by RF in order to improve reliability. Devices form a mesh topology where all of them act as two-way repeaters in a peer to peer network, without the need for either network supervision or routing tables. The very small footprint of its implementation is feasible to be implemented in very small devices such as light switches, sensors, remote controls, etc.

nanoNET[22] is an emerging technology for loss protection of animal and people, which uses the standard 802.15.4a (chirp spread spectrum) in order to achieve maximum range without loss. Typical devices for this technology are

---

[16] http://www.wibree.com.
[17] http://www.thisisant.com.
[18] http://www.dynastream.com.
[19] http://www.isa.org.
[20] http://www.z-wavealliance.com.
[21] http://www.insteon.net.
[22] http://www.nanotron.com.

location tags (mobile), anchor nodes (fixed), location gateways, and location severs. This technology is supported by a small group of manufacturers interested in the development of location systems.

9.2.2   Concluding Remarks

Certainly, there is no single solution for implementing the network structure in a pervasive system (with the sole exception of having a uniform class of devices with the same level of functionality in all of them). Considerations about network service availability and type of device to employ must be the main priority when choosing a network technology. Then, one thing should be clear: only network technologies with short ranges are capable of giving real time guarantees or location information (except GSM). For this reason, when control restrictions are present in our system or when sensing devices are a part of our system, one of these technologies must be used. In the remaining cases, the range of technologies to be used is more flexible and traffic intensity must be considered as a first priority issue followed by the type of device adopted.

One thing to note is that UMTS promises to be a technology with a wide usability (because it tries to have a permanent connection through the use of other network technologies such as GPRS, Bluetooth, WiFi, satellite, etc.) and could be a good solution except when sensor devices are part of the system. In that case, the alternative is to use a gateway node to join both networks and connect the entire system.

Coexistence problems must receive special attention in each case. There is no universal solution for avoiding coexistence problems and only the correct choice and configuration of networks would appear to provide a firm guarantee for coexistence.

Additionally, pervasive applications require interactions among systems at different levels. The ability to interact at the internetworking level, including all the lower level functions (approximately equivalent to the transport, network, data link and physical OSI layers [33]), implies that communications should be *asynchronous* because of the necessary mobility of nodes. In other words, this interaction (called *interconnectivity* in a previous work [34]) cannot be guaranteed at all times. Mobility requires wireless connections that may suffer frequent connection losses, which forces asynchronous communications. Clients asking for a service and devices offering it may not be connected at the same time. The communication paradigm should be *connectionless* (vs. connection oriented), well suited for intermittent connections.

On the other hand, heterogeneity may make a Nomadic system better that an Ad-Hoc system: a backbone fixed infrastructure plus a number of mobile devices connected through wireless links is sometimes a good solution. A simple example is a backbone network based on the IP protocol, which has demonstrated its success in the interconnection of heterogeneous devices. Most devices can be connected through this IP network while secondary, maybe simpler, devices (e.g. sensors) may be connected using non-IP communications. In this case a gateway may be used to interconnect IP and non-IP sub-networks. As explained in the first study case of Section 9.4, there are scenarios where the loss of connectivity introduces huge delays and therefore applications are designed to be delay tolerant. For example, pervasive systems deployed in Mars are designed to be delay tolerant because the delay introduced by distant.

Nomadic systems present additional advantages. They simplify connection establishment and roaming. Mobile devices usually have limited resources (computing power, bandwidth, memory, etc.) so some of the complexity required or wanted for some functions can be placed in fixed systems. Also, the infrastructure can maintain knowledge about device characteristics and manage coherent device interactions. However, the required mobility of typical pervasive applications makes ad-hoc communications more appropriate, although there are still open issues to be solved.

9.3   Middleware

Middleware is a software layer designed to overcome the complexity and heterogeneity of distributed systems. It is placed between applications and the operating system and offers a common programming interface for all devices in a distributed system, regardless of which hardware or network technology is used in each of them (Figure 9.4). This capability makes middleware the glue that joins all the pieces of a pervasive /ubiquitous system. An approach to the inclusion of middleware in this type of system can be found in [35].

This way, the problem of heterogeneity of networks, operating systems, devices, services and applications is solved. If one application wants to use an external service, it can use the same as any other application

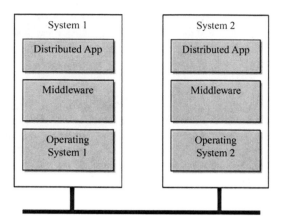

Figure 9.4   Middleware in distributed systems.

running in another device connected to the same distributed system by the same middleware, regardless of the hardware, network or operating system. Note that we consider that interaction among devices at the middleware level includes dynamic service discovering [12] (periodically or triggered by determined events [36]), service description (including actions that may be performed, properties that may be useful, even devices for which connection was not planned), and service control (actions and modifications of state or attributes of a service in a sub-network from another device connected to a different sub-network). There are four classes of middleware:

- **Distributed Tuples:** this offers abstraction for distributed tuples. Examples: L2IMBO, LINDA.
- **Remote Procedure Call:** abstraction of external procedure invocation: Java Remote Method Invocation, Modula-3, XML-RPC, .NET Remoting.
- **Message-Oriented Middleware:** message queue abstraction for messages between distributed applications. Examples: Advanced Message Queuing Protocol, Java Message Service.
- **Object Request Broker:** remote object management abstraction allows using remote objects as if they were in the user's own space. Examples: DCOM, CORBA, COM.

These examples were thought up for large systems with powerful computers. But no network heterogeneity was considered and a permanent, high speed and reliable networking is needed. Things change when smart devices and wireless networks are used to build pervasive or distributed applications. Middleware implementations have to deal with high heterogeneity of devices, networks and operating systems, and with no reliable network connections. Moreover, due to the small amount of memory in these devices, implementations of middleware stack must be small-footprint and resource-saving. For these reasons, none of these middleware classes are suitable (in our opinion) for pervasive applications taking into account the requirements mentioned before.

A number of available architectures can support these functions, and we will now revise some of them taking into account the specific characteristics of the environments and applications for distributed and pervasive systems.

HAVi[23] specification allows registration of new devices as they add to the network, so the rest of devices know what new functions they may perform. Most common A/V (Audio/Video) functions have some standard APIs (Application Programming Interfaces) to enable other devices to use them. The system allows the installation of applications and user interface software in each device in an automatic way. These APIs are specified in a generic C-like language named IDL (*Interface Definition Language*). Specification is independent of language although they are usually implemented in Java [37]. A disadvantage with HAVi is that it assumes the communication channel is IEEE1394, a wired connection with enough bandwidth to transmit audio and video, and so its architecture may show limitations when used in some applications [38].

Figure 9.5 shows the basic structure of HAVi architecture. 1394 Media Manager allows synchronous or isochronous communications between system nodes. The Message System is responsible for passing messages between all HAVi software components. The registry serves as a location directory service that allows other software

---
[23] http://www.havi.org/.

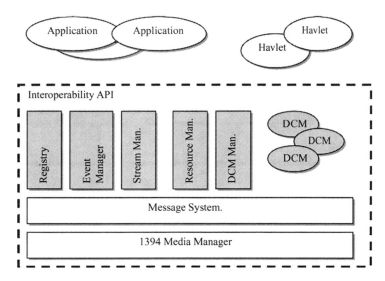

Figure 9.5   Basic structure of HAVi architecture.

elements on the network to be found. Event manager is responsible for sending status messages between software elements, or sending HAVi network configuration messages. Resource manager abstracts shared resources such as DCM or Device Control Modules, which are the central point of HAVi. They are the API defined for one device and can be installed or removed from the system as required if one device is connected or disconnected from the network by DCM Manager. This is a distributed architecture that allows: service (device) discovery on the fly; messaging; event subscription services; and auto-configuration of network devices, streaming and reserve system resources to schedule actions (such as initiate a recording on a video). With all of these features HAVi tries to make it possible to control devices through a user interface that can reside in the DCM. Applications can be written in Java, as Havlets that are applications extracted from a DCM or executed on request from a DCM. DCM and Havlets can be implemented in Java but also in a byte-code of a specific processor.

HAVi classifies its devices in four classes (Figure 9.6):

- **Full AV**: contains a complete HAVi architecture. They can run Java Byte-code, so they can upload and execute Java bytecode from other devices, providing extended control capabilities to this kind of device. An example of Full AV could be a residential gateway controlling all devices at home.

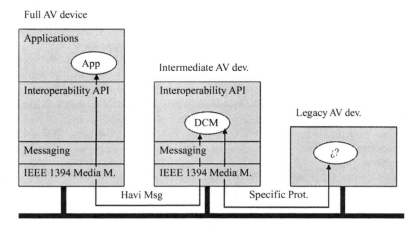

Figure 9.6   Devices in HAVi architecture.

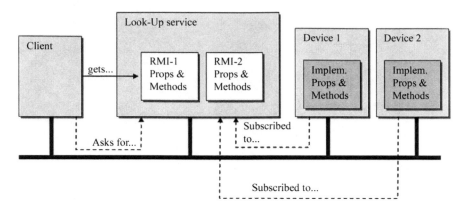

Figure 9.7 Components of Jini.

- **Intermediate AV**: they don't provide Java bytecode execution and cannot be controllers for any kind of devices but can control specific devices.
- **Base AV**: These devices only have a DCM in their memory that can export its code to other devices for being controlled through an AV controlling protocol.
- **Legacy AV**: They don't implement HAVi architecture, and use proprietary protocols to control functionality.

Another *middleware* described in the literature is Jini.[24] Jini gives support for all the described structure concerning service discovery and sharing of services by different clients. Mascolo *et al.* point to its heavy dependency on Java and its assumption of the existence of a fixed infrastructure as its main drawbacks [33]. On the other hand, some authors think that Jini is not very suitable for limited computational resources devices and, mainly, for mobile devices [39]. Nevertheless, the recent release of very small Java virtual machines has made it possible to include very small embedded devices that previously could not be integrated.

Basically, Jini is a Java-based platform consisting of a device federation in which devices and Look-up services coexist (see Figure 9.7). Look-up services are the key component of Jini. They have to be running continuously to allow device discovery services, register devices and give operation licences to their connected devices. Jini allows device announcement through multicast or unicast frames to look-up services. Applications can discover devices on network by asking to look-up services for devices. Devices can be registered in several look-up services that store information about device functionality, a Java bytecode needed to interact with the device and a set of properties and methods. Devices are controlled using RMI (Remote Method Invocation). Auto-configuration of network and security devices is not implemented.

In contrast, there are many devices on the market based on UPnP (Universal Plug and Play),[25] including all Windows XP-based systems. UPnP supports all mentioned functions including the dynamic connection of a device to a network, service offering and discovery, everything based on a unified description of functions and attributes of services through XML (*eXtended Mark-up Language*) documents. There are two factors that make UPnP especially attractive from our point of view: the first is the use of IP protocols at the lowest level; the second is the use of open and standard protocols.

As IP is used for the lowest level, the first phase of interaction among UPnP systems is *Addressing*. This is the mechanism that devices have to obtain an address from that makes them visible in all the system. At first, all UPnP devices incorporate a DHCP (*Dynamic Host Configuration Protocol*) client to obtain an IP address. If there is no answer from a DHCP server, it chooses an address and checks it wasn't used through ARP (*Address Resolution Protocol*). In any case, once the device has an IP address, it can proceed with following phases related to discovery functions and service control, that are performed using another open and simple protocol named SSDP (*Simple Service Discovery Protocol*).

IP based communications are made using HTTP protocol or any of its versions. HTTP is message based (asynchronous communication), which is especially adequate for pervasive systems, that may suffer unpredictable

[24] http://www.jini.org.
[25] http://www.upnp.org.

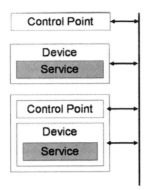

Figure 9.8 Devices in UPnP.

disconnections. Particularly, UPnP considers the usage of HTTPU (or HTTPMU for *multicast* messages, very useful when offering services to the whole system), that is HTTP version over UDP. UDP gives functionality equivalent to the transport (TCP) but is simpler, and, most importantly, non-connection oriented. So, the use of the set HTTPU/UDP/IP is, in our opinion, an adequate solution for asynchronous communication in pervasive systems.

Development of UPnP devices (see Figure 9.8) has been made much easier thanks to tools like *Intel Service Author, Device Builder and Device Validator* [40]. Siemens also has some tools, and there are even some freeware developments for Linux. A disadvantage of UPnP may be that simple devices find it difficult to interpret XML descriptions of devices and services. Moreover, these devices may also have trouble supporting all the pile of UPnP protocol. A feasible solution for simple devices and/or slow connections is the use of *Simple Control Protocol –* SCP. This is a non TCP/IP based protocol, but it uses the same schemes and models as UPnP, allowing easy interoperation among UPnP and SCP devices via simple bridges [41].

Finally, it may be worth mentioning other interoperability solutions, not considered here because they are limited to a set of applications, such as *Salutation*, or Bonjour in office environments(moreover, dissolved in 2005), or limited to specific functions or technologies, such as SDP (*Service Discovery Protocol)* of Bluetooth. They are all technologies for service discovery in local area networks (case of Salutation and Bonjour) or piconets (case of SDP).

URCC[26] (Universal Remote Console Control Protocol) is another example, focused on the control of household appliances such as TV, VCR, clocks, etc., using PDAs, laptops or whatever remote controller is suited to embedding this protocol. In URCC the main idea is to have a device capable of talking with all kinds of devices (that implement URCC) and offer a user interface suitable for the device selected by the user. Components in this architecture are the controller, the target device (controlled) and the network they use to communicate. URC was focused initially on IRDa but it can now use any other RF or wired network technology. One thing to note is that URCC has a provision for implementing natural language user interfaces, and also multilingual interaction. For these reasons, URCC is suggested for use in assistive technologies.

As there are many technologies available at *middleware* level, and they are not compatible with each other, the heterogeneity problem remains a major difficulty for pervasive and distributed systems. There are already descriptions in the literature of efforts to overcome heterogeneity at *middleware* level [42, 43], such as OSGi (Open Services Gateway specification) initiative [44]. This provides a common structure at application level that is independent of the middleware technology used. Actually what OSGi offers is a specification for a somewhat centralized gateway that allows the use of services of different technologies, such as Jini and UPnP (although in spec. R4, Jini has been dropped due to the lack of interest in it).

In OSGi architecture, bundles are OSGi components developed in Java or C (Figure 9.9). Bundles consist of a code for server (implementation) and client (interface) side, plus a manifest file that allows specifying dependences on other bundles, versioning and a basic description of capabilities. Services are connected and executed by bundles that dynamically must register the services that they want to use. Dependencies between bundles and services are managed by the Life Cycle layer. There are services included by default in the architecture (XML parser, HTTP,

---

[26] http://myurc.org.

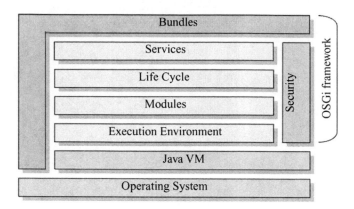

Figure 9.9   OSGi architecture.

user admin, IO connector, configuration, log, URL handlers, etc.) and others that the developers can add to it. Services and bundles can be deleted, installed and reinitiated, by the Modules layer using functionalities of the Life Cycle layer. Execution environment defines which Java classes lay in a specific platform.

All the architecture is executed by a Java Virtual Machine that resides over the operating system. Bearing this in mind, it is hard to believe that an OSGi implementation with UPnP or Jini embedded could fit into a PDA or a low cost device. It is obvious that OSGi was designed to be implemented in more powerful devices (like gateways).

OSGi has the following features:

- A free component framework for industry, service providers and developers.
- A powerful model to execute safely components/applications in Java Virtual Machine.
- A flexible API to control the life cycle of applications/components, allowing updating of and easy development of components without restarting applications.
- A collaborative model to use and share services between applications /devices.
- A set of optional services to fit the framework footprint to device requirements.

Basically, these middleware architectures are integrated in OSGi by creating services to be used by bundles that act as bridges between information generated by these services. This way, services and clients (applications) of both architectures can coexist and communicate between themselves.

In Figure 9.10, Bundle A in the gateway device uses the Jini service (in the form of a bundle, as Jini was dropped from the last OSGi specification) and the UPnP service already included in the OSGi framework to communicate with UPnP and Jini devices and view them all together as the same set of services (as an example). This bundle

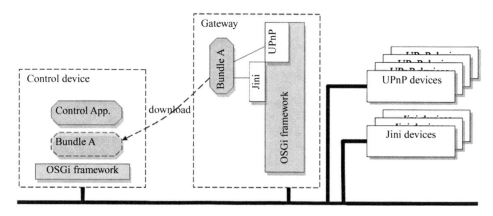

Figure 9.10   An example of OSGi.

can be imported by a control device and be used by a control application that exposes a GUI for controlling Jini and UPnP devices.

Something similar has been proposed using techniques already used in the Internet, through the concept of *Virtual Overlay Networks* [45]. Generally speaking, this idea of a gateway is placed over the level of middleware, closer to the application level, so it differs slightly from the former interoperability concept.

Whatever solution is chosen, several design considerations have become clear from the previous sections. If nomadic systems are used at the interconnectivity level, then the fixed infrastructure may support some interoperability functions. For instance, a central unit (*Residential Gateway*) may gather and distribute some services. Furthermore, the context captured by fixed elements (for instance the closest service access point/provider in public environments such as airports) can approach that of the mobile client. Mobile devices could therefore obtain an IP address; connect to a fixed system and collect/send services and context information in order to interoperate. This interoperation also requires abstract context representation as well as generic, common protocols, including common formats for multimedia content distribution.

On the other hand, a fully distributed solution based on an ad-hoc network would still be possible, with devices advertising services (multicast) and collecting context information, but this would require more powerful devices and complex protocols.

9.3.1   Future Trends: Beyond the Middleware

Interaction among subsystems at the middleware level allows the services to be discovered and shared without considering their 'meaning'. In the near future, a semantic interaction between applications could be needed, similar to those described in the literature for other fields [46, 47, 48]. This need is emphasized by the ISTAG[27] as being one of the key issues in future applications [1]. Several languages have already been proposed for semantic interaction, such as RDF (Resource Description Language), OWL (Ontology Web Language) and initiatives that emphasize semantic interactions are already present in other technological niches.

From our point of view, semantic interactions would add two main values to subsystem interactions:

- Semantic descriptions allow us to pre-select the services previewed as useful for the applications of a particular subsystem.
- They would enable the adaptation of existing applications to new services. New applications may even become available based on the new services.

And they will present us with new challenges such as:

- Mixing services from incompatible technologies.
- The use of new functions with little or no advance planning. It may be difficult to foresee potential uses of services that were not considered at the design stage.
- New services should offer different complexity descriptions and/or interfaces.

Whatever solution is considered to solve these problems, semantic interactions represent the next frontier to be crossed by the next generation of systems. More challenges and more difficulties will arise on the way to achieving this target but, in the end, the overall objective is to make machine interactions as similar to human relations as possible. Will applications be capable of defining the next frontier for their interactions by themselves?

9.4   Case Studies

The following case studies represent very different application areas of Pervasive Computing. The first one, the Hiker's PDA, is an example of pervasive computing in remote areas above the Arctic Polar Circle in north Scandinavia. The second one, MyHealthService, is an example of Pervasive Healthcare, where heterogeneous technologies are put together for providing personalized health services to patients with chronic diseases.

---

[27] Information Society Technologies Advisory Group.

9.4.1 Pervasive Computing in Extreme Areas; The Hiker's Personal Digital Assistant

Hiker's PDA is an example of a pervasive system developed in the project N4C[28]. N4C is exploring the use of Delay Tolerant Networks (DTN) [49, 50, 51] to provide connectivity and services in Communications Challenged Regions (CCR), where there is no infrastructure for connection to Internet and no mobile phone coverage. Delay Tolerant Networking is concerned with interconnecting highly heterogeneous networks together even if the end-to-end connectivity may never be available. Examples of application include networking in space (e.g. the provision of Internet during inter-planetary travels), underwater, and some forms of ad-hoc sensor/actuator networks [51]. In N4C a set of pervasive applications are provided using a DTN infrastructure in remote areas of Swedish Lapland and the Slovenian forests.

As stated earlier, the communication system is fundamental to any pervasive system. A prerequisite for pervasive applications is connection to the Internet, usually provided by wireless communication. Since TCP/IP is an end-to-end protocol it tends to time out when communication is delayed or disrupted. However, many types of applications are inherently delay tolerant and could perfectly well have survived such disruptions, if it were not for the behavior of the underlying network service. In order to extend the realm of pervasive systems to include delay tolerant applications, one must use a delay tolerant protocol. The N4C paradigm emphasizes the *nomadic* nature of both users and by implication the network as a whole – In N4C the user *is* the network, in contrast to the conventional Internet where the network and the user, with the end-node, are functionally and physically separated.

9.4.1.1 Users and Applications

The users of the N4C's Hiker's PDA are reindeer herders, tourist guides, nature park rangers, police and tourists in the vast wilderness areas in the Arctic area of Scandinavia and Slovenia. On top of the DTN infrastructure they have access to the following applications: Email, Map service with GPS (for Location Based Services), NotSoIM (Not So Instant Messaging) for communicating with other people in the field, Web Service (e.g. to access cached web pages and RSS) and the GeoBlog. The GeoBlog is a special blog containing field observations where pictures and GPS localization may be added. When random users are within communication range, their PDAs will synchronize geoblogs so that the registrations are exchanged. The GeoBlogs will be visible locally in the map application Maemo Mapper, and eventually, when the first hiker reaches a gateway with Internet connection, at the N4C GeoBlog on the Internet (see Figure 9.11).

9.4.1.2 System Architecture

These applications are provided in small and low-cost computers, such as Ultra Mobile PCs and Netbooks. Two different types of computers are tested in the current version, Asus Eee PC[29] as gateway nodes and the UMPC Nokia N810 and N900[30] as the Hiker's PDA (end-user node). The nodes are interconnected using WiFi (802.11) with a DTN protocol whenever they are within WiFi range. The Asus has no internal GPS, so an external USB-GPS from Holux is used. The Nokia N810 has built-in GPS, but it is a SiRF star III based GPS, which makes its performance suffer, compared to the Holux.

The Software architecture is divided in four different tiers; see Figure 9.12. The Application layer called Hiker's PDA contains the applications. The middleware layer relays with the DTN layer which may provide communication to the Internet or to different CCRs. The DTN is on top of the Link layer which is managing the different communications (e.g. WiFi, WiMax and Bluetooth). The implementation of the software platform is based on Python. The hybrid synchronization service in the middleware layer uses XML-RPC for synchronization of meta-data, and http for synchronization of user data. The auto-discovery function uses IPv6 and multicast.

---

[28] Networking for Communications Challenged Communities: Architecture, Test Beds and Innovative Alliances project. http://www.n4c.eu.
[29] Asus Eee PC: http://eeepc.asus.com/global/index.html.
[30] Nokia 810, http://www.forum.nokia.com/devices/N810.

Color legend:

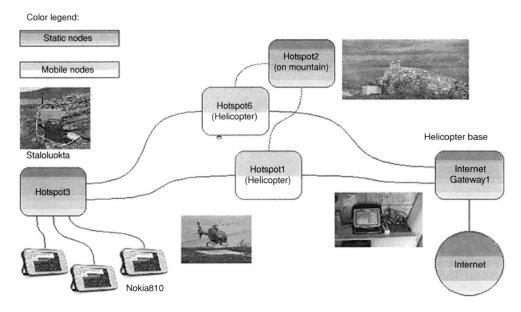

Figure 9.11   Topology for summer test in a remote Lappish village using 3 Nokia810. There are 4 hotspots: one in a tent, 2 in helicopters and one on a mountain. The gateway to Internet resides in the helicopter base station in Ritsem.

9.4.1.3   Current Status and Future Work

The described functionality of the N4C system has been developed and tested in remote areas of Norway, Sweden and Slovenia. The most advanced prototypes feature both automatic and manual synchronization. Additionally, the next version will include improvements in the user interfaces and the integration of new DTN technologies (e.g. DTN2 and/or PRoPHET).

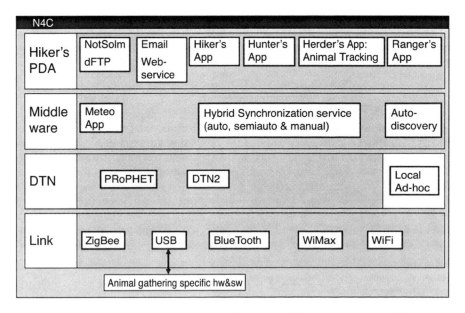

Figure 9.12   N4C software model with applications, middleware, DTN and link layers.

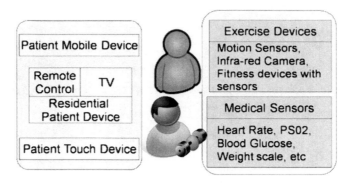

Figure 9.13   Devices and sensors in MyHealthService.

Beyond the scope of the project we are working on the integration of DTN with Unmanned Aerial Vehicles (UAVs) that can be used as hotspots while flying in remote areas and synchronize with wireless sensor networks on the ground can transmit information flying UAVs.

9.4.2   Pervasive Computing in Personal Health Systems; The MyHealthService Approach

Pervasive computing has also entered healthcare [52], and has the potential of being applied to a broad range of areas, from supporting healthcare personnel at work to personal health systems for patients at home. The example addressed in this case study, MyHealthService, is an integrated system for personal health and chronic disease management, based on Free Open Source software and low-cost consumer electronics (see Figure 9.13). The focus is on home-based services for self management and following-up [53,54]. MyHealthService service is being developed within the Tromsø Telemedicine Laboratory.[31]

9.4.2.1   Services

The services in the MyHealthService environment can be grouped into four categories; health education, health diary, exercising, and community [54]. Each of these categories consists of one or more service components, and is implemented on various patient devices. When orchestrating more complex services such as a home-based rehabilitation program for Chronic Obstructive Pulmonary Disease (COPD), the required services components are selected from a service category, and integrated in the overall system.

9.4.2.2   Devices and Communication

A core component in the MyHealthService environment is the Residential Patient Device, implemented on a small low-cost PC. This device gives users access to the personal health services at home, and provides secure communication between the patient at home, healthcare professionals, and peers. It is connected to a standard TV and has a dedicated IR remote control, low-cost video camera and a set of Bluetooth sensors for monitoring vital signs and movements.

The prototype of the residential patient device has been implemented on a Barebone PC, a Mac mini and latest on an ASUS Eee Box,[32] which is a mini desktop computer. All of them are Linux based and support wireless communication like Wi-Fi, Bluetooth, and IR. Open source software components are used extensively in the implementation, including Ekiga[33] for the videoconferencing, encrypted Postgress database for data storage, and

---

[31] Tromsø Telemedicine Laboratory, http://www.telemed.no/ttl.

[32] Asus Eee PC: http://eeepc.asus.com/.

[33] Ekiga: http://www.ekiga.org.

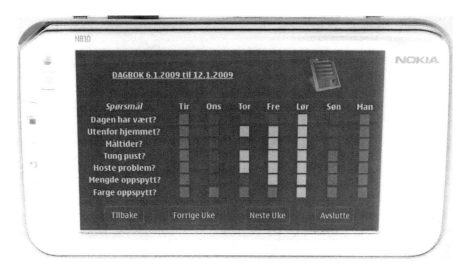

Figure 9.14   The Mobile Patient Device (Nokia 810) with the Patient Diary application.

BlueZ[34] for the Bluetooth communication with the sensors. The user interface is implemented using a standard web browser displayed in full screen mode on the TV. The menu's items are all numbered, which enables easy navigation using the dedicated remote control.

Limiting the accessibility of the personal health services to the home environment could discourage the users to take part in activities outside their home, go travelling, etc. Therefore a Mobile Patient Device has been implemented in order to support access to services while on the move, and for flexible access at home. It is currently implemented on a Nokia 810,[35] see Figure 9.14, which is a Linux-based touch sensitive Ultra Mobile PC. It has a camera, microphone, GPS, keyboard, Wi-Fi, and Bluetooth. The Wi-Fi connectivity is used to synchronize the data (e.g. questionnaires, educational materials) with the Residential Patient Device.

Other Patients devices include a touch-screen PC (Asus Eee Top) and a Notebook (Asus Eee), which allows the user easy access to services at the point of need, such as next to a home fitness training device or in the kitchen for nutritional advices.

Various sensors are integrated into the system, including movement sensors and medical sensors. Two wireless pulse oximeters, measuring oxygen saturation and heart rate using infrared, are integrated in the system, providing valuable information about the cardiopulmonary situation of the users for following-up and while exercising. These wireless Bluetooth pulse oximeters provided by Nonin[36], Nonin 4100 Bluetooth for the wrist and Nonin Onyx II 9560 wireless fingertip, have open communication protocol making integration easier.

The game-based control Wii Remote[37] is integrated providing motion sensing capabilities with its 3D accelerometer [55]. The motion information will be used to enhance services for motivation and self management for the patients, and for following-up by healthcare professionals. These Bluetooth enabled sensors can be held or worn by the user, or embedded in consumer fitness devices (Figure 9.15). Sensors from Inertia Technologies[38] which contain accelerometers, gyroscopes and magnetoscopes, are also being integrated. These sensors form a wireless sensor network using IEEE 802.15.4. Other sensors for motion tracking integrated are the WiiFit device[39], which is a board with pressure sensors.

---

[34] BlueZ: http://www.bluez.org.
[35] Nokia 810, http://www.forum.nokia.com/devices/N810.
[36] Nonin, http://www.nonin.com.
[37] WiiRemote, http://en.wikipedia.org/wiki/Wii_Remote.
[38] Inertia Technologies, http://www.inertia-technology.com.
[39] WiiFit, http://en.wikipedia.org/wiki/Wii_Fit.

Figure 9.15   Wii Remote attached to consumer fitness device.

9.4.2.3   Trials and Evaluation

Parts of the MyHealthService system and services have been field trailed and evaluated in a home-based rehabilitation program for COPD patients, and for home-based education and following-up of people with diabetes. The Residential Patient Device, TV, remote control and some medical and activity sensors were used in the trials. The user acceptance was high and the patient found the technology non-intrusive and easy to use [53].

9.4.2.4   Future Work

Future work in MyHealthService includes the development of new and improved personal health services, increased context-awareness and personalization, and to utilize innovative features of new low-cost consumer devices. For example, we are exploring the use of context-aware techniques for building a recommender system of health educational videos adapted to the changes in the patient's health [56].

One of the main barriers to develop personalized health applications is the lack of medical devices with open communication protocols. To overcome this barrier many initiatives are emerging, such as Continua Alliance.[40]

References

[1] ISTAG Draft Consolidated Report: 'Ambient Intelligence: from vision to reality'. Sept. 2003. Archived at http://www.webcitation.org/5r3ae3sQP.

[2] A. Greenfield (2006) *Everyware: the dawning age of ubiquitous computing.* New Riders.

[3] C.A. da Costa, A.C. Yamin, and C.F. Resin Geyer (2008) 'Toward a General Software Infrastructure for Ubiquitous Computing' *IEEE Pervasive Computing*, pp. 64–73, Jan–March.

[4] J.L. Sevillano *et al.* (2004) *On the Design of Ambient Intelligent Systems in the Context of Assistive Technologies.* K. Miesenberger *et al.* (eds): ICCHP 2004, LNCS 3118, pp. 914–21.

[5] D. Saha and A. Mukherjee (2003) 'Pervasive computing: a paradigm for the 21st century,' *Computer* **36**(3): 25–31.

[6] C. Fetzer and K. Högstedt (2003) *Challenges in Making Pervasive Systems Dependable. Future Directions in Distributed Computing*, LNCS2584, Springer Verlag.

[7] M.S. Obaidat and N.A. Boudriga (2007) *Security of E-systems and Computer Networks*. Cambridge University Press.

[8] A.K. Dey (2001) 'Understanding and using context,' *Pers. Ubiquit. Comput.* **5**: 20–4.

[9] K.L. Park, U.H. Yoon, and S.D. Kim (2008) 'Personalized service discovery in ubiquitous computing environments,' *IEEE Pervasive Computing*, pp. 58–65. January–March.

[10] J. Krumm (2009) 'A survey of computational location privacy,' *Pers. Ubiquit. Comput.* **13**: 391–9.

---

[40] http:// www.continuaalliance.org/.

[11] W.H. Widen (2008) 'Smart cameras and the right to privacy,' In *Proceedings of the IEEE* **96**(10): 1688–97, October.

[12] F. Zhu, M.W. Mutka, and L.M. Ni (2005) 'Service discovery in pervasive computing environments,' *IEEE Pervasive Computing* **4**(4): 81–90, Oct.–Dec.

[13] N. Golmie (2006) *Coexistence in Wireless Networks*. Cambridge University Press.

[14] A. Karnik, A. Iyer, and C. Rosenberg (2009) 'What is the right model for wireless channel interference?,' *IEEE Transactions on Wireless Communications* **8**(5): 2662–71.

[15] A.V. Kini, N. Singhal, and S. Weber (2008) 'Wireless; performance of a WSN in the presence of channel variations and interference,' *IEEE Communications and Networking Conference, 2008. WCNC 2008*, March 31–April 3: 2899–2904.

[16] L. Lo Bello, E. Toscano (2009) 'Coexistence issues of multiple co-located IEEE 802.15.4/ZigBee networks running on adjacent radio channels in industrial environments,' *IEEE Transactions on Industrial Informatics* **5**(2): 157–67.

[17] K. Premkumar, S.H. Srinivasan (2005) 'Diversity techniques for interference mitigation between IEEE 802.11 WLANs and Bluetooth,' *IEEE 16th International Symposium on Personal, Indoor and Mobile Radio Communications, 2005. PIMRC 2005* 3: 1468–72.

[18] K. Huang, V.K. Lau, and Y. Chen (2009) 'Spectrum sharing between cellular and mobile ad hoc networks: transmission-capacity trade-off,' *IEEE JSAC. 27, 7 (Sep. 2009)*, 1256–67. DOI=http://dx.doi.org/10.1109/JSAC.2009.090921.

[19] W. Stallings (2002) *Wireless Communications and Networks*. Prentice Hall.

[20] P. Nicopolitidis, A.S. Pomportsis, G.I. Papadimitriou, and M.S. Obaidat (2003) *Wireless Networks*. John Wiley & Sons, Inc.

[21] J. Yick, B. Mukherjee, and D. Ghosal (2008) 'Wireless sensor network survey,' *Computer Networks* **52**: 12 (Aug. 2008), 2292–2330. DOI=http://dx.doi.org/10.1016/j.comnet.2008.04.002.

[22] Ts 101 369 - V06.03.00 – Digital Cellular Telecommunications System (phase 2+); Terminal Equipment To Mobile Station (te-ms) Multiplexer Protocol. GSM Special Mobile Group. ETSI http://www.etsi.org/. March 1999.

[23] A.R. Noerpel (1996) 'PACS: personal access communications systems an alternative technology for PCS,'. IEEE Communications Magazine **34**(10) :138–50.

[24] K. Fazel, C. Decanis, J. Klein, G. Licitra, L. Lindh, and Y.Y. Lebret (2002) 'An overview of the ETSI-BRAN HIPERACCESS physical layer air interface specification,' *The 13th IEEE International Symposium on Personal, Indoor and Mobile Radio Communications, 2002* 1: 49–53.

[25] E.P. Vasilakopoulou, G.E. Karastergios, and G.D. Papadopoulos (2003) 'Design and implementation of the HiperLan/2 protocol,' *SIGMOBILE Mob. Comput. Commun. Rev.* **7** (2): 20–32.

[26] G.V. Merrett, A.S. Weddell, L. Berti, N.R. Harris, N.M. White, and B.M. Al-Hashimi, (2008) 'A wireless sensor network for cleanroom monitoring,' In *Eurosensors 2008*, 07-11 September, Dresden, Germany.

[27] S. Tennina, M. Di Renzo, F. Graziosi, and F. Santucci (2008) 'Locating zigbee® nodes using the ti®s cc2431 location engine: a testbed platform and new solutions for positioning estimation of wsns in dynamic indoor environments,' In *Proceedings of the First ACM international Workshop on Mobile Entity Localization and Tracking in Gps-Less Environments* (San Francisco, California, USA). MELT '08. ACM, New York, NY, 37–42.

[28] J.L. Sevillano *et al.* (2009) 'A real-time wireless sensor network for wheelchair navigation,' In *Proceedings of the 7th ACS/IEEE Intl. Conf. on Computer Systems and Applications*, pp. 103–8.

[29] I.F. Akyildiz, T. Melodia, and K.R. Chowdhury (2007) 'A survey on wireless multimedia sensor networks,' *Journal on Computer Networks* **51**: 921–60.

[30] Xin Ma, Wei Luo (2008) 'The analysis of 6LowPAN technology,' *IEEE Pacific-Asia Workshop on Computational Intelligence and Industrial Application*, pp. 963–6, 2008.

[31] J.W. Hui,, D.E. Culler (2008) 'Extending IP to low-power, wireless personal area networks,' *IEEE Internet Computing* **12**(4): 37–45.

[32] ITU Internet Reports 2005: Executive Summary. The Internet of Things. Archived at http://www.webcitation.org/5r3bKSPtr.

[33] C. Mascolo, L. Capra, and W. Emmerich (2002) 'Mobile computing middleware,' Tutorial in IEEE MWCN'02. Stockholm, Sweden.

[34] J. Abascal, JL Sevillano, A. Civit-Balcells, G. Jimenez, and J. Falcó (2005) 'Assistive environments: integration of heterogeneous networks to support ambient intelligence,' *Home-Oriented Informatics and Telematics*. New York, EE.UU. Springer-Verlag, pp. 321–35.

[35] Z. Artola, M. Larrea, D. Cascado, JL Sevillano, R. Casas, and A. Marco (2007) 'A middleware-based approach for context-aware computing,' *Upgrade: The European Online Magazine for the It Professional* **8**(4): 24–30.

[36] A. Wils A. *et al.* (2002) 'Device discovery via residential gateways,' *IEEE Trans. Consumer Electronics* **48**(3): 478–83.

[37] R. Lea, S. Gibbs, and A. Dara-Abrams, and E. Eytchison (2000) 'Networking home entertainment devices with HAVi,' *IEEE Computer* **33**: 35–43.

[38] T. Nakajima T. *et al.* (2002) 'A virtual overlay network for integrating home appliances,' In *Proceedings 2002 Symp. Applications and the Internet*, Nara, Japan, pp. 246–53.

[39] S. Helal (2002) 'Standards for service discovery and delivery,' *IEEE Pervasive Computing* **1**: 95–100.

[40] http://www.intel.com/technology/upnp/index.htm.

[41] M. Jeronimo, J. Weast (2003) *UPnP* Design by Example*. Intel Press.

[42] P. Grace P. (2004) 'Overcoming middleware heterogeneity in mobile computing applications,' Ph D. Thesis. Computing Departament. Lancaster University. March 2004.

[43] L. Capra, G.S. Blair, C. Mascolo, W. Emmerich, and P. Grace (2002) 'Exploiting reflection in mobile computing middleware,' *SIGMOBILE Mob. Comput. Commun. Rev.* **6**(4): 34–44.

[44] P. Dobrev, D. Famolari, C. Kurzke, and B.A. Miller (2002) 'Device and service discovery in home networks with OSGi,' *IEEE Communications Magazine* **40**: 86–92.

[45] T. Nakajima T. *et al.* (2002) 'A virtual overlay network for integrating home appliances,' In *Proceedings 2002 Symp. Applications and the Internet*, Nara, Japan. pp. 246–53.

[46] C. Stephanidis, A. Savidis (2001) 'Universal access in the information society: methods, tools, and interaction technologies,' *Universal Access in the Information Society* June, **1**: 40–55.

[47] M. Paolucci, K. Sycara (2003) 'Autonomous semantic web services,' *IEEE Internet Computing* 7(**5**): 34–41.

[48] Special issue on 'Web Services,' *Computer* **36**, Oct. 2003.

[49] Delay-Tolerant Networking Research Group, http://www.dtnrg.org/wiki, accessed 22 Aug. 2009.

[50] S. Farrell, V. Cahill, V. (2006) *Delay- and Disruption-Tolerant Networking.* Artech House.

[51] R. Beck *et al.* (2007) 'GPSDTN: predictive velocity-enabled delay-tolerant networks for arctic research and sustainability,' *The First International Workshop on Tracking Computing Technologies, IARIA-IEEE Track 2007*, July 1-6, 2007 – Silicon Valley, USA.

[52] C. Orwat, A. Graefe, and T. Faulwasser (2008) 'Towards pervasive computing in health care – a literature review,' *BMC Medical Informatics and Decision Making* **8**: 26.

[53] T.M. Burkow, L.K. Vognild, T. Krogstad *et al.* (2008) 'An easy to use and affordable home-based personal eHealth system for chronic disease management based on free open source software,' *Studies in Health Technology and Informatics* **136**: 83–8, 2008.

[54] L.K. Vognild, T.M. Burkow, and L. Fernandez-Luque (2009) 'The MyHealthService approach for chronic disease management based on free open source software and low cost components,' *31th Annual International Conference of IEEE Engineering in Medicine and Biology Society*, Sep. 2–6, Minneapolis, MN, USA.

[55] S. Romero, L. Fernandez-Luque, J.L. Sevillano Ramos, and L.K. Vognild (2010) 'Open source virtual worlds and low cost sensors for physical rehab of patients with chronic diseases (eHealth 2009),' *Lecture Notes of the Institute for Computer Sciences, Social-Informatics and Telecommunications Engineering (LNICST)* **27**: 84–7.

[56] L. Fernandez-Luque (2009) 'Personalization in the Age of Health 2.0: MyHealthEducator,' In *Proceedings of International Workshop on Adaptation and Personalization for Web 2.0 (AP-WEB 2.0 2009), UMAP 2009*, Trento, Italy, June 22, pp. 139–42, CEUR Workshop Proceedings, ISSN 1613-0073, online http://ceur-ws.org/Vol-485/paper1.pdf.

Part Two

Pervasive Networking Security

10

Security and Privacy in Pervasive Networks

Tarik Guelzim and Mohammad S. Obaidat

Dept. of Computer Science and Software Engineering, Monmouth University, W. Long Branch, NJ 07764, USA.

10.1 Introduction

The technological advances in recent years have revolutionized the area of pervasive computer networking and pervasive computing. This is thought to be a natural evolution of the expansion of computing devices into every aspect of our daily life from bread toasters to nano-sized chips. With all the advantages that this computing paradigm brings, there are also numerous disadvantages when it comes to protecting privacy of individuals and users. Many systems try to bridge this gap by finding a compromise between pervasive computing and privacy, a balance that can only be improved by adding the dimensions of context and locality to this complex equation. Beyond classical security concepts such as perimeter security and access control, other complex security mechanisms, including the use of logical schemes such as contract based security, will become the norm for software download 'handshake' protocols. The newer methods rely on RFID based authentication and identification in pervasive systems. In these systems, a personal tag device is used to define rights of access. Other statistical methods based on neural networks have also been implemented as a means of controlling systems security. In this type of method, an *a priori* distribution is constructed and used as the base knowledge for all requests. Although this method has a very high rate of success, it has the drawback of not being able to evolve when the environment changes.

Privacy is also a big issue when dealing with pervasive networks. Pervasive networks cannot proliferate unless privacy is preserved. As will be explained later in more detail, privacy needs to be seamless and not obstructive. This is important in order to maintain the user's trust as well as a low overhead. One of the proposed solutions was embedding privacy software inside the devices instead of handling this task by way of the server. Other privacy schemes relied on concepts of data ownership and data collection. These needed to be controlled by the users only. In recent years, these rules have not been respected by ISPs who on numerous occasions have leaked their subscribers' private information. Nonetheless, privacy is not easy to maintain for several reasons. First, because of the nature of pervasive networks, users are expected to be moving in different cells. In the case where these cells have different privacy policies, it is usually difficult in practice to maintain customized policies for different users seamlessly with no overhead. Another issue, also inherent in the nature of pervasive networks, is the aspect of sharing with this type of network. Although mobile devices are personal, these devices might be shared between different family members or friends each with specific expectations of privacy. Solving problems such as these requires the addition of contextual data and the use of more flexible systems.

Pervasive Computing and Networking, First Edition.   Edited by Mohammad S. Obaidat, Mieso Denko, and Isaac Woungang.
© 2011 John Wiley & Sons, Ltd. Published 2011 by John Wiley & Sons, Ltd.

In the following sections we will present these issues in detail as well as prototype systems that attempt to resolve them.

10.2   Security Classics

10.2.1   Perimeter Security

In any computer network, firewalls are used to protect devices and users against any intrusion or misbehavior. Nodes are authorized to work with only well defined interconnections as well as trusted computers. Many application level protocols such as the Lightweight Directory Access Protocol (LDAP) or Active Directory (AD) are used to control resources and to effect rights to every component that wishes to connect to the network or utilize its resources. In Wireless Networks (WNs), perimeter security must accommodate the dynamic nature of access and control such as in Roaming for Global Subscriber Mobile (GSM) or MobileIP for 802.11 networks [1].

10.2.2   Access Control

Access control (AC) in pervasive networks usually relies on the right management matrices, which define roles, rights and context. The most predominant scheme is Role Based Access Control (RBAC). Since the early days of computing, it has been recognized that a reference monitor is needed in order to manage users and devices and in order to distinguish their concerns. Although this scheme is coherent, it is unsuitable for today's wireless networks owing to to the lack of a central authority that names all of the possible actors [1]. Most of the recent research in ACs focuses on context as a criterion for privacy instead of single dimension mapping between privacy enforcement and access control.

10.3   Hardening Pervasive Networks

10.3.1   Pervasive Computational Paradigms

Over the last decade, we have witnessed a revolution in computing by way of smart and energy efficient personalized gadgets. A broad range of technologies is all mixed in together to form these devices. Looking back at the history of computing, we can identify three major milestones in the computing revolution [2]:

- *First wave*: A computing resource shared among many users through mainframes.
- *Second wave*: A computing resource per user; a personal PC.
- *Third wave*: Many computing resources per user; pervasive computing.

The third wave allows millions of computers embedded in an environment all coming together to put technology at the service of our 'new' every day's needs. Figure 10.1 shows a pervasive system model stack.

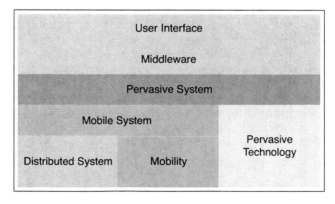

Figure 10.1   Pervasive computing from a system perspective.

10.3.2 Pervasive Hardware

Breakthroughs in the microprocessor industry and nanotechnology by which we now can cram billions of transistors into chips enabled the expansion of small and smart devices that can carry out many operations that were once a task for big computers and thus make computing popular. Examples of today's powerful devices are handheld PDAs, smart phones, GPS receivers, small laptops, etc. These devices come in many different forms and sizes [2] and are expected in the near future to communicate with each other intelligently.

To further understand pervasive computing, we break down the devices into these three categories:

1. *Sensor devices*: These detect environmental changes, user behavior, remote commands, amongst others.
2. *Processors*: These are normally manifested as nano-sized electronic chips that analyze and interpret input data.
3. *Actuators*: These receive input data and update or alter environmental variables using electronic or mechanical means (e.g. temperature control).

The current trend with chip makers is to produce chips as small as a grain of sand that function independently in terms of networking and power supply. These chips can be distributed throughout an environment so as to form a heterogeneous and complementary network of helper devices.

10.3.3 Pervasive Networking and Middleware

As an extension to hardware, pervasive computing will rely on connecting heterogeneous devices together within the same network. We can achieve this today using Bluetooth, ZigBee or WiFi protocols. The success of future pervasive networks relies on the need to make all of these technologies communicate through the use of middleware.

Amongst other functions, middleware enables different networking kernels to communicate with each other in order to provide a transparent and unified interface for users. This middleware consists of firmware and software packages distributed on the server side or client side level. These user interfaces are capable of acquiring information about users through sensors, exploiting and processing this information using processors and displaying or starting an action using actuators. Research today is examining more elaborate ways to solve the human-machine interaction by focusing on input interfaces that provide visual interactions. iPhone[1] is a current example of how well this interaction has been understood and represents only the beginning of the next generation of state of the art gadgets.

10.3.4 Pervasive Applications

From control of machinery and home automation to user base interaction relying on biometric features, applications in pervasive systems are expanding at an exponential rate due primarily to advances in the hardware that supports them. Heart implants are a striking example of how we can harvest the benefits of pervasive applications. These little devices can communicate wirelessly through a local network so as to notify a computer designed to detect anomalies, raise the alarm and save a life.

In this ever growing field, the vision expressed by manufacturers and researchers alike is to enhance two fundamental characteristics:

1. **Invisibility**: From a user's perspective, pervasive computing technology should be transparent. This fosters the concept of self-organized machine learning systems that are learning continuously from the environment.
2. **Scalability**: As 'Smart space' [2] grows in sophistication, so do the number of the devices that supports it. This heterogeneous domain obliges software companies to write code for many platforms, processors and devices. From a technical point of view, it becomes very hard to manage all of these devices in one environment. The short-term solution is to write applications in a managed platform such as Microsoft's .Net or Sun's Java. But with the growth of scripting languages such as python, it will soon be possible for them to be the front runners in this race.

---

[1] iPhone is a trademark of Apple.

10.3.5   Pervasive Distributed Application

Tomorrow's pervasive applications will use the Internet to a greater extent in order to form a mesh network in a distributed manner. Here are some examples:

1. Cell phones will ask land line phones for their numbers in order to route or forward calls.
2. Kitchen appliances will communicate in order to inform the supermarket as to the quantity of products contained in the refrigerator.
3. Cars will connect to the Internet so as to acquire advance notice of the number of available places in their preferred restaurant's parking lot.
4. More futuristic examples propose a kitchen stove that communicates with the fridge and proposes recipes based on the available items.

There is an ongoing effort to expand pervasive computing. However, this raises the debate over privacy and safety [2]. As an example, the Class-C Mercedes has an electronic braking system that detects when a car is braking in front of it and will thus initiate its own deceleration. Now, the question is what happens when these systems are breached and controlled by malware. Privacy, on the other hand, is the privilege accorded to users to determine for themselves when, where and how their private information can be shared with others. By users we denote individuals, groups or organizations. Issues also arise with sharing information between 'other groups'. For instance, financial data are often saved on remote servers by credit card companies and merchants. This information is also linked to personal data so that if one side is breached, the chances are that the other will be as well.

Issues with privacy and security in pervasive networks can be especially acute in that the connections will be made transparently and automatically leading to the breach of data protection with or without a user's consent.

10.3.6   Logic Based Level Security

10.3.6.1   Security by Contract

Today's handheld devices have more computing power than a PC did 15 years ago [3]; this is how far computing has come. There are many scenarios in which a user, or a mobile, can exploit the available pervasive network. A tourist in an airport can start downloading the city guide in order to facilitate the journey in a newly visited city. This city guide can direct a rented car to a preferred site that a person wants to visit by interacting automatically with the surrounding navigation systems.

10.3.6.2   Un-Trusted Code

If we want to delve into this concept, which is actually not too far from reality, this interaction invokes downloading un-trusted code from local area networks via Bluetooth, WiFi or even WiMAX networks. Current programming language, the dominant ones such as Java and .Net, cannot function in this model because their code signature will not give any trust information unless we use some sort of web service security.

10.3.6.3   Massive Multiplayer Online Role Playing Games (MMORPG)

These emerging multi-million dollar servers are based on Peer to Peer (P2P) networks so as to connect millions of online game users. However, these environments are too bloated and are not suited for the network model of pervasive games. One solution to this problem is to create virtual environments in which mobile users in a lounge, for example, can play or compete against each other in a collaborative environment.

Security By Contract [3] is a scheme proposed in order to take advantage of four parties: mobile operators, software developers, mobile users and security providers. Each application in this model can be described by:

- A *contract*: This defines the overall permitted or granted behavior of an application.
- A *policy*: This defines the framework of the application's perimeters of action.

Let us imagine a scenario in which an application joins a pervasive mobile network. The mobile device must either download or transmit a policy to the requesting party. Based on whether this policy is accepted or not, a contract is 'signed' and is enforced during the load and execution phases of the software. This behavior must be accomplished in a fast and transparent manner in order to keep the overhead light.

10.3.7 Deterministic Access Models

10.3.7.1 RFID Based Identification and Authentication

Radio Frequency Identification (RFID) enables automatic detection of tagged objects using RF signals [4]. The advantage of RFID over other technologies is that it has the properties of storage and contact-less and self-computing. RFID is destined to replace barcode. It is found everywhere, from manufacturing and supply management to inventory control.

Advances in the field of RFID have given birth to a mobile version that enables a new array of functionalities. As an example, in the US market, new credit card companies are now introducing RFID based credit cards that allow a buyer to make a payment by simply holding the card up to a reader.

In order to fully understand the benefits of this technology, imagine a customer who enters a supermarket, picks up the items and walks out of the store without passing by the clerk. Of course, the customer had to pay for the purchases somehow. By passing through specialized gates, a reader scans the items in the cart using the RFID chip, adds up the amount of each item, scans the customer credit card's RFID chip, deducts the amount of the purchases; and all of this happens in a fraction of a second.

Let us look at a typical Mobile RFID as proposed by [4] in Figure 10.2.

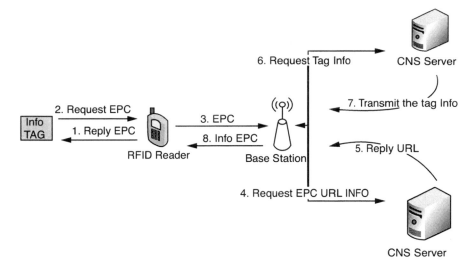

Figure 10.2 Mobile RFID network architecture.

In general, a mobile RFID network is composed of a tag, an RFID reader and network servers. Mobile RFID is a good technology; however it presents many security disadvantages such as information leakage, traceability and impersonation which are inherent in the properties of the wireless medium. Here is a summary of each of these issues:

- **Impersonation**: A tag communicates with a legitimate mobile reader. A misbehaving node can collect the tag information being transmitted and create a clone. In a second pass when the mobile reader communicates with the RFID tag, the clone responds and behaves as the legitimate one.
- **Leakage of information**: By incorporating various tag objects in a pervasive environment, some objects might contain sensitive information that is private and should be kept confidential to its owner only. The design of mobile RFID should take into consideration the security constraints of their respective users.
- **Traceability**: This is an extension to leakage of information in which determining the owner of a tag may lead to tracing the location of the person by linking a response to a tag.

In an attempt to solve these issues, [4] proposes the following method described in Figure 10.3:

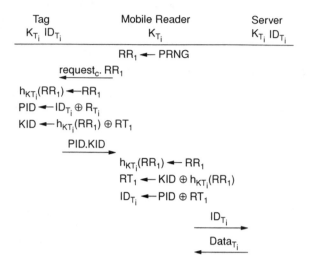

Figure 10.3 Privacy protection in mobile RFID.

In a nutshell, this is how this algorithm works:

- The mobile node generates a random number RR_i and sends $request_C$ to a tag with the random value RR.
- Upon receiving the query ($request_C$, RR_i), The tag computes as follows:
  - compute h_{KTi} which is a a key of a i-th tag(Ti).
  - (RR_i) using the received RR_i and a key of the tag T_i.
  - Generate a random value R_{Ti}.
  - Compute PID = ID_{Ti} R_{Ti} and KID = h_{KTi} (RR_i) R_{Ti}.
    The tag sends PID, KID to the mobile node.
- The node computes:
  - h_{Kj} (RR_i), $j\{1,..,N\}$ where N is the number of keys the mobile node holds.
  - Then RT_i = KID h_{Kj} (RR_i).
  - And ID_{Ti} = PID R_{Ti}.
  - Finally ID_{Ti} using possessed keys.
    The server sends $Data_{Ti}$ with ID_{Ti} to the mobile node.

To go back to how this scheme can be useful, let us look at how it resolves each of the mobile RFID problems.

- **Information leakage**: The tag applies random values which work by sending a PID tag and KID using a keyed hash function and random values from the mobile reader. Extracting this ID requires an attacker to guess a 96 bits value.
- **Traceability**: By traceability, an adversary can obtain the PID and KID. These two values are refreshed by both the mobile reader and the tag on each session. Even though a malicious adversary transmits the same random values to different tags, the tags themselves respond by different random numbers, which makes it hard to track which PIDs belong to which tags.

10.3.8   Predictive Statistical Schemes

10.3.8.1   Location Aware Neuro-Computing Based Authentication

The IEEE 802.11 standard is considered to be one of the most popular and profitable network topologies in use today and is considered to be a serious contender along with the Bluetooth and ZigBee networks in the field of home automation. The scalability of Wireless Local Area Networks comes with the burden of ensuring integrity, confidentiality and trust in the network. By integrity we mean the need to develop a mechanism by which only authorized users can gain access to the network's resources. Confidentiality implies that all data transmitted by each user stays known only to the communication parties. The two characteristics above can then enforce a trust environment in which all wireless nodes and users are authorized and secure. In [5], the authors proposed a scheme that relies on a neural networks decision engine to restrict network access to mobile nodes whose physical location is within a threshold distance from the wireless access point or the controller of the network.

Another problem solved by this neural net based security scheme is context based security. Let us look at these scenarios:

Scenario 1: Consider two parties: a user who wants to use the already existing free and public wireless network while on the source's premises and a controlling authority (CA) who wants to offer this free service to only those clients who are trying to connect from within its business area. We emphasize the fact that the CA wants to offer the free service to its clients only. Although this can still be implemented using an Authorization, Authentication and Accounting server (AAA) such as FreeRadius, it does not offer the flexibility required by such a dynamic environment where in the case of a coffee shop, customers come in, check their emails and leave as quickly as possible. Using such a server would be overkill. Instead, we need a solution that will ensure that only the customers of that hotspot zone can receive access to the system and that this functionality is accomplished transparently for both server and client, i.e. no password or subscription is needed.

Scenario 2: In this second scenario, we consider a network setup in which we have a public wireless file server that serves files to its clients. Under normal circumstances, this scenario does not provide any deployment difficulties; however, if we have a larger number of users that are trying to retrieve data all at the same time, we can see clearly that we might need to add mirror servers to the current setup in order to distribute the load on the single server.

Scenario 3: Often, in big environments, users are given the right to access corporate data remotely. This can pose a serious threat if security measures are breached. One way to strengthen security is to use some mechanism by which we can add a challenge to a user in addition to the already used public key schemes (PK) and cryptographic techniques. Using our work, for example, we can restrict access to resources to only those users that are within a physical threshold area. To make this clearer, we can restrict usage of the network to only those users that are within 'x' meters of the source where 0<x<maximum threshold [6]. This method is illustrated in Figure 10.4.

10.3.8.2   Context Based Security

Discretionary Access Control (DAC) and Mandatory Access Control (MAC) along with Role Based Access Control (RBAC) were the first access schemes used in computer security. These schemes are well adapted for

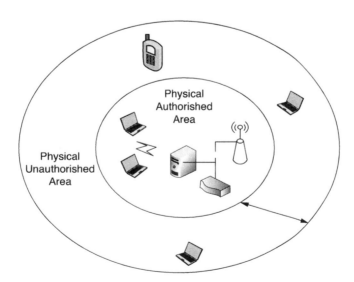

Figure 10.4   Neural Network based access scheme.

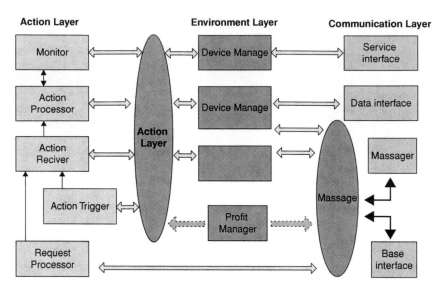

Figure 10.5   PerSE layered architecture.

many organizations such as hospitals, administrations and universities, amongst others. In today's networks, these types of security schemes are obsolete. Setting a user role and access per user is time consuming and would present a burden to the administrator. In [7], the authors present the implementation of a system that relies on context in order to secure pervasive networks. Their system is called PerSE. A brief description of PerSE is given next.

10.3.8.3   PerSE (Personal Security)

PerSE [8] is a user-oriented pervasive environment, in which the user can access resources that exist on Peer to Peer (P2P) surrounding devices. Moreover, this platform is proactive and non-intrusive [8]. As a part of the PerSE environment, each device has to run a base to share its local resources. The PerSE Base service is in charge of communications with other Bases [7], in order to run distributed services in a smart way. It also consists of many independent Bases, able to discover each other, and to send and receive messages through different communication channels including wired LANs, Wireless LANS and Bluetooth PANs. The PsaQL language [9] allows the user to choose a partial action describing the services that the user wants to use and their possible location. The PerSE Base has then to interpret this intention into a connected graph of services that are meant to be executed. This architecture is composed of three modules: communication, environment and action layers as illustrated in Figure 10.5:

- The communication layer is the lowest level layer and it is task is to enable communication amongst the users of the system.
- The environment layer manages local knowledge and available services. The context manager manages both local and distant access.
- The action layer is intended to gather requests from users and/or applications.

10.3.8.4   Ubiquitous Computing Context-Based Security Middleware (UbiCOSM)

UbiCOSM is an access control scheme for securing agent-to-agent and agent-to-environment interactions in a context aware environment. In particular, UbiCOSM focuses on three specific aspects: context-centric access control, active context view provisioning to middleware components and support for disclosure of the security properties of UbiCOSM agents and resources to interested entities [10]. UbiCOSM, as illustrated in Figure 10.6, depends on dynamic context variables, such as sensors, in addition to more traditional attributes.

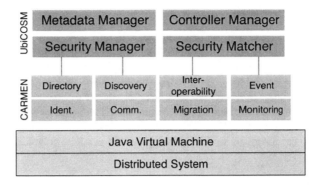

Figure 10.6   UbiCOSM pervasive model.

10.3.8.5   Context-Aware Middleware for Resource Management in the Wireless Internet (CARMEN)

CARMEN is system that adds accessibility in mobile terminals in the Web [11]. CARMEN is a distributed system that deploys active middleware proxies over the fixed network to support service provisioning to portable devices. CARMEN provides any portable device with a companion middleware proxy (shadow proxy) that acts autonomously on its behal, negotiates service tailoring to fit user/device characteristics and follows user/device movements among network localities. CARMEN implements hidden proxies and also provides execution environments, called places or nodes. These nodes can be grouped into domains that correspond to network localities, e.g. either IEEE 802.3 LANs or IEEE 802.11b-based wireless LANs.

Carmen uses Resource Description Framework (RDF) or Extensible Markup Language (XML) notation to define the profiles and policies of the system.

Here is an example of sample service declaration as described in [10] and [11]:

```
<cesa:CESAEntertainmentService rdf:ID = 'Pervasive Network Movie'>
       <cesa:typeOfService rdf:resource = '&entertainment;Movie' />
       <cesa:servbiceOfferLocation rdf:resource = '&airport;TerminalB' />
</cesa:CESAEntertainmentService>
```

10.4   Privacy in Pervasive Networks

Privacy appears as a major issue for pervasive computing applications. Many models have been proposed in order to address privacy issues. Successful schemes require awareness of the technology's users and that their desires and concerns are understood [12]. This is difficult as there is little statistical research about the number of potential pervasive users that designers can accommodate. Complicating design further is the fact that pervasive systems are typically embedded or invisible, making it hard for users to know when these devices are present and collecting data.

Today, users have a more or less limited understanding of the technology, thus several privacy and safety issues are raised [12].

10.4.1   Problem Definition

The problem with privacy began with the trend in Web applications and pervasive networks in general that enabled users to collect, communicate, save, organize and reuse information [14]. Pervasive services are provided through many devices embedded in the user such as cell phones or personal computers. This continuous information collection exposes personal behavior, habits, preferences and associations [15]. Privacy and security of this information has not been protected enough and many security breaches have resulted in exposing information about individuals and putting in jeopardy their personal life and financial situation. In addition, pervasive computing applications rely heavily on mobile and wireless communications that bring with them new privacy issues.

10.4.2   Challenges to Privacy Protection

Pervasive computing need not be obtrusive. This is why technology is embedded into many objects that transmit and receive data. Embedding reduces the visibility of the pervasive computing environment surrounding the user and makes it more friendly and acceptable [16]. Ironically, the same characteristics make it possible to invade the privacy of the user without the user realizing it. This leaves users with limited control over their own privacy and also adds a responsibility that they do not intrude on others' privacy. This invasion and responsibility cannot be managed or imposed through social and organizational controls [14]. There is a need to find a balance between privacy and pervasive sharing of identity. Traditionally, networks required a verbose input from the user. Today, pervasive networks need to be transparent when requiring private data whilst providing the maximum protection.

10.4.3   Location Dependency

Pervasive computing applications use location information to provide usable local services such as maps, local searches and local matches, amongst others. To utilize these location-based services, users have to reveal their location to service providers. Nonetheless, access to location information about a user can be misused by legitimate and illegitimate requestors for identity theft, spam or marketing mails. There is also the need for the services to be 'smart' enough to update different location policies based on the situation. For instance, a user might want location privacy, but may have to amend this need in the case of an emergency to point to the 'right' location [17].

10.4.4   Data Collection

Pervasive computing implementation relies on the quality and accuracy of the generated data. The sheer amount of data collected and processed leads to users frequently ignoring or being deprived of allowing their personal data to be released. In addition, pervasive computing environments are characterized by having many mobile devices. These factors limit the creation of protocols and schemes for privacy protection that might depend on extensive use of these pervasive devices.

10.4.5   Internet Service Provider (ISP) Role

ISPs play a major role in shaping privacy action laws. Their role is to maintain and preserve all private information, sensitive or not, about their users. There are numerous opportunities for misuse of data passing through the devices of the service provider. The Platform for Privacy Preferences (P3P) of the World Wide Web Consortium (W3C) defines the specification that can be used to ensure that each information request by service providers also specifies the purpose, retention and recipients of the data [18]. In reality, enforcing this law is very difficult to achieve.

10.4.6   Data Ownership

Resources in today's computing systems need to have owners and access control policies. On the other hand, pervasive computing networks 'permit loose and more dynamic couplings between people and resources, thereby invalidating the usual approaches to ownership and control of resources' [18]. It is difficult and challenging to implement privacy and control when data ownership cannot be determined.

10.4.7   Private Systems

Several projects and models have tried to address the concerns of privacy protection in pervasive computing. Some of this work is discussed below [19].

10.4.7.1   PSIUM

The Privacy Sensitive Information Diluting Mechanism (PSIUM) is a scheme that tries to thwart the misuse of private data by ISPs. The model realizes this by eliminating the usefulness of the data collected by the service providers without affecting the service provided to the users. A PSIUM-enabled device sends multiple location-based service request messages to the service provider and only one of these messages contains the true location of the user. The device knows how to utilize the correct information and make it available to the user. PSIUM maintains false data that appears realistic so that it is difficult to distinguish it from the true data. This false data is created from locations used previously by the user. The strength of PSIUM lies in preventing misuse of users' data by a service provider without reducing the quality of the service.

10.4.7.2   Mix Networks and Mix Nodes

This type of store and forward network offers anonymous communication [20]. In this model, a mix node collects n equal-length packets as input and reorders them by some metric before forwarding them. This scheme eliminates any link between incoming and outgoing messages and provides protection even in compromised environments. This scheme works best when the number of sends is large to allow a larger array of permutations; however, its weakness lies in the vulnerability of communication between user device and the service provider.

10.4.7.3   Pseudonyms and Mixed Nodes

This is the same as the technique above with the addition of 'nicknames' to represent identities in the system. Although pseudonyms can be tracked down, this is harder to accomplish when paired in connected spatial regions in which the designated users have not registered. This model is very complex to implement in practice.

10.4.7.4   Information Space Model

Information spaces define different boundaries on data by using 'owners' and 'permissions' [21]. With this in mind, information is channeled through context-aware gates and authorizations routers. Privacy is preserved in this model by defining signals and triggers. Once a privacy issue arises, privacy gates are notified and software components stop the information leakage.

10.4.7.5   PawS

The Privacy awareness System (PawS) protects personal privacy and helps others respect that privacy [22]. In contrast with other systems that put forward rigorous technical protection, PawS ensures respect for privacy by emitting beacons for each service that is requested. A proxy checks user privacy against the service that is demanded. Once they agree, the user can use the service in question.

10.4.8   Quality of Privacy (QoP)

QoP enables balancing between the data that can be conceded and the value the service in question brings up to the user [23]. QoP is based on the following elements:

1. *Location*: This is the segment of the network the user is connecting from.
2. *Identity*: This is a 1 to n identification process, such as employing a user name.
3. *Access*: This is a 1:1 identification process, such as providing a password.
4. *Activity*: This is the intended used services of the pervasive network.
5. *Persistence*: This is to indicate whether the user is using a session to hold QoS setting information.

The user can demand a level of quality of privacy in the pervasive environment and as in QoS, QoP adjusts the necessary parameters to respect such a request. The highest level of privacy in this scheme is the 'Anonymous mode'. This mode is mapped in the system using a context that satisfies the level of QoP that both the user and the application have agreed upon.

10.4.9   Open Issues in Privacy of Systems

There are many challenges in protecting the privacy of systems and individuals. Although there are many issues to be resolved, a partial list is given below.

1. *Insufficient privacy response*: The current problem is to model the response of the user to the level of privacy.
2. *Scalability*: Most of the research that has been done up to now has been somewhat experimental. The models developed in this research have to prove their concept when applied in larger population systems.
3. *Changing environment*: Naturally, users will change their location and thus their privacy context information, especially when we are dealing with mobile nodes. It is a real design challenge to create devices that support these models and scenarios with transparent privacy preservation for the user.

10.4.10   'Sharing' in Personal Networks

One of the most difficult problems in personal network PNs is to look at the social aspect. Today's architectural model promotes a strong relationship between the device and the device owner. Although this preserves privacy to a certain degree, having such a model poses flexibility issues when it comes to people who have various affiliations in different communities [24]. With this in mind, sharing physically a gadget or a personal device becomes more difficult if the device does not support profiles and contextual usage. This is currently a major topic in many research laboratories.

10.5   Conclusions

Pervasive computing provides a good picture of how computing and services will evolve in the future. There is no need to use this resource in the old-fashioned way of sitting 'immobile' in front of a screen to surf the Web and use the Internet, but today we are presented with many 'smaller' versions of computers such PDAs, setup boxes, embedded controllers, etc, that enable us to process information based on our location as well as context of use. Today we can connect anything, anytime, anywhere, transparently and in a seamless dynamic manner in order to accomplish our tasks. This technology is rapidly finding its way into our everyday life and is also changing our everyday activities as well as the way we do them. Nonetheless, with greater advantages come some drawbacks. In fact, pervasive networking raises fundamental questions and concerns about the privacy of users. Finding a reliable privacy-preserving technology is a prime issue that goes hand in hand with ubiquitous computing and traditional privacy preserving techniques such as access control and anonymization are no longer sufficient. Robust schemes rely on contract based privacy assurance between service providers and users. This QoP is adjusted dynamically based on context and the level of confidentiality.

References

[1] A. Zugenmaier, T. Walter (2007) ' Security in pervasive computing calling for new security principles,' *IEEE International Conference on Pervasive Services*, pp. 96–9.
[2] V. Dhingra, A. Arora (2008) 'Pervasive computing: paradigm for new era computing,' *First International Conference on Emerging Trends in Engineering and Technology (ICETET 2008)*, pp. 349–54.
[3] N. Dragoni, F. Massacci, C. Schaefer, T. Walter, and E. Vetillard (2007) 'A security-by-contract architecture for pervasive services,' *Third International Workshop on Security, Privacy and Trust in Pervasive and Ubiquitous Computing (SecPerU 2007)*, pp. 49–54.
[4] Il Jung Kim, Eun Young Choi, and Dong Hoon Lee (2007) 'Secure mobile RFID system against privacy and security problems,' *Third International Workshop on Security, Privacy and Trust in Pervasive and Ubiquitous Computing (SecPerU 2007)*, pp. 67–72.
[5] C.-E. Pigeot, Y. Gripay, M. Scuturici, and Jean-Marc Pierson (2007) 'Context-sensitive security framework for pervasive environments,' *Fourth European Conference on Universal Multiservice Networks (ECUMN'07)*, pp. 391–400.
[6] T. Guelzim, M.S. Obaidat (2008) 'A novel neurocomputing-based scheme to authenticate WLAN users employing distance proximity threshold,' In *Proceedings of 2008 IEEE International Conference on Security and Cryptography, SECRYPT 2008*, pp. 145–53, Porto, Portugal.
[7] G. Zhang, M. Parashar (2004) 'Context-aware dynamic access control for pervasive applications,' In *Proceedings of the Communication Networks and Distributed Systems Modeling and Simulation Conference (CNDS 2004)*.

[8] Y. Gripay, J.-M. Pierson, C.-E. Pigeot, and V.-M. Scuturici (2006) 'Une architecture pervasive sécurisée : PerSE,' In *Proceedings of 3e Journées Francophones Mobilité et Ubiquité (UbiMob'06)*, pp. 147–50.

[9] P. Bihler, V.-M. Scuturici, and L. Brunie (2005) 'Expressing and interpreting user intention in pervasive service environments,' In *Proceedings of the 1st International Conference on Signal-Image Technology and Internet-Based Systems, SITIS*, pp. 183–8.

[10] A. Corradi, R. Montanari, D. Tibaldi, and A. Toninelli (2005) 'A context-centric security middleware for service provisioning in pervasive computing,' In *Proceedings of the Symposium on Applications and the Internet*, pp. 421–9.

[11] P. Bellavista, A. Corradi, R. Montanari, and C. Stefanelli (2003) 'Context-aware middleware for resource management in the wireless internet,' *IEEE Transactions on Software Engineering* **29**(12): 1086–99.

[12] N. Damianou, N. Dulay, E. Lupu, and M. Sloman (2001) 'The Ponder Policy specification language,' In *Proceedings. of Policy 2001*, Springer-Verlag, LNCS 1995, pp. 18–39, Bristol.

[13] J. Seigneur, C.D. Jensen (2004) 'Trust enhanced ubiquitous payment without too much privacy loss,' In *Proceedings of the ACM symposium on Applied computing*, pp. 1593–9, Cyprus.

[14] V. Bellotti, A. Sellen (1993) 'Design for privacy in ubiquitous computing environments,' In *Proceedings of 3rd European Conference on Computer Supported Cooperative Work*, pp. 77–92.

[15] J. Cas (2005) 'Privacy in pervasive computing environments – a contradiction in terms?', *Technology and Society Magazine, IEEE* **24**(1): 24–33.

[16] D. Hong, M. Yuan, and V. Y. Shen (2005) 'Social communication: dynamic privacy management: a plug-in service for the middleware in pervasive computing,' In *Proceedings of the 7th International Conference on Human Computer Interaction With Mobile Devices & Services*.

[17] D. L. Lee, J. Xu, B. Zheng, and W.C. Lee (2002) 'Data management in location-dependent information services,' *Pervasive Computing*, pp. 65–72.

[18] J. Cas (2005) 'Privacy in pervasive computing environments – a contradiction in terms?,' *Technology and Society Magazine, IEEE* **24**(1)..

[19] K. Henricksen, R. Wishart, T. McFadden, and J. Indulska (2005) 'Extending context models for privacy in pervasive computing environments,' *Third IEEE International Conference on Pervasive Computing and Communications Workshops*, pp. 20–4.

[20] P. Bhaskar, S.I. Ahamed (2007) 'Privacy in pervasive computing and open issues,' In *Proceedings of the Second International Conference on Availability, Reliability and Security, ARES*, pp. 147–154, Vienna.

[21] A.R. Beresford, F. Stajano (2003) 'Location privacy in pervasive computing,' *Pervasive Computing, IEEE* **2**(1): 46–55.

[22] J. Xiaodong, J.A. Landay (2002) 'Modeling privacy control in context-aware systems,' *Pervasive Computing, IEEE* **1**(3): 59–63.

[23] M. Langheinrich (2002) 'A privacy awareness system for ubiquitous computing environments,' bicomp, *Lecture Notes in Computer Science* **2498**: 237–45, Springer.

[24] M. Tentori, J. Favela, M.D. Rodriguez, and V.M. Gonzalez (2005) 'Supporting Quality of Privacy QoP) in pervasive computing,' (2005) *Sixth Mexican International Conference on Computer Science*, pp. 58–67.

[25] Roshan K. Thomas, and R. Sandhu (2004) 'Models, protocols, and architectures for secure pervasive computing: Challenges and research directions', *First IEEE International Workshop on Pervasive Computing and Communication Security (PerSec)*, Orlando, USA.

11

Understanding Wormhole Attacks in Pervasive Networks[1]

Isaac Woungang,[1] Sanjay Kumar Dhurandher,[2] and Abhishek Gupta[2]

[1]Department of Computer Science, Ryerson University, Toronto, Ontario, Canada.
[2]CAITFS, Division of Information Technology, Netaji Subas Institute of Technology (NSIT), University of Delhi, New Delhi, India.

11.1   Introduction

Wireless ad hoc and sensor networks are comprised of nodes that must cooperate to establish routes dynamically using wireless links without the intervention of centralized access points or base stations. Routes may involve multiple hops with each node acting as a router. Nodes present several security problems such as easy to hear, modification, impersonating, etc. Since ad hoc and sensor networks typically work in such an open untrustworthy environment with little physical security, they are subject to a number of unique security attacks. Further, Wireless ad hoc and sensor networks have varied applications such as environmental monitoring, structural health monitoring and traffic analysis. They can be easily deployed in hostile environments that require a high degree of security such as battlefields in war and disaster rescue operations. Because of these reasons, safe transmission of packets and especially establishing a method for safe routing are most important challenges for ad hoc networks.

In order to achieve secure communication in such networks, understanding the likely security attacks on wireless ad hoc and sensor networks is an important task. Such networks suffer from a variety of security attacks and threats such as Denial of Service (DoS), flooding attack, impersonation attack, selfish node misbehaving, routing table overflow attack, wormhole attack, blackhole attack and so forth. Wireless ad hoc and sensor networks are vulnerable because of their basic characteristics, such as having no source of network management, vigorous topology changes, resource restriction, no certification authority or centralized authority, to mention but a few [1, 2, 3]. One of the most severe attacks that can be launched against wireless ad hoc and sensor networks is the Wormhole Attack. A detailed analysis of the Wormhole Attack is provided in this chapter.

11.2   A Wormhole Attack

A Wormhole Attack is a type of a *collaborative attack* in which the attacker provides two choke-points of a malicious nature, which are used to degrade the network or analyze the network traffic [4]. These two choke points constitute the end points of a wormhole. The end points are connected via a high speed link (Figure 11.1) of some sort or a tunnel. Packets are captured from one end point and are tunneled to the other malicious end point in some

---

[1] This research was supported in part by the New Opportunity Fund, CFI/OIT Project #9662 of the Canadian Foundation of Innovation & Ontario Innovation Trust.

Pervasive Computing and Networking, First Edition.   Edited by Mohammad S. Obaidat, Mieso Denko, and Isaac Woungang.
© 2011 John Wiley & Sons, Ltd. Published 2011 by John Wiley & Sons, Ltd.

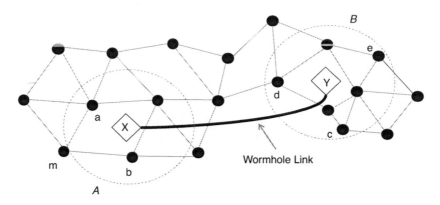

Figure 11.1 X and Y are the end points of the wormhole with a communication link between them known as the wormhole link. a and b are in the transmission range of X where as d, e and c are in the transmission range of Y (courtesy of [17]).

other part of the network, where they are replayed, typically without modification. The above figure illustrates a network topology affected by a wormhole.

11.3   Severity of a Wormhole Attack

Wormhole Attack is considered to be one of the most severe attacks on ad hoc networks. A Wormhole Attack is severe against on demand as well as proactive routing mechanisms. Firstly, in on demand routing mechanisms, a wormhole is capable of attracting a significant percentage of network traffic. This is because most of the on demand routing protocols are shortest path routing mechanisms using hop count as a metric for route selection. The link between the two adversarial nodes of the wormhole is a fast link with a small number of hops, in most cases a single hop. Data forwarded via the wormhole thus reaches the destination sooner or with smaller number of hops as compared to data forwarded by the genuine nodes using multiple hops for transmission. To understand this, let us divide the network in two *Partitions A* and *B* (Figure 11.2), each containing one of the end points of the wormhole.

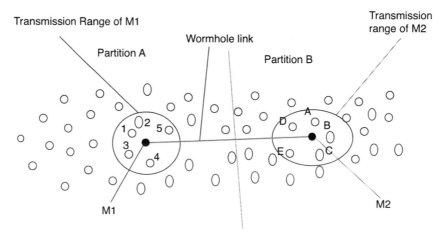

Figure 11.2 M1 and M2 are the malicious nodes of the Wormhole. Nodes 1, 2, 3, 4 and 5 are in the transmission range of M1. Nodes A, B, C and D are in the transmission range of M2. The network is divided into two Partitions A and B. The wormhole will handle a significant amount of routing between Partition A and Partition B. Also nodes 1, 2, 3, 4 and 5 will consider nodes A, B, C, D and E as their immediate neighbors due to the presence of the wormhole.

If packets are to be transmitted from a node in *Partition A* to *Partition B*, most of the routes discovered will include the wormhole owing to the presentation of a shorter path. Therefore, most of the data transfer carried out between these two partitions is affected by the wormhole. In another situation a wormhole can tunnel a *ROUTE REQUEST* packet direct to its destination. When a destination node's neighbor hears the *ROUTE REQUEST* packet it will follow the normal routing procedure to rebroadcast it and then discard all of the other *ROUTE REQUEST* packets originating from the same Route Discovery. Any route other than the one that is affected by a wormhole is thus prevented from being discovered. If the wormhole is near the originator of the *ROUTE REQUEST* packet routes more than two hops can be prevented from being identified.

Using wormholes an attacker can also break any protocol that relies directly or indirectly on *geographic proximity*. For example, target tracking applications in sensor networks can be paralyzed easily if the network is affected by wormholes. Similarly, wormholes will affect connectivity based localization algorithms, as two neighboring nodes are localized nearby and the wormhole links essentially *'fold'* the entire network. This can have a major impact as location is a useful service in many protocols and applications, and often out-of-band location systems such as *GPS* are considered to be expensive or unusable because of the environment. Wormholes are equally dangerous even in the case when they are simply relaying the packets without any modifications or disruptions, on some level providing a communication service to the network. With a wormhole in place, the affected network nodes do not have a true picture of the network, which may disrupt localization-based schemes, and hence lead to wrong decisions. A wormhole can also be used to simply aggregate a large number of network packets for the purpose of traffic analysis or compromise of encryption.

After the wormhole has become a significant part of routing the possible way of exploiting it may be to use it to analyze the routing traffic. The critical points of the network such as the sender node or the destination node may be identified and the attack may then be launched against them. The adversarial nodes of the wormhole may drop the packets instead of forwarding them all, thus creating a permanent *Denial of Service Attack*. In this case, this attack would be more detrimental as the wormhole is handling a significant amount of the routing of the network. The wormhole may also discard the packets selectively such as the control packets in the on demand routing mechanisms or modify them.

In the case of pro-active routing mechanisms which employ neighbor discovery procedures a wormhole attack is equally dangerous. These protocols use *HELLO PACKETS* for neighbor discovery. If the *HELLO PACKETS* of A are tunneled across via a wormhole and are transmitted to B then A will consider B as its neighbor. The routing will become disrupted when A tries to communicate with B as its one hop neighbor and is not able to, as they are not within transmission range. In Figure 11.2 nodes *1, 2, 3, 4* and *5* will take nodes *A, B, C, D* and *E* as their immediate neighbors.

The severity of wormholes is also reflected in the fact that they are not easily detectable. *Cryptographic techniques are not useful in detecting wormholes* as in most cases they only relay the encrypted or authenticated packets. Suppose the attacker places two transceivers at two critical positions in the network and initiates a fast link between the two. These transceivers will simply pick up packets from the network and tunnel them across. These transceivers need not be part of the network for performing this task as they will be simply smuggling on the packets transmitted by the neighbor nodes. Cryptographic techniques will be useless in this case. The nature of wireless communication allows the attacker to design such transceivers. It is also possible for the attacker to transmit each bit instead of waiting for the whole packet thus decreasing the delay of transmission. If the attacker does the tunneling non-maliciously then the wormhole can be very useful in routing as it provides a fast route with a smaller number of hops. However, in most scenarios this is not the case.

A Severity Analysis of a Wormhole Attack has been carried out by Mazid Khabbazian and his co-workers in [5]. It has been shown that in shortest path routing protocols, two strategically located malicious nodes can disrupt on average 32% of all communications across the network, when the nodes of the network are distributed uniformly. When the wormhole targets a particular node in the network, it can disrupt on average 30% to 90% (based on the location of the target) of all communication between the target node and all other nodes in the network. In a network of grid topology it has been shown that 40% to 50% of all communication can be disrupted if the wormhole is placed along the diagonal of the grid. The study above illustrates the severity of wormhole attacks in wireless ad hoc networks.

11.4   Background

A Wormhole Attack is considered to be one of the most severe attacks against Wireless Ad Hoc Networks and is detrimental against both On Demand and Pro-active routing protocols, as discussed in the previous section. A wide variety of wormhole attack mitigation techniques have been proposed for specific kinds of networks: sensor networks, MANETs and static networks.

Hu *et al.* [6], who introduced the discussion of wormhole attacks in ad hoc networks, suggested the use of geographical or temporal packet leashes to detect wormholes. A geographical leash requires each node to know its own location and all nodes to have loosely time synchronized clocks. The nodes need to exchange location information securely. A sender node can then ensure that the receiver is within a certain distance and detect discrepancies therein. With temporal leashes, all nodes must have tightly synchronized clocks. The receiver will compare the receiving time with the sending time attached to the packet. It can determine if the packet has traveled too far in too little time and detect the wormhole attack.

Hu and Vans [7] proposed a solution to wormhole attacks for ad hoc networks in which all nodes are equipped with directional antennae. In this technique nodes use specific 'sectors' of their antennae to communicate with each other. Each couple of nodes has to examine the direction of received signals from its neighbor. Hence, the neighbor relation is set only if the directions of both pairs match. This extra bit of information achieves wormhole discovery and introduces substantial inconsistencies in the network, which can easily be detected.

Wang and Bhargava [8] introduce an approach in which network visualization is used for discovery of wormhole attacks in stationary sensor networks. In their approach, each sensor estimates the distance to its neighbors using the received signal strength. All sensors send this distance information to the central controller, which calculates the network's physical topology based on individual sensor distance measurements. With no wormholes present, the network topology should be more or less flat, while a wormhole would be seen as a 'string' pulling different ends of the network together.

Lazos *et al.* [9] proposed a 'graph-theoretical' approach to wormhole attack prevention based on the use of Location-Aware 'Guard' Nodes (LAGNs). Lazos uses 'local broadcast keys' – keys valid only between one-hop neighbors – to defy wormhole attackers: a message encrypted with a local key at one end of the network cannot be decrypted at another end. Lazos proposes the use of hashed messages from LAGNs to detect wormholes during key establishment. A node can detect certain inconsistencies in messages from different LAGNs if a wormhole is present. Without a wormhole, a node should not be able to hear two LAGNs that are far from each other, and should not be able to hear the same message from one guard twice.

Khalil *et al.* [10] propose a protocol for wormhole attack discovery in static networks which they call LiteWorp. In LiteWorp, once deployed, nodes obtain full two-hop routing information from their neighbors. While in a standard ad hoc routing protocol nodes usually keep track of where their neighbors are, in LiteWorp they also know who the neighbor's neighbors are that is they can take advantage of two-hop, rather than one-hop, neighbor information. This information can be exploited to detect wormhole attacks. Also, nodes observe their neighbor's behavior to determine whether data packets are being forwarded properly by the neighbor.

Song *et al.* [11] proposes a wormhole discovery mechanism based on statistical analysis of multipath routing. Song observes that a link created by a wormhole is very attractive in a routing sense, and will be selected and requested with unnaturally high frequency as it only uses routing data already available to a node. These factors allow for easy integration of this method into intrusion detection systems to routing protocols that are both on-demand and multipath.

Thaier *et al.* [12] proposed 'De Worm', a protocol that uses routing discrepancies between neighbors along the path from the source to the destination to detect a wormhole. 'De Worm' is based on the observation that for a wormhole to have a successful impact on the network it must attract a significant amount of network traffic towards itself and routes going through a wormhole must be shorter than the alternate routes going through the valid network nodes.

Marianne *et al.* [13] proposed a statistical based approach to defend against Wormhole Attacks in which each node is assigned a cost function whose value depends on the contributions of that node to the networking tasks. As a route involving a wormhole is used more frequently than the other routes the cost value associated with this route will increase and cross the threshold value and this path will be prevented.

11.5 Classification of Wormholes

Various classifications have been made in the literature regarding wormholes. Some of them are as follows.

11.5.1   In Band and Out Band Wormholes

In Band Wormholes: This method creates a *secret overlay tunnel* within the active wireless medium. This method is *more dangerous* than the out band method of creating a wormhole. It *needs no extra hardware* and uses the existing wireless medium for transmission.

    Out Band Wormhole: The attacker creates a *direct communication link* between the two choke points. The link can be wired or use any other mode of communication. Data is received at one end and transmitted to the other, thus providing room for huge amount of data to be transmitted.

11.5.2   Hidden and Participation Mode Wormholes

Hidden Mode: Attackers do not use their identities so they remain hidden from the legitimate nodes. The attackers create a virtual link to transmit data. The attackers act as two simple transceivers that capture data at one end and forward it to the other end. There is no need for the possession of cryptographic keys by the attackers.

    Participation Mode: The attackers possess cryptographic keys. The attackers do not create any virtual link and take part in routing as legitimate nodes using the wormhole to deliver packets in less time or with smaller hops. These nodes, like hidden nodes, can drop packets selectively or wholly.

11.5.3   Physical Layer Wormhole

In this type of Wormhole Attack, one end of the wormhole captures bits of the waveforms and transmits them to the other using high speed links. Transmission and replay can start before the receipt of the entire packet. This decreases the delay that would otherwise occur when the node waits for the entire packet to be received before transmission.

11.6   Wormhole Attack Modes

A Wormhole Attack can be carried out in various modes [14]. Some of them are as follows.

11.6.1   Using the Existing Wireless Medium

This mode of Wormhole Attack uses the technique of *data encapsulation*. In this mode, when the malicious node at one end of the wormhole receives a *RREQ* packet it first encapsulates the packet and then tunnels it towards the other end of the wormhole using multiple hops of legitimate nodes. These hops are not added in the *RREQ* packet due to the encapsulation done. When the other end of the wormhole receives this packet, it demarshals it and forwards it to the neighbors. Now, the neighbors following their normal routing procedures will discard any other legitimate *RREQ* with same sender-destination pair that may arrive through the legitimate routing paths. Also, the cost or the hops associated with the route in the encapsulated packet will be much lower than any of the arriving *RREQ* since the hops between the two ends of the wormhole are not added into the packet due to the encapsulation. In Figure 11.3 node *A* transmits a *RREQ* packet to find a route to node *B*. On receiving the *RREQ* packet malicious node *M1* encapsulates it and transmits it to the next node which is legitimate. This packet travels through nodes *N1*, *N2* and *N3* and reaches node *M2*. Due to the encapsulation hops *M1-N1*, *N1-N2*, *N2-N3* and *N3-M2* are not included in the packet. Node *M2* demarshals the packet and forwards it to destination node *B*. *RREQ* generated at *A* reaches node *B* using alternate route *A*, *P1*, *P2*, *P3* and *B*. *B* will select the route affected by the wormhole as it presents a lower number of hops *(3 hops)* as compared to that of the legitimate route *(4 hops)*. In reality the selected route is *6 hops* long. Any routing protocol that selects the routes using the number of hops as a metric is vulnerable to this mode of wormhole attack.

    This mode of wormhole attack is easy to launch as it does not need any specialized hardware or high power transmission capabilities because it uses the existing wireless medium and legitimate nodes as hops between the

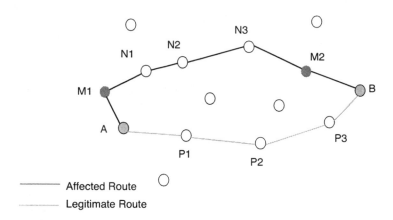

Figure 11.3 M1 and M2 are the two malicious end points of the wormhole. A is the sender node and node B is the receiver. The route A-M1-N1-N2-N3-M2-B is the affected route. The route A-P1-P2-P3-B is the legitimate route. M1 encapsulates the packets sent by node A and the addresses of hops M1-N1, N1-N2, N2-N3 and N3-M2 are not included in the packet header. Therefore, the affected route will be selected over the legitimate route as it appears to have lower number of hops (3 hops) than in the legitimate route (4 hops).

ends of a wormhole. Also, there is no need for the two ends of the wormhole to have any cryptographic information as they are already part of the network.

11.6.2   Using the Out-of- Band Channel

This mode can be achieved by using *long range directional wireless links or high speed wired links* between the two ends of the wormhole, i.e. the two ends of the wormhole communicate directly in this case. Consider the case depicted in Figure 11.4. A sends the *RREQ* packet and B receives it through two routes. They are *A-N1-N2-N3-B* which is the valid route of *4 hops* and *A-M1-M2-B* which is a malicious node of *2 hops*. *M1*, one of the endpoints of the wormhole, receives the *RREQ* from A and tunnels it directly to *M2* using a high speed long distance Out of Band link such as an optical fiber wired link or an Out of Band long distance wireless link. *Node B* will select the malicious route as it has lower number of hops. This mode is more difficult to achieve since there is a need for specialized hardware. This mode creates an entirely different picture for network nodes.

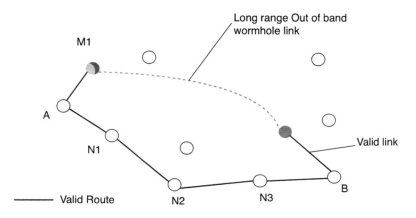

Figure 11.4 A is the sender node and B is the destination. The route A-N1-N2-N3-B is the valid route of 4 hops whereas A-M1-M2-B is the malicious route affected by a wormhole. M1 and M2 are the end points of this wormhole link. The malicious route is of 2 hops.

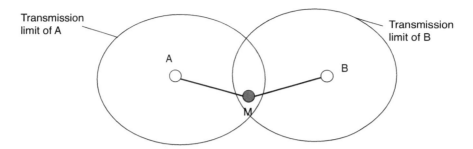

Figure 11.5   A and B are two legitimate nodes of the network that are not in transmission range. M is a malicious node that links A and B using packet relays.

11.6.3   Using High Transmission Power

In this mode a malicious node, on receiving a RREQ, broadcasts it at *high power level*, a capability which is not available to the other nodes in the network. The nodes receiving this packet forward it towards the destination. By doing this the malicious node increases its chance of being in the route.

11.6.4   Using Packet Relays

A wormhole using *Packet Relay* is another mode of Wormhole Attack in which a malicious node relays packets between two distant nodes that are not in transmission range in order to convince them that they are neighbors. This attack mode can be carried out even by single malicious node. It is performed by an intruder node M located within transmission range of legitimate nodes *A* and *B* (Figure 11.5), where *A* and *B* are not themselves within transmission range of each other. Intruder node *M* merely relays the packets between *A* and *B* (and vice versa), without the modification presumed by the routing protocol, e.g. without stating its address as the source in the packet's header so that *M* is virtually invisible. This results in the creation of a link between two legitimate nodes which otherwise are not in transmission range. This link is under the influence of a malicious node. *Node M* can afterwards drop or selectively drop relayed packets or break this link at will. Two intruder nodes connected by a wireless or wired private medium can also collude to create a longer (and more harmful) wormhole.

11.7   Mitigating Wormhole Attacks

The solutions provided to counter Wormhole Attacks can be divided into different categories. Broadly they can be classified under [5]:

1. Pro-active countermeasures.
2. Reactive countermeasures.

11.7.1   Pro-Active Counter Measures

These measures attempt to prevent wormhole formation by the use of *specialized hardware* or by using *time synchronization* techniques, i.e. every transmission done by nodes is checked for the presence of wormholes. The drawback of pro-active solutions is that they fail to detect wormholes running in *participation mode*. This is due to the fact that these countermeasures often attempt to prevent wormholes between legitimate nodes; however, in participation mode the wormhole is formed between malicious nodes participating in routing.

Some of the solutions proposed that fall under the category of pro-active countermeasures are as follows:

- *Timing based solutions*: These solutions use *time synchronization* between the network nodes in order to estimate the distance between the legitimate nodes and then compare the computed travel time to check whether the packet has traveled too far. A time synchronization technique using the concept of leashes [6] is discussed later.

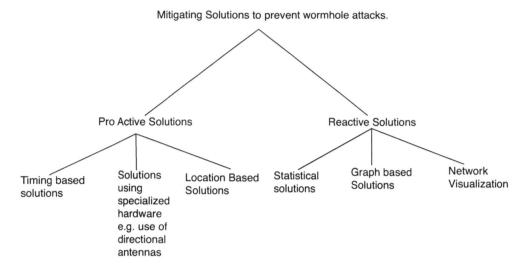

Figure 11.6   Tree representation of the classification of various mitigating solutions to prevent wormhole attacks.

- Another method for mitigating Wormhole Attacks is to use special nodes known as *guard nodes* [11] dispersed uniformly in the network and monitor the other nodes for detecting the presence of wormholes.
- *Directional antennae* [9] have also been used to mitigate Wormhole Attacks. In this method the receiver shares directional information with the sender. The receiver can prevent a wormhole end point being used as a false neighbor by detecting the direction of the received signal and comparing it with the information of the sender.

11.7.2   Reactive Countermeasures

Reactive countermeasures, on the other hand, use no specialized hardware and do not stop wormhole formation. These measures use simple authentication techniques to avoid Wormhole Attack. The drawback of these counter-measures is that these are not useful when the wormhole acts as a *passive attacker* or is used to simply analyze the traffic.

Other mitigating solutions include *statistical methods* and *graph based solutions*. Many *neighbor monitoring* schemes have also been proposed.

11.8   Discussion of Some Mitigating Solutions to Avoid Wormhole Attacks

11.8.1   Immuning Routing Protocols Against Wormhole Attacks

For a Wormhole Attack to be effective, it has to attract a significant amount of network traffic. This is a characteristic of a wormhole. This is the basic idea behind the proposed solution in [13]. The authors suggest a modification to the *AODV* [16] routing protocol. They have assigned to each node a cost function whose value depends on the contribution of that node to the networking tasks. Doing this will not allow the wormhole to attract all of the traffic all of the time. The cost function as referred in [13] is:

$$C(i)_{new} = 2^n + c(i)_{old}$$

Where, $C(i)$ = cost of node i, and initially $c(i) = 0$.

n = no. of times node i has participated in routing towards a particular destination.

As we can see the function used for cost is an exponential function. The cost will thus increase rapidly each time the node is used for forwarding a packet. Therefore, cost associated with a route affected by a wormhole will be very large due to the fact that a wormhole attracts a significant percentage of traffic by advertising shorter paths, with a smaller number of hops. This route will then be discarded by the end node.

To implement the idea above various modifications have to be made to the existing *AODV* protocol. These are:

- Firstly, in the basic *AODV* protocol if a node receives a *ROUTE REQUEST* packet for a sender destination pair which it has already processed it discards this packet. This step has been modified in this protocol. An intermediate node processes all the *ROUTE REQUEST* packets it receives for the same sender destination pair so as to generate multiple routes between sender and destination.
- A new cost field has to be added to *RREQs, RREPs* and to the routing tables of the node.

Every time a signaling packet is forwarded by an intermediate node the node adds its cost to the packet. When this packet reaches its destination, the destination calculates the total cost associated with the route that was followed by that particular signaling packet. After this, the cost calculated is compared with other routes. A route with minimum cost is selected for routing.

The suggested method is simple and completely *decentralized*, which is suitable for the distributed nature of ad hoc network. There is no resource requirement, except that only one extra field is required in the signaling packets and the routing table of table of the nodes. The solution above designed to mitigate Wormhole Attack also performs the *load balancing function* for the network by ensuring that no particular node uses a large percentage of its resources in one go.

The visible shortcomings of the solution above are as follows:

- During the initial phases of the network non malicious nodes or a route free of malicious nodes may also be prevented from taking part in the routing.
- Some of the data may become *compromised* as the wormhole may not be identified immediately.
- The function chosen for the cost is an exponential function that will increase the value of the cost significantly each time it takes part in routing so as to detect wormholes as soon as possible. The authors say that their proposed scheme also carries out load balancing. However, due to the nature of the function that is used some nodes will have a very high value of cost associated with them (non malicious nodes in the route containing the wormhole) and they may be excluded from taking part in routing. Therefore, the load balancing function of the protocol above will not be effective.
- The malicious nodes in the wormhole may modify the cost function to suit their own needs.
- A wormhole may not be detected in a case where the wormhole participates in routing only for certain time period and remains inactive for some time. In this case the cost associated with the route including the wormhole may not increase during the dormant phase, and it may be selected for routing when it becomes active. But the time for which it remains inactive must be chosen carefully as there is a time set out for route entry in the route table.

11.8.2   De Worm: A Simple Protocol to Detect Wormhole Attacks in Wireless Ad Hoc Networks

The De Worm protocol [12] designed to detect Wormhole Attacks is based on the observation that for a wormhole to impact on the network it must attract a significant amount of network traffic towards itself and routes going through wormhole must be shorter than the alternate routes going through the valid network nodes.

Most of the work done in providing mitigating solutions to prevent Wormhole Attacks uses specific information regarding the location of the nodes, time of packet transmission or time synchronization between the nodes, or use of specialized hardware(*e.g. directional antennae*) [8]. *GPS* is required in the nodes, where there is need for precise location information, which may not provide accurate information in indoor and urban areas. Defence mechanisms that rely on time measurements cannot always detect a *Physical Layer Wormhole*. Protocols that use specialized hardware are costly to set up and add complexity to the set up.

De Worm is a simple and localized protocol that can be applied on demand (i.e. when there is a need to verify the existence of a wormhole in the route selected for routing). It does not need any kind of time synchronization between the nodes nor is there a need for any kind of specialized hardware. De Worm uses routing discrepancies between the neighbors along the path from the source to the destination in order to detect a wormhole. The following is the procedure De Worm follows to detect a wormhole (algorithm for On Demand ad hoc routing protocol):

11.8.2.1   Algorithm

1. Sender node **S** sets the target node *T* to be the node *two* hops away from *S* on the route selected by the routing protocol.
2. *S* discovers all its one hop neighbors by broadcasting a *HELLO* packet. These nodes will reply back to *S*. *S* will store these addresses in the neighbor list.
3. Let *N1* be the next node in the route after *S* as selected by the routing protocol. This node is identified by *S*.
4. **S** broadcasts the list as well as the address of *T* to all the one hop neighbors and asks every node except *N1* to find a route to *T* such that the route does not include any other node in the *neighbor list* (this list also known as the *forbidden list*). Each one hop neighbor node except node *N1* will then run the routing algorithm to find the route to target node **T** and will reply to **S** with length *L*, in number of hops, of the routes found out by the nodes.
5. The sender will then pick a *'selected route'* and determine its length. For example, it may select the route with the maximum number of hops. The wormhole will be detected by comparing this length with a fixed parameter *X*. For, e.g. in the case above if *L-2>X* then the presence of a wormhole is assumed. *X* is the *tunable sensitivity parameter*.
6. If no wormhole is detected then one extends the procedure to the next hop along the route.
7. In short, De Worm is an algorithm that runs when a route has been identified by the routing protocol of the network and this route needs to be checked for the presence of a wormhole. Therefore, it does not prevent the formation of wormhole. It can thus be placed in the category of Reactive Solutions designed to prevent wormhole formation. This algorithm checks for the existence of a wormhole hop by hop in the route selected by the routing protocol using parallel routing operations from the neighbors of the nodes in the main route.

The advantages of the solution above are as follows:

- This method designed to mitigate Wormhole Attack does not need any *time measurement or time synchronization* between the nodes.
- Capable of detecting *physical layer* Wormhole Attack.
- There is no need for *specialized hardware*.
- No need for accurate *location information*.
- *Less overhead* associated with the protocol.
- Will be effective even when the wormhole operates in *participation mode*.
- Has a *high wormhole detection rate* and a *few false positives*.

Some of the drawbacks associated with the De Worm protocol designed to avoid wormhole attack are as follows:

- Not effective in mobile environments.
- The protocol is not energy efficient. The reason is that at each and every hop the routing protocol runs to find the path to the target node not involving the nodes present in the Forbidden List. This task is energy consuming as this step is performed by all the neighbors of the present node. The node's energy resources will deplete at a rapid rate.
- Consider the case of a large scale network in which the length of wormhole is also very large. In this scenario, when the present node is the node one hop away from one of the ends of the wormhole, then all the one hop neighbors of the present node will try and find a route to the target node which in this case will be on the other side of the wormhole. Now, this route finding technique will be both time and energy consuming as a large number of hops will be involved, due to the length of wormhole. Therefore in the case of long wormholes this protocol will be very time as well as energy consuming.
- Large number of control packets involved.
- This protocol will not be effective in the case of *short wormholes*.
- This protocol does not prevent the formation of wormhole link.

11.8.3 Wormhole Attacks in Wireless Ad Hoc Networks – Using Packet Leashes to Prevent Wormhole Attacks

In [6], the authors provide a new and general mechanism for detecting and thus defending against Wormhole Attacks. They introduce the general mechanism of *packet leashes* to detect Wormhole Attacks, and propose two types of leashes: *geographic leashes* and *temporal leashes*.

11.8.3.1 Leash

The notion of a *packet leash* has been introduced as a general mechanism for detecting and thus preventing Wormhole Attacks. A leash is any information that is added to a packet designed to restrict the packet's maximum allowed transmission distance. Leashes are designed to protect against wormholes over a single wireless transmission. When packets are sent over multiple hops, each transmission requires the use of a new leash. The authors have distinguished between *geographical leashes* and *temporal leashes*. A geographical leash ensures that the recipient of the packet is within a certain distance from the sender. A temporal leash ensures that the packet has an upper bound on its lifetime, which restricts the maximum travel distance, since the packet can travel at most at the speed of light. Either type of leash can prevent a Wormhole Attack, because it allows the receiver of a packet to detect if the packet traveled further than the leash allows.

11.8.3.2 Geographical Leashes

To construct a geographical leash, in general, each node must know its own location, and all nodes must have loosely synchronized clocks. When sending a packet, the sending node includes in the packet its own location, p_s, and the time at which it sent the packet, t_s; when receiving a packet, the receiving node compares these values to its own location, p_r, and the time at which it received the packet, t_r. If the clocks of the sender and receiver are synchronized to within $\pm\Delta$, and v is an upper bound on the velocity of any node, then the receiver can compute an upper bound on the distance between the sender and itself, d_{sr}. Specifically, based on the timestamp ts in the packet, the local receive time tr, the maximum relative error in location information δ, and the locations of the receiver p_r and the sender p_s, then d_{sr} can be bounded by the following equation as referred in [6]:

$$d_{sr} \leq ||p_s - p_r|| + 2v(t_r - t_s + \Delta) + \delta$$

A standard digital signature scheme or other authentication technique can be used to enable a receiver to authenticate the location and timestamp in the received packet. In certain circumstances, bounding the distance between the sender and receiver, d_{sr} cannot prevent Wormhole Attacks; for example, when obstacles prevent communication between two nodes that would otherwise be in transmission range, a distance-based scheme would still allow wormholes between the sender and receiver. A network that uses location information to create a geographical leash could control even these kinds of wormholes. To accomplish this, each node would have a radio propagation model. A receiver could verify that every possible location of the sender *(a δ + v. ($t_r - t_s$ + 2Δ) radius around ps)* can reach every possible location of the receiver *(a δ + v. ($t_r - t_s$ + 2Δ) radius around pr)* [6].

11.8.3.3 Temporal Leashes

To construct a temporal leash, in general, all nodes must have tightly synchronized clocks, such that maximum difference between any two node's clocks is Δ. The value of the parameter Δ must be known by all nodes in the network, and for temporal leashes, must generally be in the order of a *few microseconds or even hundreds of nanoseconds*. This level of time synchronization can be achieved with the use of specialized hardware. The use of such specialized hardware is currently not common in wireless nodes. But as the cost, size and weight of the future systems are reducing these specialized hardware will get incorporated in the systems being developed. Although the general requirement for time synchronization is indeed a restriction on the applicability of temporal leashes, for applications that require a defense against the wormhole attack, this requirement is justified due to the seriousness of the attack and its potential disruption of the intended functioning of the network. To use temporal leashes, when sending a packet, the sending node includes in the packet the time at which it sent the packet, t_s;

when receiving a packet, the receiving node compares this value to the time at which it received the packet, t_r. The receiver is thus able to detect if the packet traveled too far, based on the claimed transmission time and the speed of light. Alternatively, a temporal leash can be constructed by including instead in the packet an expiration time, after which the receiver should not accept the packet; based on the allowed maximum transmission distance and the speed of light, the sender sets this expiration time in the packet as an offset from the time at which it sends the packet. As with a geographical leash, a regular digital signature scheme or other authentication technique can be used to allow a receiver to authenticate a timestamp or expiration time in the received packet.

An advantage of geographical leashes over temporal leashes is that *time synchronization* can be much looser. Another advantage of using geographical leashes in conjunction with a signature scheme (i.e. a signature providing non-repudiation), is that an attacker can be caught if it pretends to reside at multiple locations. When a legitimate node overhears the attacker claiming to be in different locations that would only be possible if the attacker could travel at a velocity above the maximum node velocity v, the legitimate node can use the signed locations to convince other legitimate nodes that the attacker is malicious.

Advantages:

• These solutions can be categorized within pro-active mitigating solutions for Wormhole Attacks and thus prevent wormhole formation.
• Is reliable and has a *high detection rate*.
• A wormhole is detected between two nodes by checking whether the packet has traveled too far by comparing the traveling time (calculated with the help of leashes) and thus detection can be achieved using any packet that is to be transmitted by a node. The network can be checked for the presence of a wormhole in the *RREQ* phase of routing (*in On-Demand routing protocols*) itself. There is no need for a separate detection phase for the detection of wormholes.

The disadvantages of the proposed solution are:

• The nodes require tightly synchronized clocks.
• Each node requires one to one communication with each of its neighbors.
• Each node requires predicting the sending time and computing signature while having to timestamp the message with its transmission time.
• Special hardware is needed to achieve tight time synchronization between the nodes which makes the set up more complex and costly.
• These solutions cannot always detect a physical layer wormhole.
• Overheads may increase as the size of the packet is increased due to the use of a packet leash.

11.9   Conclusion and Future Work

In this chapter, a detailed analysis of Wormhole Attacks is provided. The Functioning of a wormhole is discussed along with analysis of its severity and its various modes of operation. Classification of Wormhole Attacks is achieved on the basis of whether the wormhole nodes are hidden or are participating in the routing of the network as do any other genuine nodes. Another classification is made based on how the data is transmitted in the wormhole. If the wormhole uses the existing wireless medium to transmit the data it is placed in the category of In band Wormhole. If it uses a medium which differs from the existing wireless medium, it is put in the category of Out Band Wormhole. Mitigating solutions designed to prevent Wormhole Attacks have been classified broadly as Pro Active and Reactive. Various solutions to prevent Wormhole Attacks are discussed and their respective advantages and drawbacks have been analyzed.

A significant amount of research has been devoted to study Wormhole Attacks as well as the countermeasures to prevent these. However, there is still much research work that needs to be done. The underlying rationale is that existing security solutions to prevent Wormhole Attacks are well-matched to specific scenarios with various assumptions and these solutions have proven to be useful in defending against scenarios that match these assumptions. The development of a general mitigation plan capable of defending against the various forms of Wormhole Attacks would be considered to be an important direction for future work.

References

[1] H. Deng, W. Li, D.P. Agrawal (2002) 'Routing security in wireless ad hoc networks,' *IEEE Communications Magazine* **40**: 70–5.

[2] H. Yang, H. Luo, F. Ye, S. Lu, and L. Zhang (2004) 'Security in mobile ad hoc networks: challenges and solutions,' *IEEE Wireless Communications* **11**: 38–47.

[3] L. Peters, F. De Turck, I. Moerman, B. Dhoedt, P. Demeester, and A.A. Lazar (2006) 'Network layer solutions for wireless shadow networks,' In *Proceeding of the Intl. Conference on Networking, Intl. Conference on Systems and Intl. Conference on Mobile Communications and Learning Technologies, IC/ICONS/MCL'06*, Mauritius, pp. 138–62.

[4] C.H. Vu, A. Soneye (2009) 'An analysis collaborative attacks on mobile ad networks' Master Thesis, School of Computing, Blekinge Institute of Technology, Sweden, June.

[5] M. Khabbazian, H. Mercier, and V.K. Bhargava (2009) 'Severity analysis and countermeasures for the wormhole attack in wireless ad hoc networks', *IEEE Trans. on Wireless Communications* **8**(2).

[6] Y.C. Hu, A. Perrig, and D.B. Johnson (2003) 'Packet leashes: a defense against wormhole attacks in wireless networks', In *Proceedings of IEEE Intl. Conference on Computer Communications (INFOCOM 2003)*, San Franscisco, CA, USA.

[7] L. Hu, D. Evans (2004) 'Using directional antennas to prevent wormhole attacks', In *Proceedings of Network and Distributed Security Symposium (NDSS'04)*, San Diego, CA, USA.

[8] W. Wang, B. Bhargava (2004) 'Visualization of wormholes in sensor networks', In *Proceedings of ACM Workshop on Wireless Security*, Philadelphia, PA, USA, pp. 51–60.

[9] L. Lazos, R. Poovendran, C. Meadows, P. Syverson, and L.W. Chang (2005) 'Preventing wormhole attacks on wireless ad hoc networks: a graph theoretic approach,' In *Proceedings of IEEE Wireless Communications and Networking (WCNC'05)*, New Orleans, LA, USA.

[10] I. Khalil, S. Bagchi, and N. B. Shroff (2007) 'Liteworp: detection and isolation of the wormhole attack in static multihop wireless networks,' *Computer Networks* **51**(13): 3750–72.

[11] N. Song, L. Qian, X. Li (2005) 'Wormhole attack detection in wireless ad hoc networks: a statistical analysis approach,' In *Proceedings of the 19th Intl. Parallel & Distributed Processing Symposium (IPDPS'05)*, Denver, Colorado, USA.

[12] T. Hayajneh, P. Krishnamurthy, and D. Tipper (2009) 'DeWorm: a simple protocol to detect wormhole attacks in wireless ad hoc networks,' In *Proceedings of 3rd Intl. Conference on Network and System Security (NSS'09)*, Gold Coast, Australia.

[13] M. Azer, S.M. El-Kassas, and M.S. El-Soudani (2009) 'Immuning routing protocols from the wormhole attack in wireless ad hoc networks,' In *Proceedings of 4th Intl. Conference on Systems and Networks Communications (ICSNC'09)*, Porto, Portugal.

[14] M. Azer, S. El-Kassas, and M. El-Soudani (2009) 'A full image of the wormhole attacks – towards introducing complex wormhole attacks in wireless ad hoc networks,' *Intl. Journal of Computer Science and Information Security (IJCSIS)* **1**(1): 41–52.

[15] S. M. Safi, A. Movaghar, and M. Mohammadizadeh (2009) 'A novel approach for avoiding wormhole attacks in VANET,' In *Proceedings Of the 2nd Intl. Workshop on Computer Science and Engineering (WCSE'09)*, Quingdao, China, pp. 160–5.

[16] C.E. Perkins and E.M. Royer (2000) 'The ad hoc on-demand distance vector protocol', In *Ad hoc Networking*, C. E. Perkins (ed.), Addison-Wesley, pp. 173–219.

[17] R. Maheshwari (2007) 'Detecting wormhole attacks in wireless networks using connectivity information,' In *Proceedings Of the 26th IEEE Intl. Conference on Computer Communications (INFOCOM 2007)*, Anchorage, AK, USA, pp. 107–15.

12

An Experimental Comparison of Collaborative Defense Strategies for Network Security

Hao Chen and Yu Chen

*Department of Electrical and Computer Engineering, The State University New York
(SUNY) – Binghamton, Binghamton, NY 13902, USA.*

12.1   Introduction

As the number of Internet users grew from 361 million in 2000 to 1.8 billion in 2009 [12], the Internet became even more attractive to hackers. The Information Security Forum (ISF) report, entitled *Threat Horizon 2010* [6], predicted an increase in Web 2.0 vulnerability, mobile malware, industrial espionage and attacks from organized crime. Malicious behavior such as Distributed Denial-of-Service (DDoS) attacks, turbo worms, e-mail spam, phishing and viruses have been identified as the primary challenge to an ISP that is asked to satisfy certain requirements with regard to Quality-of-Service (QoS).

Security of network infrastructure has become a major concern. Ideally, a comprehensive solution for issues of security is expected to cover the entire fabric of a network's infrastructure. However, this is not feasible owing to scalability and complexity. It is more reasonable to deploy infrastructure security applications within a limited range. For example, to protect a subset of targets that is under serious threat, such as government agencies, financial institutions, health care facilities, etc.

An efficient defense against distributed attacks has been a hot topic in the network security community in recent years. Instead of completely brand-new, dedicated systems, the collaboration of existing, widely applied single-point network security applications for defense would be more feasible. Through collaboration, a security shield that covers the infrastructure of multiple network domains could be built without significant changes. Collaboration offers individual security applications a wider view of situations that otherwise might not be observed. It improves their resilience and confidence to handle sophisticated security problems with optimized strategies.

Existing collaborative schemes for a distributed defense can be divided into two major categories: centralized or decentralized. The most significant example of a centralized collaboration scheme lies in the use of a coordination daemon, as shown in Figure 12.1(a). It could be either a powerful multi-role server, or a dedicated server. Most likely, this server is located at one of the participating network domains. Its major responsibilities include information collection, processing, analyzing and distribution from or to each individual node. This server may also conduct decision-making, either making optimal final decisions for all of the participating nodes, or providing suggestions to those nodes for reference, depending on the detailed mechanism of collaboration. The major advantages of centralized schemes are high accuracy and efficiency. When considering overhead and scalability, the centralized approaches have distinct limitations.

Pervasive Computing and Networking, First Edition.   Edited by Mohammad S. Obaidat, Mieso Denko, and Isaac Woungang.
© 2011 John Wiley & Sons, Ltd. Published 2011 by John Wiley & Sons, Ltd.

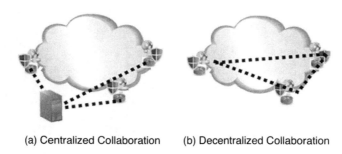

(a) Centralized Collaboration   (b) Decentralized Collaboration

Figure 12.1   Centralized and decentralized collaboration.

In contrast, a decentralized collaboration scheme is much more flexible. It behaves in a similar manner to that of Peer-to-Peer (P2P) networks. This is due to the fact that most decentralized schemes are developed on top of P2P network protocols. The P2P collaborative architecture provides decentralized schemes with good scalability. Theoretically, any network node featuring compatible collaboration protocols could participate. The boundary of a covered network could be loose. Rather than having a collaborative server, each participating node takes responsibility for collaboration, as shown in Figure 12.1(b).

Individual nodes usually have greater flexibility for self-management by comparison with network nodes in centralized scheme. In addition, the cost of application is relatively low, since it does not require any modification to the network beyond the installation of software.

It is important to have a full understanding of the behavior of variant collaborative schemes. A deeper insight is critical for the designers of network security systems so as to adopt proper strategies that can match their requirements. However, there is little reported work that has studied the collaborative behavior of different schemes at a system level. Instead, most of the research has focused on application-specific solutions.

One major challenge in conducting such a behavioral study lies in the lack of methodology that is capable of presenting networks at an abstract level. In practice, many technical and/or non-technical issues make this task more complicated. Fortunately, this challenge may be handled through modeling technology. With the help of modeling, a virtual environment that mimics a simplified world could be set up [10]. After abstracting and scaling-down original problems, it is feasible to conduct further research on substantial behavior.

In this chapter, a comparison study of two major collaborative schemes for distributed defense is described. A three-layered network model including two network layers and one application layer has been developed. The network environment layer at the bottom and the overlay network layer in the middle take advantage of a small-world network model for setup, while the application layer at the top is used for the modeling of centralized and decentralized collaborative schemes. Through adjusting the parameters of the model, different scenarios have been created for collaboration schemes and for evaluating their impact on network infrastructure security.

The chapter is organized as follows. Section 12.2 provides a brief review of the reported work on distributed defense schemes for network infrastructure security. A developed three-layered modeling platform is introduced in Section 12.3. Section 12.4 focuses on a description of our modeled worm-based attack and defense. Section 5 describes the operational detail of how the simulation experiments have been conducted. We also analyze the results for the overall performance evaluation of applying multi-points collaboration for distributed defense. Section 12.6 concludes our work.

12.2   Background

In network security systems, it is common that multiple end-hosts work collaboratively against attacks [15, 16, 19, 23]. A blacklist is exchanged among the potential victims to mitigate the threat. Usually a two-stage operation is conducted for distributed defense, which includes local detection and global collaboration.

There are two popular collaboration schemes. Schnackenberg *et al.* have proposed a centralized coordinative scheme called CITRA [22] for network intrusion detection. A central coordinator takes action for coordinating countermeasures based on a complete view of the network. Janakiraman *et al.* [13] have introduced a decentralized defense scheme for prevention of network intrusion. Information is shared among trusted peers in order to guard the network against intrusion. Subscription-based group communication is conducted over a P2P architecture, which brings with it excellent scalability.

For collaborative detection at the victim's end, more advanced techniques have been developed. Beyond focusing on certain detectable facts in the same domain, the emergence of cross-class detection [21] and multi-domain alter correlation [33] are able to link these detectable facts to certain essentials for further analysis. With the help of cross-class detection, hosts can monitor and share information on different attacks. Meanwhile, multi-domain alter correlation can aggregate alters that possess common values. Instead of focusing only on traffic volume, researchers have extended the anomaly detection scheme to the frequency domain, in which traffic distribution has been considered as random signals, and their energy distribution in different frequency band has been analyzed [4, 28, 36].

Collaboration should not be confined to the victim's end. Deploying network security systems in the fabric of a network increases the initiative of defense system. Gamer *et al.* [9] have extended their research to achieve coordinated collaboration among independent systems for anomaly-based attack detection. Their approach combines an in-network deployment of neighboring detection systems with information exchanging. Working in a self-organized manner, however, each node makes decisions independently.

Considering the advantages of the P2P network, researchers have attempted to address the major challenges in large scale collaboration: the scalability and avoidance of a central point of failure [32]. They have merged multi-dimensional correlation for collaborative intrusion detection [33] and have developed a self-protecting and self-healing collaborative intrusion detection architecture for the trace-back of fast-flux phishing domains [34].

A Distributed Change-point Detection (DCD) scheme has been proposed to detect DDoS attack over multiple network domains [29, 30]. Distributed information is collected through a Change Aggregation Tree (CAT) for centralized analysis and decision making [27]. Another collaborative approach has been designed to detect and stop DDoS attacks at the intermediate network [31]. To achieve this purpose, detection nodes are deployed at both the victim's and source's end for collaborative detection [35]. In a more ambitious approach based on the DefCOM [18] scheme, collaborative nodes are deployed all over the network. Not only at the victim's end and the source's end, the intermediate network is also included [20].

The Internet can be considered as a complex system. All network activities, including attack and defense, may be treated as subsets of it. From this perspective, it is practical to study network security using a complex system model. The earliest attempts at using a complex system for the modeling of distributed network defense schemes were reported in 2001 [8]. However, little similar research has been conducted since then, to the best of our knowledge. Differing from that proposal, which consisted of only describing a preliminary agent-based model without concrete experimental results, this chapter presents our work on specific evaluation based on a three-layered network model.

12.3 Small-World Network Based Modeling Platform

Many network security applications are based on traffic monitoring. The more traffic information obtained, the more confident security applications are. In order to achieve best performance, it is preferable that security applications be deployed at the gateway of the intended network. In practice, it has been the trend to integrate traffic monitoring functions into routers for a simple solution. Nowadays, many advanced commercial-available routers are security enhanced. It is not only software applications that are implemented, hardware based applications are also embedded for advanced security improvement.

Essentially, security applications for collaborative defense are set up on top of the cooperation of corresponding gateway routers. The upper-level application chooses the countermeasure, while the lower-level agent supports its execution. In this section, taking advantage of the small-world network theory, an abstract three-layer platform is built for this modeling study.

12.3.1 Small World Network

The Internet is a scale-free network [2]. A scale-free network is a network whose degree distribution follows approximately the power law [3]. Most nodes in a scale-free network have only one or two links, while only a few nodes have a large numbers of links. This small portion of nodes acts as hubs responding for the connection of the whole network.

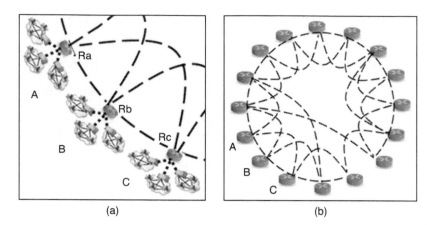

(a)   (b)

Figure 12.2   Illustrations of logic networks (a) A segment of Internet in logic. (b) A simplified network representation.

Figure 12.2(a) represents a segment of the Internet in logic. With the connection to router R_a, R_b, and R_c, end hosts belonging to different network domain A, B, and C can communicate with each other across domains. In addition, multiple logic links among R_a, R_b, and R_c make communication more efficient by choosing optimal paths. For example, by forwarding packets directly from domain A to C through the shortest links between R_a and R_c without passing R_b. Therefore, routers R_a, R_b, and R_c play critical roles as gateway hubs bringing local network A, B and C together to form a larger network.

The theory of a small-world network describes the scale-free network well. A small-world network is a network that is mostly local-connected and with a few global connections. In fact, many real-world networks might be described accurately by using small-world network models, such as cells [14], social networks [25], the World-Wide Web [11] and the Internet [1]. The Watts and Strogatz model is the most famous small-world network model. It is a random graph generation model that produces graphs with short average path lengths and high clustering [26]. It is also the foundation of our model.

Figure 12.2(b) illustrates a classic small-world network. This modeled network consists of finite numbers of nodes. Each node represents a network domain. Let us assume that each network domain has only one outlet to the Internet. Actually, this is true in most cases. The dashed lines represent logic links among different network domains over the Internet in general.

The graph in Figure 12.3 shows an example generated by taking the specific approach that we employed in order to construct a small-world network. The entire network contains 50 nodes. Each node is connected to two nearest neighbors and has a probability of 0.175 to add another edge. Small-world networks set up the basic topology for our intended network layers.

12.3.2   Structure of Three-Layered Modeling Platform

As demonstrated in Figure 12.4, the platform that we developed for the modeling of collaborative schemes consists of three layers. From the bottom to the top, they are the Internet layer, overlay-network layer and application layer, respectively. The development of the bottom two layers was inspired by the Watts and Strogatz small-world network model. The Internet layer models an abstract Interment environment in general, while the overlay network layer sets up a dedicated environment for the operation of different collaborative schemes. Finally, the application layer focuses on the description of defense schemes.

The physical Internet layer is shown as the bottom layer in Figure 12.4. Each solid node represents a network domain participating in collaborative defense and the hollow nodes are network domains that do not participate in the collaboration. The dashed lines represent the physical network topology. The middle layer overlay network consists only of network domains that participate in collaborative defense. The solid lines are the logic connections among these nodes.

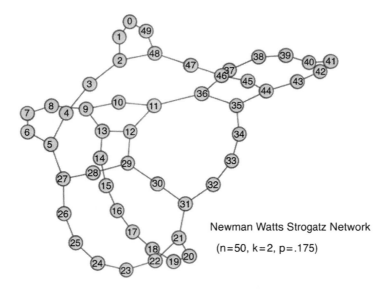

Figure 12.3   An example of small world based network.

In this model, link weight is adopted as the metric of distance between nodes. One hop is the minimal distance between any neighboring network domains. For example, in Figure 12.4, nodes A and B, nodes B and C are adjacent respectively, so that the weight of $Link_{ab}$ and $Link_{bc}$ is one. Meanwhile, the weight of $Link_{ac}$ is two.

However, their link weights may not remain the same when corresponding nodes are mapped to the bottom Internet layer. Though Node A' and B' still appear to be adjacent, Node B' and C' are four hops away from each other, so the weight of $Link_{b'c'}$ is four. As illustrated by Figure 12.4, there are three other network domains on the path from B' to C'.

In the Internet layer, the six-degree-of-separation theory [17] indicates that the average width of a large scale network is six. Specific to the Internet, relevant research [29] has also verified statistically that more than 99% of network domains in the Internet can be reached within six hops. In our work, the weight assigned to the associated links follows normal distribution.

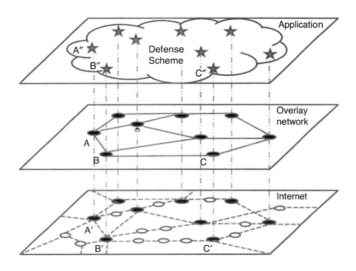

Figure 12.4   Structure of three-layered network model.

The top layer is the application layer, which is a conceptual layer where we model defense schemes. This layer focuses on the behavior modeling of participant network domains in an abstract manner.

As shown in Figure 12.4, star-nodes represent individual network domains with security applications deployed. The cloud generalizes the defense schemes organizing these applications for reaction.

Three types of defense schemes are described in this layer: the single-point defense scheme, as well as centralized and decentralized collaborative defense schemes. All three defense schemes are applied to the same constellation of security applications, but with different concentrations of network coverage. The first single-point scheme concentrates on the protection of individual network domains. Each security application works independently. The other two collaborative schemes, concentrate on the protection of a widely ranging network area. Through multi-points collaboration, valuable information is shared among individual security applications for the improvement of their overall performance.

12.4   Internet Worm Attack and Defense

As well as the network model introduced above, a comparison study has been conducted in order to investigate the performance of three different defense schemes against a typical Internet worm attack. This section describes the following: modeling a worm attack and modeling corresponding defense schemes.

12.4.1   Modeling a Worm Attack

Featuring self-duplication and automatic propagation, Internet worms are truly autonomous during an attack. They are able to spread across the network, breaking into end hosts and replicating. It is extremely challenging to prevent Zero-day worms. In addition, the worms themselves are good carriers for other malicious attack tools. Some sophisticated worm attacks are intended to propagate stealthily so as to survive for further action, such as remote controlling infected hosts in order to launch DDoS attacks.

Various strategies have been adopted to achieve fast propagation, such as exploring security holes, or increasing scanning rate with different scanning schemes [7, 24]. The propagation patterns of most Internet worms have shown a certain similarity. After a short period of modest increase, the number of infected network domains grows exponentially. Once it has achieved maximum infection, the corresponding curve trends to be flat, if there is no efficient way to defeat it. At that time, the wave of worm attack enters a saturated state.

We modeled the worm attack as follows. Assume that there is only one type of worm in the entire process. The infection of network domains follows a simple classical epidemic model (SI model) as shown in Figure 12.5 (left hand side). Participating nodes in the space concerned have two states: susceptible (S) and infected (I). They are all initialized to be susceptible to the attack. One of the participant nodes is selected randomly as the first infected node. Worms propagate to all the neighbor nodes from the current infected node. This propagation follows the network topology at the Internet layer as shown in Figure 12.4.

The SI model only considers the attack under a pure infection mode. With the engagement of all kinds of defensive efforts, worms may be detected and contained. Consequently, network domains may be immune from this attack, regardless of their current status either in susceptible or infected states. The adapted SIR model depicts such an infection and recovery, as shown in Figure 12.5 (right hand side). A removed (R) state is introduced. Once it has entered the 'Removed' state, it represents that the network domain has recovered and become invulnerable

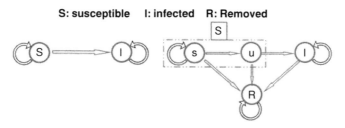

Figure 12.5   Infection/defense modes (a) Pure Infection mode (SI model). (b) Adapted Infection /Recovery Mode (SIR model).

to the worm. We have also adjusted the Susceptible (S)' state in the original SI model to those of 'susceptible (s) and 'under attack (u)' states.

12.4.2   Modeling Defense Schemes

The adaptive SIR model also represents the basic behavior of individual security applications. We assume that the immunity of network domains to worm attacks resulted from the reaction of security applications. Each security application plays its part as an agent for reaction. Although individual defense behavior may vary, collective behavior determines overall efficiency.

A single-point defense scheme is inefficient when facing fast propagation, widespread and stealth Internet worms. For example, consider the security applications modeled in the application layer of Figure 12.4, network domains A', B' and C' adopt different security applications A'', B'' and C'', respectively. While B'' has detected the signature of a worm in network B', neither A'' nor C'' has detected that network A' has been infected. Since all of the applications work individually, successful detection at B'' does not imply that other network domains will benefit. Their collective behavior shows that the overall reaction efficiency of security applications running under a single-point defense scheme is low. It depends greatly on the performance of each individual.

In contrast, a collaborative defense can improve effectiveness significantly. Benefiting from the earlier alarm and assistance from B'', C'' can start and optimize its defense in advance, and A'' will realize what happened and be able to contain any maliciousness so as to minimize the negative effect.

This on-the-fly collaboration runs seamlessly without human intervention. As a result, the corresponding network domains are all saved from attack. The overall reaction efficiency of participants running under a collaborative scheme is higher. Their performance is correlated and impacted significantly by the first agent reacting to the attack. Theoretically, the larger the number of network domains involved and the wider the area they span for co-defense, the higher the probability of prompt detection.

Owing to the different collaborative strategies, centralized and decentralized schemes have been proposed based on the improvement of the single-point defense scheme. The individual behavior of the participating agents remains the same, but obviously their collective behavior is changed. In this study, we will focus on the different mechanisms for collaboration.

The collaborative defense of participating agents running under a decentralized scheme behaves in a similar manner to social networking activities. Besides defending individually, they interact with peers for information sharing and decision making. This type of collaboration is flat, no agent is dominant. We assume that all of the agents collaborate only with their neighbors and all neighbors have the same significance for each other. On one hand, each agent still makes decisions individually, but with reference from its neighbors under consideration, such as issuing an alarm for worm attack. On the other hand, each agent acts as a relay that passes the proper information efficiently to others, such as spreading the issued worm alarm to its peers. Individual agents are highly flexible in collaboration.

In the centralized scheme, all of the collaborative activities operate as a root-leaves structure. A centralizer acts as the root that is in charge of the whole collaboration. It may locate in any of the network domains. This centralizer has reliable communications with all of the participants. Security applications act as leaves, collecting and pre-screening useful information to the root. Through the analysis of gathered information, corresponding feedback is returned from the root to every leaf. Obviously, the overall efficiency of agents running under this scheme is more consistent than that which the decentralized scheme can achieve.

12.5   Experiments and Performance Evaluation

This section presents simulation results and performance evaluation. The simulation experiment is discussed in detail, including the basic assumptions, parameters and attack-and-defense operations. Through the analysis of simulation results, the performance of centralized and decentralized collaboration schemes for distributed defense is evaluated. The single-point scheme is also discussed for reference.

12.5.1   Experimental Setup

In the simulation, the operation of attack and defense are relatively independent from each other. According to the platform shown in Figure 12.4, the simulation of a worm attack is conducted in the Internet layer. The propagation of worms spreads through the paths defined in this layer. Their targets are those susceptible interested network domains (solid nodes). The propagation does not stop as long as any susceptible node has not been compromised in the concerned network space.

The simulation of defense is carried out in the overlay network layer. The entire constellation of security applications defined in this layer acts as agents. They take the defense schemes from the upper layer to protect the corresponding network domains mapped in the lower layer. The collaboration among agents follows the topology defined in this layer. The defense countermeasures will not stop until either all of the solid nodes in the Internet layer are alarmed or all of the nodes are immune from the worm, depending on the detailed setup for simulation.

The appearance of the first infected node triggers the worm attack aimed at the relevant network area. This node was selected randomly from the relevant network domains in the Internet layer at the beginning of the simulation. After that, the worm propagates following the approach described by the model.

Meanwhile, the location of the centralizer was also assigned randomly to one of the participant network domains. Through its association with its related defense agents, it records all of the shortest paths from the related agents to all of the other agents as defined in the overlay network layer. The structure for centralized collaboration was set up during the initialization of the simulation.

The initial alarm for worm detection from a defense agent triggers the entire defensive reaction. This triggering event is associated with the progress of worm propagation. With the propagation of worms through the network space, the probability of being detected also grows. All of the worm packets are detectable once the signature has been identified [5].

Basically, our simulation process follows the adaptive SIR model as shown in Figure 12.5(b). Those event-triggered activities could be well managed under a Finite State Machine (FSM). The simulation is executed with a discrete time scale. The execution time of each activity is scaled to one or multiple time slots. The complete worm propagation procedure consists of three phases: online probe, data transmission and local infection. For simplicity, we assume that the time delay for one node infection is one unit time slot.

Consider that transferring 100 k data in 100 M/bps takes 1 ms, while infecting a node takes a few seconds. It is obvious that the infection time is dominant in worm propagation. From the perspective of the defenders, the time for spreading the alarm is expected to be short, but the overhead that results from collaboration among different agents is non-trivial.

To describe the variable security vulnerabilities that network domains possess, we assigned the resistant time of each network domain to worm attack randomly from one to three unit time slots following normal distribution. It means that the most vulnerable network domains would be infected in one unit time and the least vulnerable network domains would also be infected in three unit time, if there is no available defense.

At the defenders' side, we assume that the time delay of information exchange for collaboration between adjacent agents in a decentralized scheme or between agents and the centralizer in centralized scheme is one unit time. The collaborative activities include alarm spreading and advanced knowledge sharing, which includes the updated signatures. It is also assumed that the execution time for all of the local activities is one unit time, including the issuing of the alarm, advanced knowledge issuing, relay determination and knowledge updating. With the exception of the original agent which issued the alarm or other advanced knowledge, all of the other agents have to receive the updated knowledge and finish the update processing in order to contain the worm in their network domains.

Although setting the spreading of the knowledge update just one unit time after the worm alarm is issued may be debatable, it does not affect relative defending trends after the knowledge update process is finished by local agents. Instead of the trends, the difference rests in the reaction time.

12.5.2 Experimental Results and Discussion

As described above, we conducted extensive simulation experiments on worm attack-and-defense on the modeling platform. The sample of the first simulation consisted of 200 security applications in the overlay layer. Mapping to the Internet layer, they correspond to 200 intended network domains. Taking advantage of the Watts-Strogatz network function, we generated a small-world network environment for the Internet layer with $n = 200$, $k = 2$, $p = 0.4$. The average distance between any two adjacent nodes is 5.514, which is acceptable for the number of represented unintended network domains in between them along the way. Since the nodes in the overlay network are usually tightly connected in logic, the average distance between any pair of adjacent nodes should be shorter. Thus, we increase the link probability for setting up the topology of Overlay network layer, as $n = 200$, $k = 2$, $p = 0.6$.

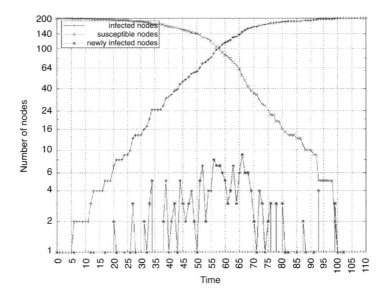

Figure 12.6 Pure worm infections without containment.

12.5.3 Simple SI Model

Figure 12.6 demonstrates the modeled worm propagation without any defense. The X axis represents the time intervals.

The Y axis represents the number of network domains (nodes). The line with vertical bars represents the increasing trend of worm infection over all of the network domains. Those vertical bars record the total number of infected nodes. The line with crosses represents the decreasing of susceptible nodes due to the increase of infected nodes from the left-top to the right-bottom. The star spots in the lower part of the plot indicate the number of newly infected nodes in each time unit.

In order to present a clear view of their variation, we used a log scale of the representation of values in the Y axis. The number of infection nodes stays small at the both ends, but is large in the middle. This is because the exponential increase of worm propagation usually happens in the middle with respect to the whole process.

Figure 12.7 shows the alarm spreading after the worm attack has been detected. The line across diagonally is the referred pure infection line, which represents the number of infected nodes under pure infection. It is identical to the line with vertical bars in Figure 12.6. The shape is different because the Y axis is scaled normally, which represents the number of alarmed nodes. One difference between the centralized and the de-centralized collaborative schemes is that the knowledge update or alarm generated by the central server is sent to each agent to contain the worms, not by itself.

In this simulation, it is assumed that when 60% of the network domains have been infected, the worm is detected and the first alarm is generated. In Figure 12.7, the cubic box on the pure infection line marks this point. Referring to the X axis, this is the time for issuing the first alarm from that agent. The line with vertical bars represents the agents working under a single-point defense scheme and the line with stars and the line with crosses represent agents working under centralized and decentralized defense schemes, respectively.

As expected, the line with vertical bars is almost flat during the entire simulation period since none of the peer agents is expected to be able to share the alarm. For centralized and decentralized schemes, the alarm spreads quickly to all of the defense agents through the topology built into the overlay network layer. This topology models the paths for collaboration. It is obvious that a centralized scheme is more efficient for alarm spreading with the same set of collaborative nodes and the same network topology.

12.5.4 The SIR Model

For further insight into the impact of different defense schemes, we simulated the defeat scenario. Once a worm attack has been detected and an alarm has been issued, the security agent continues to update its knowledge base

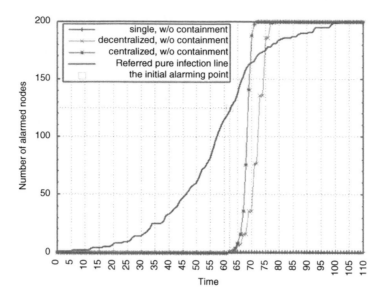

Figure 12.7   Alarming for worm attack.

and spread the newly generated signatures to peers for worm containment. Through sharing the signatures database, other agents can prevent the attack effectively.

Figure 12.8 presents the scenario of worm containment. The curves reflect the number of infected nodes along the time span. The referred pure infection line and the initial alarming point remain the same as the previous examples for consistency. The difference is that the initial update point is introduced, which represents the time point when the first agent has finished knowledge updating and is ready for worm containment. The trend of the line with vertical bars that represents the defense running under a single-point scheme almost remains the same as the pure infection line. Since only one node is alarmed and become immune from attack, all other nodes are still vulnerable and become infected.

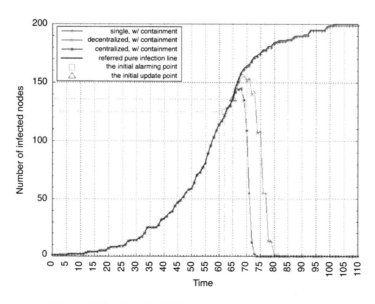

Figure 12.8   Trends of infected nodes with containment.

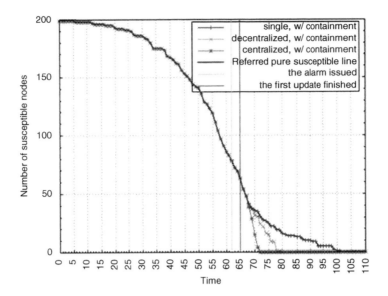

Figure 12.9   Trends of susceptible nodes with containment.

Two collaborative defense schemes demonstrate much greater containing efficiency. As observed, the trends of the line with stars and the line with crosses turn down sharply after a small delay. The line with stars represents the number of infected nodes under the centralized scheme, while the line with crosses represents that under the decentralized scheme. Finally, in this case the former touches the ground at the 74th tick, and the latter touches down at the 82nd tick.

From another perspective, the decreasing trend of susceptible nodes along the time span also supports the observation that collaborative schemes are efficient in defense.

As illustrated in Figure 12.9, the first and second vertical lines from the left to the right represent the initial alarm time line and the initial update time line, as described in the previous example. The trend lines regarding different defense schemes overlap most of the time. The interesting point is that the two lines with respect to the centralized and decentralized schemes divert shortly after they pass the initial update line, and diminish quickly to zero at the 74th and 82nd ticks, respectively. This rapid diminution is due to the efficient signature database updating of both collaborative schemes. Before being attacked by the worm, they have already been immunized.

To show the overall efficiency of different defense schemes against worm attack, we have explored the ratio between the immune and infected nodes in our simulation.

Figure 12.10 provides an alternative view of this exploration. The immunity ratio is defined as:

$$\frac{number\_of\_immune\_nodes - number\_of\_infected\_nodes}{total\_number\_of\_nodes}$$

At the beginning of the simulation, none of the nodes is infected or immune from the worm, so the immunity ratio is zero. Without effective containing measures, the rapidly propagating worms bring the immunity ratio swiftly down to negative and finally lock it into fully infected status at -1, or -0.995 in terms of the single-point defense scheme. However, this ratio could also be increased by proper countermeasures. As observed in Figure 12.10, the trend lines corresponding to two collaborative schemes rise quickly and finally enter full immune status at 1. In fact, the trends depend on two factors: the efficiency of collaboration and the delay in the detection of the worm.

The second simulation exhibits the impact of variable alarming times on overall worm defense, as shown in Figure 12.11.

The X axis lists a range of different ratios for the compromised network domains in percentage terms, at which the first alarm is issued. The Y axis represents the alarm correlated infection rate when the first alarm reaches all of the agents. For consistency, all of the experimental configurations are the same as the previous examples. This simulation studies the different compromised node ratios from 0.0 to 0.99 with an interval of 0.02. In order to

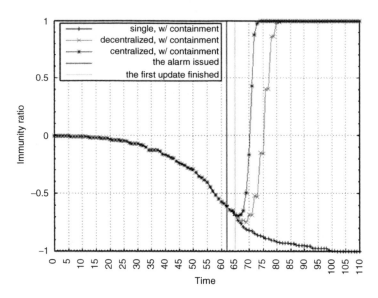

Figure 12.10   Overall defense efficiency.

achieve a refined output, every data point in Figure 12.11 is the average of five experiments. Intuitively, the earlier the alarm is issued, the lower the correlated infection rate would be, if no further containing action follows. From the top down, the first two lines represent the schemes without containment. The infection rate of decentralized scheme represented by the upper line is higher than that of centralized scheme represented by the lower line under the same alarm rate. It is obvious that both infection rates trend to full when the compromised ratio moves from 0 to 99%.

Furthermore, we compared the infection rates with the corresponding containment and the results are represented by the last three lines in Figure 12.11. They become flat even if the first alarm is being postponed continuously. This is due to the reaction of containment. After the alarm and the following update signatures are shared, more and more security agents are capable of containing the worm. They lower the infection rate of the network space.

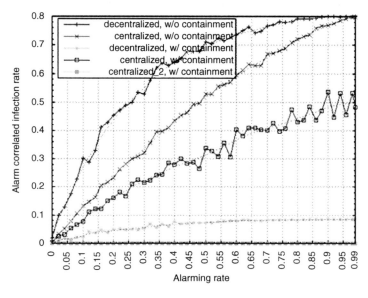

Figure 12.11   Correlated infection rate with variable-compromised node ratio.

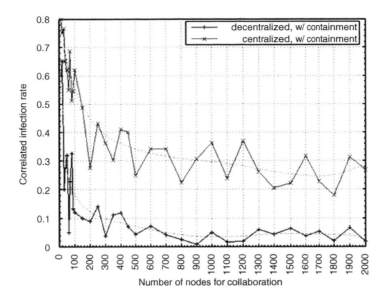

Figure 12.12   Correlated infection rate under different collaborative scales.

The scalability of collaborative defense schemes has also been evaluated. The configuration remains the same, except with a different number of collaborative nodes for operation each time. Figure 12.12 presents the simulation results. The number of nodes ranged from 10 to 2000. According to the simulation results, although the vibration of the infected rate is still obvious, it trends to flat when the collaboration scale is greater than 200 nodes. The correlated infection rate of a centralized scheme is higher than that of a decentralized scheme when both are under the same simulation conditions.

12.6   Conclusions

This chapter presented a comparison study of the characteristics of different collaborative defense schemes against Internet worm attacks. Based on the small-world network model, our experimental results have verified that both centralized and decentralized collaborative schemes can improve system performance effectively as compared with a single-point defense scheme. Based on a survey of the relevant work, most of the reported research is application-dependable and gives specific problem-based solutions. This chapter provides new insight into collaborative strategies on a higher abstract level.

References

[1] R. Albert, H. Jeong, A.-L. Barabási (1999) 'Internet: diameter of the world-wide web,' *Nature* **401**: 130–131.

[2] R. Albert, H. Jeong, A.-L. Barabási (2000) 'Error and attack tolerance of complex networks,' *Nature* **406**: 378–382.

[3] A.-L. Barabási, E. Bonabeau (2003) 'Scale-free networks,' *Scientific American* **288**: 50–9.

[4] H. Chen and Y. Chen (2008) 'A novel embedded accelerator for online detection of shrew DDoS Attacks,' In *Proceedings of the 2008 International Conference on Networking, Architecture, and Storage*, pp. 365–72.

[5] H. Chen, D.H. Summerville, and Y. Chen (2009) 'Two-stage decomposition of SNORT rules towards efficient hardware implementation' In *7th International Workshop on Design of Reliable Communication Networks, 2009 (DRCN 2009)* Washington, DC, pp. 359–66.

[6] J. Creasey (2008) 'Threat Horizon 2010,' ISF 08 04 03.

[7] R. Dantu, J. Cangussu, and A. Yelimeli (2004) 'Dynamic control of worm propagation,' In *Proceedings of the International Conference on Information Technology: Coding and Computing (ITCC'04)*, pp. 419–23.

[8] D. Frincke, E. Wilhite (2001) 'Distributed network defense,' In *IEEE Workshop on Information Assurance and Security*, West Point, NY, pp. 236–8.

[9] T. Gamer, M. Scharf, and M. Schöller (2007) 'Collaborative anomaly-based attack detection,' In *Lecture Notes in Computer Science* 4725, ed: Springer Berlin/Heidelberg, pp. 280–7.

[10] D.M. Gorman, J. Mezic, I. Mezic, and P.J. Gruenewald (2006) 'Agent-based modeling of drinking behavior: a preliminary model and potential applications to theory and practice,' *American Journal of Public Health* **96**: 2055–60.

[11] B.A. Huberman, L.A. Adamic (1999) 'Internet: growth dynamics of the world-wide web,' *Nature* **401**: 131.

[12] Internetworldstats.com. (2009) World Internet Usage and Population Statistics: http://www.internetworldstats.com/stats.htm.

[13] R. Janakiraman, M. Waldvogel, and Q. Zhang (2003) 'Indra: A peer-to-peer approach to network intrusion detection and prevention,' In *Proceedings of 12th IEEE Workshops on Enabling Technologies,Infrastructure for Collaborative Enterprises (WETICE)*, Los Alamitos.

[14] H. Jeong, B. Tombor, R. Albert, Z. N. Oltvai, and A.-L. Barabási (2000) 'The large-scale organization of metabolic networks,' *Nature* **407**: 651–4.

[15] J. Kannan, L. Subramanian, I. Stoica, and R.H. Katz (2005) 'Analyzing cooperative containment of fast scanning worms,' In Proceedings of the Steps to Reducing Unwanted Traffic on the Internet on Steps to Reducing Unwanted Traffic on the Internet Workshop, Cambridge, MA, pp. 17–23.

[16] M. Allman, E. Blanton, V. Paxson, and S. Shenker (2006) 'Fighting coordinated attackers with cross-organizational information sharing,' In *The Fifth Workshop on Hot Topics in Networks (HotNets '06)*, Irvine, CA.

[17] S. Milgram (1967) 'The small world problem,' *Psychology Today* **2**: 60–7.

[18] J. Mirkovic, M. Robinson, and P. Reiher (2003) 'Alliance formation for DDoS defense,' In *Proceedings of the 2003 Workshop On New Security Paradigms*, Ascona, Switzerland pp. 11–18.

[19] D. Moore, C. Shannon, G.M. Voelker, and S. Savage (2003) 'Internet quarantine: requirements for containing self-propagating code,' In *IEEE Infocom*, San Francisco, CA, USA.

[20] G. Oikonomou, J. Mirkovic, P. Reiher, and M. Robinson (2006) 'A Framework for a collaborative DDoS defense,' In *Computer Security Applications Conference, 2006. ACSAC '06*, Miami Beach, FL, pp. 33–42.

[21] H. Ringberg, A. Soule, and M. Caesar (2008) 'Evaluating the potential of collaborative anomaly detection,' Technical Report, Princeton University.

[22] D. Schnackenberg, H. Holliday, R. Smith, K. Djahandari, and D. Sterne (2001) 'Cooperative intrusion traceback and response architecture (CITRA),' In *Proceedings of the DARPA Information Survivability Conference and Exposition (DISCEX)*, pp. 56–8.

[23] A. Soule, H. Larsen, F. Silveira, J. Rexford, and C. Diot (2007) 'Detectability of traffic anomalies in two adjacent networks,' In *Lecture Notes in Computer Science* 4427/2007, ed: Springer Berlin/Heidelberg, pp. 22–31.

[24] A. Wagner, T. Dubendorfer, B. Plattner, and R. Hiestand (2003) 'Experiences with worm propagation simulations,' In *Proceedings of the 2003 ACM Workshop on Rapid Malcode*, Washington, DC, USA, pp. 34–41.

[25] S. Wasserman, K. Faust (1994) *Social Network Analysis: methods and applications* Cambridge: Cambridge University Press.

[26] D.J. Watts, S.H. Strogatz (1998) 'Collective dynamics of 'small-world' networks,' *Nature* **393**: 440–2.

[27] Y. Chen, K. Hwang (2006) 'Collaborative change detection of DDoS attacks on community and ISP networks,' In the *IEEE International Symposium on Collaborative Technologies and Systems (CTS'06)*, Las Vegas, NV., USA, pp. 401–10.

[28] Y. Chen, K. Hwang (2007) 'Spectral analysis of TCP flows for defense against reduction-of-quality attacks,' In the *2007 IEEE International Conference on Communications (ICC'07)*, Glasgow, Scotland, pp. 24–8.

[29] Y. Chen, K. Hwang, and W.-S. Ku (2007) 'Collaborative detection of DDoS attacks over multiple network domains,' *IEEE Transactions on Parallel and Distributed Systems* **8**: 1649–62.

[30] Y. Chen, W.-S. Ku, and K. Hwang (2007) 'Distributed change-point detection of DDoS attacks: experimental results on DETER testbed,' In *DETER Community Workshop on Cyber Security Experimentation and Test, in conjunction with USENIX Security Symposium (Security' 07)*, Boston, MA.

[31] G. Zhang, M. Parashar (2006) 'Cooperative defence against DDoS attacks' J. Res. Pract. Inf. Technol. **38**: 69–84.

[32] C.V. Zhou, S. Karunasekera, and C. Leckie (2005) 'A peer-to-peer collaborative intrusion detection system,' In *13th IEEE International Conference on Networks, Jointly held with the IEEE 7th Malaysia International Conference on Communication*, Kuala Lumpur, Malaysia, p. 6.

[33] C.V. Zhou, C. Leckie, and S. Karunasekera (2009) 'Decentralized multi-dimensional alert correlation for collaborative intrusion detection' *J. Netw. Comput. Appl.* **32**: 1106–23.

[34] C.V.L. Zhou, C. Karunasekera, and S.T. Peng (2008) 'A self-healing, self-protecting collaborative intrusion detection architecture to trace-back fast-flux phishing domains,' *IEEE NOMS Workshops 2008*, pp. 321–7.

[35] Z. Zhou, D. Xie, and W. Xiong (2009) 'A P2P-based distributed detection scheme against DDoS attack,' In *First International Workshop on Education Technology and Computer Science, 2009. ETCS '09*, Wuhan, Hubei, pp. 304–9.

[36] L. Zonglin, H. Guangmin, Y. Xingmiao, and Y. Dan (2009) 'Detecting distributed network traffic anomaly with network-wide correlation analysis,' *EURASIP J. Adv. Signal Process* **2009**, pp. 1–11.

13

Smart Devices, Systems and Intelligent Environments

Joaquín Entrialgo[1] and Mohammad S. Obaidat[2]

[1]*Department of Computer Science and Engineering, University of Oviedo, Gijón, Spain.*
[2]*Department of Computer Science and Software Engineering, Monmouth University, West Long Branch, NJ 07764, USA.*

13.1 Introduction

Mark Weiser's vision [1] for ubiquitous computing included elements such as houses with coffee machines which anticipated the needs of its inhabitants, or offices with objects that have computing capabilities that are integrated seamlessly. These places have come to be called 'intelligent environments' and rely on the availability of smart devices and systems which comprise a world of pervasive computing.

There are several technological foundations that make this vision possible [2]. Moore's law provides smaller, faster and cheaper computing, storing and communicating technologies, which enable the creation of smart devices and systems with ever increasing capabilities. New materials, such as electronic paper or semi conducting organic polymers, can be used to integrate computing elements into everyday objects. Communication technologies, especially wireless ones, make it possible to integrate computation capabilities in mobile nodes and facilitate sharing of information between them. Finally, sensors are integrated in objects, so they can obtain information from the physical world and make the objects context aware.

Smart objects are of paramount importance for ubiquitous computing. General-purpose computers, such as PCs and laptops, cannot fulfill the vision of computers that help people invisibly in their daily activities. Thus, there is a need for blending computing capabilities into every kind of object. In this regard, the use of wireless networks of sensors and of micro electromechanical systems (MEMS) can expand the capabilities of objects and environments without altering the basic properties that the user is familiar with. Some examples of these smart objects are smart chairs, smart badges or smart medicine cabinets. Of course, this proliferation of objects augmented with computing capabilities should not mean that the user has to deal with the problems of configuration that often arise in computing environments: that would defeat the purpose of invisible computing.

The major goal of pervasive computing is to adapt computers to people instead of adapting people to computers. This requires a new kind of interface that is more natural for humans than the traditional combination of a keyboard, mouse and monitor. It also calls for intelligence in many ways. Firstly, understanding how people interact naturally – by voice, gestures and using many ambiguous signals – requires artificial intelligence in order to interpret the user's input. In addition, a computing platform should use information from the context, as people do and expect constantly in a natural interaction. Furthermore, one of the key missing components in traditional computing is proactivity. A ubiquitous system should be proactive, anticipating the needs of the user and changes in the context. This requires understanding user intent, so that the system can determine which actions might help or hinder the user.

This need for intelligence is another factor that makes smart objects and intelligent environments an important issue in the field of pervasive computing. Furthermore, the field of smart objects, which was born within the domain of wireless sensor networks, focused in the beginning on aspects such as self-organizing networks and power-efficient computing. It has received very recently a great deal of attention from the artificial intelligence community. The development of intelligent environments is one of the main challenges that this community is facing. Examples of intelligent environments include houses that control their lightning scheme, learning from user habits or smart cars that warn the user when they detect fatigue.

The field of smart objects and intelligent environments, as a relatively new area, faces many challenges and still requires a considerable amount of research so as to realize the ubiquitous vision. In spite of this, a considerable amount of research has already been carried out and the foundations of the field have been established.

This chapter introduces smart devices and systems, and then intelligent environments. Next, it presents the trends, limitations and challenges for these technologies. Finally, it provides several examples of application as well as case studies that illustrate the capabilities that they can offer.

13.2   Smart Devices and Systems

13.2.1   Definition and Components

A smart device is characterized by possession of some kind of intelligence. Often, the device in isolation cannot provide this extra intelligence, but requires the help of a background infrastructure. For instance, Satyanarayanan [3] introduces the term 'cyber foraging' for describing augmenting the capabilities of a wireless device dynamically by exploiting a wired hardware infrastructure. Another possibility is using the capabilities of other more powerful smart devices in the environment, as shown in [4].

Thompson in [5] proposes that a smart device must have some or all of these capabilities:

- Communications, to send or receive messages, both from humans and from other smart devices.
- Identity and type to be able to carry out different functions.
- Memory and status tracking in order to maintain their settings and histories.
- Sensing and actuating to monitor the environment and, if necessary, change it.

Figure 13.1 shows a diagram of a system that has the components required to carry out all of these functions. Depending on the real capabilities of a given smart device, some of these components can be missing. For instance, the simplest type of a smart device would be an object with a passive tag which identifies it; this can be considered

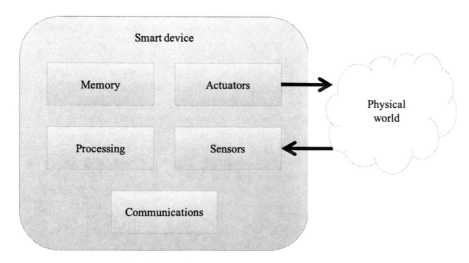

Figure 13.1   Smart device components.

as an object which only contains read-only memory, but it would be the base for a smart system that can provide intelligent behavior. A smart telephone would include at least a processing element, memory and communications components; some models might include also sensors (i.e. a GPS subsystem or an accelerometer).

13.2.2 Taxonomy

Several criteria for classification of smart devices and systems have been proposed. They are orthogonal, so a device can be classified independently according to each criterion.

Weiser in [1] used the form factor for classifying smart devices as three kinds:

- *Tabs*: A tab is basically a device the size of a post-it note (a few inches); for example, an active badge which identifies its carrier. Modern cell-phones can also be considered to be tabs according to their form factor.
- *Pads*: These devices have the size of a paperback book (a few feet); for example, a laptop. Instead of carrying the pads around, they would be in a room and their information could be accessed anywhere. They would not have individualized identity and would try to provide the functionality of real documents on a physical desktop.
- *Boards*: These are devices have the size of a blackboard (a few yards). They would be used for viewing by multiple people in large displays, such as bulletin boards, white boards or flip-flops. They are also useful for displaying large amounts of information for just one individual, as in maps, for example.

Poslad [6] noted that all of the form factors proposed by Weiser are macro-sized; they have a planar form and incorporate an output visual display. He extended the classification of possible form factors to include:

- *Smart dust*: With a size smaller than a millimeter and without display, these devices are built from micro elec-tromechanical systems (MEMS). The Smart Dust project [7] explores the technology needed for an autonomous sensing, computing and communicating device that can be packed into a cubic-millimeter mote to form the basis of integrated, massively distributed sensor networks. This research gave birth to the field of sensor motes [8].
- *Smart skin*: These are MEMS integrated into fabrics that make it possible for the development of flexible surfaces such as clothing.
- *Smart clay*: These are ensembles of MEMS arranged into arbitrary three-dimensional shapes. The Claytron-ics project [9] from Carnegie Mellon University and Intel Research-Pittsburgh explores how a collection of millimeter-scale devices can be used as a programmable system so that it reproduces moving 3D physical objects.

These six form factors, however, are not enough, as they do not include everyday objects with arbitrary forms which have been augmented with the ability to process and exchange digital information. Thus, an arbitrary form factor must be added to the classification.

Another classification can be done according to the technology used in the smart devices, as described briefly below:

- *Passive devices*: These systems do not have a power source of their own; therefore, they have to receive it from other element. The most typical configuration uses a passive tag (a passive RFID tag or a barcode) on the object, and a reader which provides the power source. Usually the tag is used just for identification. The reader is responsible for connecting to an infrastructure that offers services according to the identification of the object. As the object generally does not include sensors, properties different from its identification, such as position or temperature, have to be inferred from the properties of the reader. The background infrastructure converts the object in a smart object.
- *Active devices*: This category has a power source. Typically, active devices are more complex than passive devices as they include, in addition to sensors, computation, storage and communication subsystems. Therefore, they can collaborate with other smart devices. As the sensors are embedded in the device, the information about their properties can be more accurate. In this approach, the devices are less dependent on a background infrastructure in order to provide services.

Smart objects can also be classified according to their main purpose. A smart object, as defined in [10], would be a 'physical device equipped with a processor, memory, at least one network connection, and various sensors/actuators, some of which may also correspond to human interfaces;' this would include, for instance, a PDA and a smart umbrella. On the other hand, a smart everyday object, as defined by Siegemund [11], is 'an arbitrary object from our everyday environment, such as a chair, a hammer, a car, or an umbrella—augmented with information technology;' this would be a more restrictive category and, thus, exclude a PDA and similar objects that are conceived mainly as technological devices where their computation features are explicit from the beginning. This distinction is also fundamental for the artifacts computing model presented in [12], as it emphasizes the aim of moving processing to the background of familiar activities and objects instead of introducing new devices. Thus, two categories can be defined according to the main purpose of the object: smart technological devices and smart everyday objects.

Both of these categories can be subdivided further. Technological devices can be classified according to many characteristics, but from the point of view of pervasive computing, probably the most important is their mobility; thus, there are fixed and mobile devices. On the other hand, the best criterion for classifying smart everyday objects is their main functionality, i.e. what kind of everyday object they are.

Figure 13.2 summarizes the proposed taxonomy of smart devices. As an example, the taxonomy can be used to classify two smart objects: a smart phone and a Mediacup [12]. The smart phone would have a tab form factor,

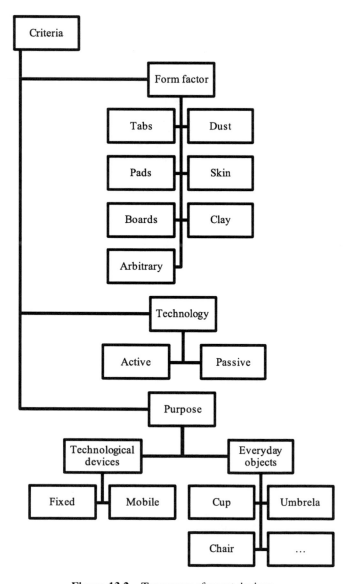

Figure 13.2   Taxonomy of smart devices.

use active technology and be a mobile technological device. A Mediacup would have an arbitrary form factor, use active technology and be a smart everyday object acting as a cup.

It should be noted that different authors use different definitions for the same term. For instance, Poslad in [6] considers that smart devices are multi-purpose information and technological devices, such as personal computers or mobile phones, and distinguishes them from the devices present in a smart environment, which he characterizes as having only one function; so he would not consider a smart everyday object to be a smart device. Baeg *et al.* [13] use another terminology: passive objects with an RFID tag are called smart items, objects with sensor or actuator capabilities are called smart objects, and electrical equipment or products having their own CPUs and communicating modules are called smart appliances. In this chapter, a broad vision has been chosen and any object that can exhibit smart capabilities is called a smart device.

13.3   Intelligent Environments

13.3.1   Definition and Components

The concept of intelligent environments is already present in Weiser's paper on ubiquitous computing. In the examples that he provided there are homes and offices which understand the needs of their inhabitants and adapt to them.

An intelligent or smart environment, as defined by Das and Cook [14], is an environment that is able to acquire and apply knowledge autonomously about the environment and adapt to its inhabitants' preferences and requirements in order to improve their experience. In this definition, there are two concepts of special significance. First, the need for acquiring knowledge about the environment, which results in a need for context information to be one of the main components in intelligent environments, and also requires some kind of instrumentation in order to gather information from the physical world. Secondly, the need to apply knowledge autonomously, which requires some kind of artificial intelligence.

Sensing is carried out by using smart devices with sensors, or by inserting instrumentation into the environment, such as, for example, using cameras to capture the situation of people or objects in a room. The information gathered gives the system the context in which it is working. There has been a great deal of work on context-aware systems which is reviewed in [15].

Sensing technology is one of the foundations of intelligent environments and, therefore, many types of sensors are used for a variety of physical situations, which can be position, light, sound, pressure, velocity, direction, presence of gas or chemicals, amongst others.

Although the first work on intelligent environments was carried out outside the artificial intelligence (AI) community, several authors have emphasized the need for AI in intelligent environments. For instance, in 1998, Coen [16] pointed out that AI-based approaches have much to offer to these environments. The term 'ambient intelligence' (AmI) was introduced in the late 1990s, and received a big push from the European Commission's Information Society Technologies Advisory Group (ISTAG) [17]. As a result, the term is used mainly in Europe, while in other regions it is more common to refer to similar developments as 'smart environments' or 'intelligent environments.' Ramos *et al.* in [18] define AmI as a digital environment that proactively, but sensibly, supports people in their daily lives, and propose that AmI is the next step for AI. They survey the fields where AI techniques can help in intelligent environments:

- *Interpreting the environment's state*: The information gathered from the sensors has to be analyzed. This includes communicating with the user through the most common interfaces for humans: sound and vision. Here, the research carried out in AI about speech recognition and synthesis, natural language processing and computer vision plays an important role.
- *Representing the information and knowledge associated with the environment*: As incomplete information is obtained, knowledge representation and reasoning is a fundamental feature. Techniques from AI research, such as Bayesian networks, fuzzy logic, data mining or ontologies can be used.
- *Modeling, simulating and representing entities in the environment*: This can be done using intelligent agents, which has been one of the main paradigms in AI.
- *Planning decisions or actions*: As the system must react to the user's preferences, deciding which action has to be taken is a required step. This has been another classical problem studied in AI. In addition, optimization problems

often arise in intelligent environments and techniques coming from the AI field, such as genetic algorithms or simulated annealing, can be used to address this problems.

- *Learning about the environment*: In order to adapt to different users, they must be observed and their preferences ascertained. Here, machine learning techniques from AI are very useful.
- *Interacting with humans*: In addition to interacting through natural language and gestures, intelligent environments must take into account social and emotional factors, such as those explored by AI research on affective computing and social computing.
- *Acting in the environment*: Research on cognitive-robotics can provide the elements required to act in the environment by means of intelligent robots.

The components of an intelligent environment [19] are depicted in Figure 13.3. Sensors gather information from the environment using physical devices. Hardware and software interfaces are defined in order to abstract the details of the devices. The communication layer sends the environment data to a database and other components process this raw data, obtaining more useful information such as interaction models, patterns, etc. This new information is presented to the decision layer, where its components make decisions according to rules. These decisions are incorporated to the information layer and transmitted by the communication layer to the actuators, which change the physical world.

13.3.2   Taxonomy

As with smart devices and systems, several orthogonal criteria will be used to define a taxonomy for intelligent environments.

First, there are two main approaches for obtaining smart environments: using special objects with instrumentation or placing the instrumentation in the environment and using everyday objects without any instrumentation in themselves. These are extreme positions and there are intermediate options according to the degree that instrumentation is used in the objects. The first approach relies on smart objects and is more in line with the vision proposed by Weiser of having computers everywhere. The second approach comes from the AI community [16] and relies on sensors, such as cameras and microphones, placed in the environment so that there can be intelligent interaction with normal objects.

Apart from instrumentation, another component of an intelligent environment that can be distributed in different ways is the required intelligence; it can be located in a smart device (or shared between smart devices as the ad-hoc smart spaces defined in [20]), in a background infrastructure, or in both.

Another classification can be made taking into account the different kinds of location awareness that an intelligent system uses. In [21], the following categories have been proposed:

- *Absolute location awareness*: The intelligent system knows the position of entities (objects or users) in a coordinated system. It can be a geographic coordinated system using, for instance, GPS.
- *Space awareness*: The intelligent system knows that an entity is in a certain space such as a room, a building, etc.
- *Proximity awareness*: The intelligent system includes objects that are aware of their proximity to other objects.
- *Transition awareness*: The intelligent system is aware of entities' transitions between spaces.

Finally, there is a great variety of intelligent environments according to the main purpose of the environment: offices, homes, classes, cars, shops, laboratories, etc. The kind of intelligence required in each one might be very different since they are guided by very different interactions and expectations from the user; let alone the different objects present in the environment.

Figure 13.4 summarizes the taxonomy proposed. As shown in the example, the intelligent room presented in [16] places all of the instrumentation in the environment. The intelligence is in the background infrastructure and the system has proximity awareness. The purpose of the system is to be used as a meeting room in an office. On the other hand, the work in [22] presents a smart kindergarten where the instrumentation is located in both the objects (toys) and the environment. The intelligence is in the background infrastructure (although the objects may have computing capabilities and therefore, some intelligence), which presents space and proximity awareness. The purpose is educational.

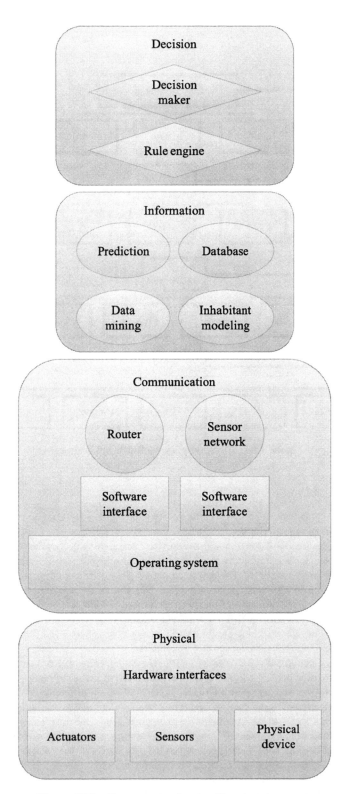

Figure 13.3   Components of an intelligent environment.

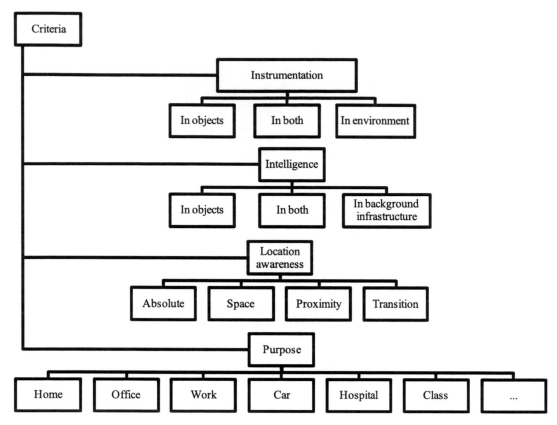

Figure 13.4 Taxonomy for intelligent environments.

13.4   Trends

There are several trends that have been observed in the field of smart devices, systems and intelligent environments; some of them moving in opposite directions.

On the one hand, there is a trend for miniaturization, with research projects related to smart motes or programmable matter [9]. On the other, ubiquitous smart devices of considerable size have emerged from the telecommunications industry: the smart phones [8]. The first smart phones provided computing, storage and communication capabilities that are in common with smart devices, but they lacked sensing capabilities. These are being added in newer devices, where it is not uncommon to find an accelerometer, a GPS, a temperature sensor or a light level sensor, while other components, such as RFID, are provided as add-ons. In addition, users are familiar with the devices. Hence, smart phones are becoming part of smart systems such as health and fitness monitoring, crisis and emergence response, or mobile environment sensing systems.

On the software front, a movement towards Service Oriented Architectures (SOA) can be observed [8, 23]. This technology, which can be used in the background infrastructure and in the smart objects themselves, enables the integration of different devices in a common intelligent system. However, different capabilities will be present in the background infrastructure and in the devices, and thus some kind of middleware would still be needed.

Intelligent environments that take into account the psychological state of the user are a necessity and an emerging reality. The necessity arises from the nature of intelligent environments, where the assumption that the user is in the correct disposition to handle a tool – as happens when the user works with a traditional computer – cannot be made. This requires that in many situations, such as for example when the user is driving a car, the smart system must accommodate to the user's needs.

The character of intelligent environments, so different from traditional computing, demands new human-computer interaction patterns. Interfaces that go beyond keyboards, mice and monitors are needed. The features that these new interfaces require [24] include: (a) adapting to the environment and to the user's intentions or situation; (b) learning from experience; and (c) the ability to respond to voice speech, gestures or other senses. All of these capabilities rely heavily on AI techniques. In addition, a trend of combining several modes of interaction in multimodal interfaces can be observed. It is a step towards a more natural interaction with intelligent computers so as to achieve Mark Weiser's vision of invisible computing.

13.5   Limitations and Challenges

Although the field of smart devices, systems and intelligent environments has seen enormous advances, there are still limitations and challenges that will have to be addressed in future research.

One of the main limitations for smart objects is energy consumption. Smart objects that use active technologies require energy to operate. In a vision where the environment is populated by smart objects, it is not reasonable to expect users to change the batteries of tens or hundreds of objects; this would defeat the purpose of pervasive computing as an invisible form of computing. Therefore, smart objects should be shipped with batteries that last an entire life time or they should be able to obtain energy from the environment.

Miniaturization, as discussed in the previous section, is one of the trends for smart devices, but it is also one of its challenges. As shown in [25], new technologies are needed to obtain a system-on-a-chip mote, i.e. a single-chip sensor mote that integrates computing, communication and sensing without requiring off-chip components and without sacrificing performance. Some of the technologies for wirelessly enabled applications include reliable data delivery, low power consumption, security and location discovery.

Economic factors have to be taken into consideration when developing smart systems. Normally, greater capabilities in the devices mean an increase in price. For instance, one of the reasons that make barcodes more common than active RFID tags is that the latter are more expensive. These economic factors have to be addressed in order to expand the presence of smart devices.

In intelligent environments such as smart homes, tracking the location of the inhabitants is a very important feature [14], because it enables offering services adapted to the user's position and preferences. An optimal strategy exists for the case with only one inhabitant, but the problem of optimal location tracking of multiple inhabitants is an NP-hard problem, that presents many limitations, such as not accounting for interaction between inhabitants and conflicting goals.

In general, increasing the intelligence of smart environments is an ongoing challenge. This expansion of intelligence can involve elements of human to computer interface, such as gesture recognition, and activities such as pattern recognition in people's actions in order to develop user models and create proactive plans designed to satisfy the user's desires.

Much of the research done in ambient intelligence has considered each environment (each house, each hospital, each classroom, etc.) as a separate system [19]. However, from the point of view of the users, they are all part of a bigger system, namely, the world in which they are living. This leads to many problems. First, different protocols, user interfaces and devices may be needed in each environment, and this can cause confusion to the user. In addition, the synergies that could be obtained from sharing information between all of these environments are lost. Therefore, integrating different intelligent environments remains a great challenge in the field of pervasive computing, a challenge that involves developing standard architectures, protocols, interfaces and devices.

However, it must be noted that a greater integration of intelligent systems and environments would increase the challenges of pervasive computing: providing services without creating privacy worries. These challenges have been present from the inception of the ubiquitous computing research, as the pervasive and invisible nature that is sought make the user's privacy vulnerable.

Finally, as several technologies, architectures and strategies are proposed by different researchers, a set of metrics that takes into account the complete intelligent environment would be desirable in order to have an easy way of comparing these different approaches. There has been work done in evaluating ubiquitous and pervasive systems [26], mainly in the area of user experience, but many questions still remain open.

13.6   Applications and Case Studies

In this section, several applications of smart devices, systems and intelligent environment are presented. There has been a great amount of work in the field since the late 1990s, so a selection had to be made. Several criteria have been used to choose the case studies presented. First, each selected case study presents a different purpose, in order to provide a general vision of the applications of intelligent environments. Some case studies have been selected because of their value as influential projects in the subject as they have been around for several years and the technology has advanced since then; other more recent case studies are also presented. This helps in illustrating the evolution and the trends that the field is experimenting with.

Applications whose main goal is related to aspects studied in other parts of this book, such as healthcare or education, have been discarded in order to avoid duplication.

13.6.1   Smart Everyday Objects

The Mediacups project [12] is an exploration of ubiquitous computing where the focus was on augmenting everyday objects with computing, communication and sensing capabilities. It incorporates all of these capabilities in a regular cup, which could then infer its context (full, moving, etc.) and communicate it to a background infrastructure using infrared communication. Other smart objects were developed in the project, such as a smart door plate that displays a message indicating that there is a meeting anytime it detects co-located hot cups in the room, or a PC-based watch that enables reading the context of the Mediacup.

The generalization of the ideas of the Mediacup project is the Smart-Its project [27], a European project that ran from 2001 to 2003. Its aim is developing a set of tools that make it easy to integrate the software and the hardware required for enhancing everyday objects easily.

To achieve this goal, small computers called Smart-Its were developed. They can be attached to any everyday object. The Smart-Its hardware was made modular in order to ensure flexibility. It consists of a core board with a wireless transceiver plus a sensor board. A standard sensor board includes sensors for light, sound, pressure, acceleration and temperature. Other sensors, such as a gas sensor or a camera, can be added if required.

Several APIs to aid application development were also developed. These include a communication API that allows different Smart-Its to share information, and a perception API that transforms low-level sensor data to higher-level concepts.

Smart-Its is aimed at creating easily prototypes as a research platform in ubiquitous computing without having to create a large amount of hardware and software from scratch; however, some sample applications were developed within the Smart-Its project. In a proactive furniture assembly system, a Smart-Its device was attached to the unassembled components, so that the next action in assembling is suggested to the user, taking into account the components' position. In a supporting avalanche report system, wearable Smart-Its sensors indicate the position and vital signs of its carrier, so that if there are several victims of an avalanche, rescue teams can use this information in order to save the most urgent first. As a final example, in a smart restaurant, Smart-Its was attached to an oyster box in a refrigerator, so that its state can be traced and its best-before date can be shown dynamically and used to propose that the oysters could be reduced in price if they were close to expiration and it was known that a box of fresh oysters was coming.

13.6.2   Smart Home

Since automation in the home environment is increasing for elements such as air conditioning, lights, multimedia or surveillance systems, homes have been one of the most active research fields for intelligent environments. One of the applications where the research has been most prominent is in the combination of smart homes with telemedicine [28] for improving the health-care of elderly and people with reduced physical functions.

From the several smart home projects that have been developed [19], the MavHome [29] has been chosen as a case study, as it focuses on the entire environment management and not just on a single area.

In MavHome architecture, the four abstract layers presented in Figure 13.3 are realized by a series of concrete layers. At the bottom is the Physical Components layer, which includes all the real devices of the system (touch screens, cameras, etc.), with the exception of the computer, which manages all of the hardware. This computer (or computers) and its associated network is the host of the rest of the layers. The hardware interfaces to physical devices (USB, Firewire, PCI, etc.); the device drivers, the operating system of the computer and the software interfaces that provide access to hardware are located in the Computer Interface layer. Next, the Logical Interface layer contains a series of basic services that act as logical proxies for each sensor and actuator in the system, providing access to their information and control interfaces via sockets and shared memory. The Middleware layer provides services to facilitate communication, process mobility and service discovery. The Service layer aggregates information obtained by the middleware layer and provides the aggregated information to the Applications layer.

The Service layer includes a database, user interfaces, prediction components, data mining components and logical proxy aggregators. Finally, the Applications layer encompasses all of the components for learning, decision making and arbitration.

One of the most important features of the MavHome is intelligent lightning control. For this, it uses lamps and appliance modules based on the X-10 home-automation standard. There are more than 60 controlled devices.

A sensor network that includes light, switches, humidity, temperature, smoke, gas and motion sensors is deployed. It is used for a key element in intelligent environments: inhabitant location.

The intelligent aspect of the MavHome is realized via a prediction module. It predicts the next action of its inhabitants using previously observed interactions of the inhabitants with the different devices. The goal is to predict accurately in real-time the repetitive actions of its inhabitants, so that their comfort is maximized and the consumption of resources is minimized, while preserving safety and security. In experimentation with one inhabitant during one month, it has been shown that 18 lighting interactions can be reduced on average to just five interactions.

13.6.3   Smart Office

The office environment has been one of the first to receive the benefits of traditional computing and it has also been identified as a perfect target for pervasive computing systems. The goal is obtaining an increase in productivity by handling the large amounts of information and interactions that feature in an office.

Bagci et al. present in [30] a mobile agent paradigm implemented in a smart office environment. Touch screens are mounted beside each door within an office building. These smart doorplates replace traditional doorplates and provide users (employees and visitors) with several services. One of these is a navigation system: the user asks a doorplate for directions to an office; the system shows an arrow pointing in the correct direction; as the user approaches other doors, new arrows indicate the direction to the destination. Another service makes it possible for users to be notified in locations separate from their PCs when e-mails that fulfill a specified filter are received. In a similar fashion, a file-reader agent reads out a file which the employee selects and shows it on a doorplate, and a file-fetcher agent brings the specified file to the doorplate and leaves it in a public folder.

In order to carry out these functions, the paradigm of mobile agents is used, defining two types of agents. The user-agents contain information about the user (name, office room, etc.), security data (public and private keys, user names, passwords, etc.) and context-information. The service-agents fulfill the actions indicated by the user-agents.

The users carry a tag with infrared sensors and a radio chip in order to track their position. The user-agents reside in the environment, i.e. in the computer system formed by the smart doorplates, which are realized with PC technology connected by an Ethernet network. User-agents can receive location messages and migrate to a host nearer to the user.

The mobile agent system is implemented in the UbiMAS framework, which has three layers. The base layer consists of the operating system and the network, the second layer is the middleware layer and the top layer is the services layer. Each node starts at least an UbiMAST service in order to host agents. The middleware is in charge of communications between the agents and the host and between agents.

Security is an important feature of the UbiMAS framework, realized by encrypted messages and a thread based security manager.

This system demonstrates how the concept of ubiquitous computing by moving data to the location of the user in a seamless way can be realized.

13.6.4   Smart Room

One of the first intelligent environments to be developed was the Intelligent Room [16] at the MIT Artificial Intelligence Laboratory. It was created as a platform for Human-Computer Interaction research that connects the real-world to computers by means of cameras and microphones.

In a more recent example of a smart room [31] at the Technical University of Catalonia (UPC), Spain, the room was created in order to serve as an experimentation environment for multimodal interaction and to collect data for further research. The technologies that can be tested give an idea of what can be done with these systems. Audio technologies include Automatic Speech Recognition, Speaker Identification, Speech Activity Detection, Acoustic Source Localization, Acoustic Event Detection and Speech Synthesis. Video is used for Multi-Camera Localization and Tracking, Face Detection, Face ID, Body Analysis and Head Pose Estimation, Gesture Recognition, Object Detection and Analysis, Text Detection and Global Activity Detection. Moreover, multimodal approaches with video and audio provide Person Identification and Person Location and Tracking.

The room is equipped with a network of sensors consisting of 85 microphones and eight cameras of different capabilities distributed so that they can be used for different functions. The room also includes a speaker system so as to interact with its users.

In addition to the aforementioned recognition, tracking and detection technologies, artificial intelligence is present in the form of a Question and Answer engine that provides answers provided by a database that can be adapted to different domains.

An application developed specifically for this room is the Memory Jog. It helps people in the room by providing background information and memory assistance. It can be used, for instance, in educational scenarios to help students in solving problems. The information can be provided on request or proactively; the latter requires knowing when it can be given without disturbing the user, which requires IA in the form of context-awareness.

Smart rooms show how multimodal interfaces can work in helping users and are also an example of introducing technology in the background infrastructure without requiring smart objects.

13.6.5   Supply Chain Management

In supply chain management, adding RFID tags to products can improve the cost effectiveness of the process involved, from manufacturing to disposal, including distribution and warehousing.

In [32], the benefits that an intelligent product driven supply management can offer are analyzed. An intelligent product is defined by five characteristics: (a) unique identity, (b) communication with the environment, (c) data stored about it, (d) displaying information about itself, and (e) participating in decisions relevant to its own destiny.

Depending on how many of these characteristics a product has, two levels of intelligence are defined. The first level covers characteristics (a) to (c), so that the product is capable of sharing its identity and data about itself with the environment. This can be carried out by means of RFID tags and results in an information oriented product. The second level covers all of the characteristics; results in a decision oriented product, and requires, in addition to a RFID tag, software agents to implement characteristics (d) and (e). The software agent corresponding to an intelligent product is located in the background infrastructure, so that it is able to communicate with other software agents and with the intelligent product to gather information and to send decisions.

Functionalities that can be fulfilled with a level 1 intelligent product include obtaining product status information, product tracking and product history access. A level 2 intelligent product has more functionality, such as triggering its replenishment when inventory levels reach a certain minimum point, modifying the picking planning based on status and product tracking, or collaborative inventory management between different distribution centers.

These functionalities impact on all the levels of supply chain management, providing new applications, such as authentication of returned products, stock rotation based on dynamic due-by-dates, automated proof of delivery and improved quality assurance levels or self-organizing production schedules.

Apart from many other functionalities and applications, [32] also proposes strategic changes in supply chain management that can be achieved with intelligent products, such as enabling companies to change from selling a product to selling a service.

A more recent approach, using service oriented architectures, is shown in [23], where web services are integrated on the floor shop in order to communicate floor shop devices and enterprise applications.

13.6.6   Smart Car

People spend a great amount of time in transportation systems and, in some parts of the world, cars are the most important means of transport. Modern cars are highly computerized systems, with CPUs, sensors, software and communication. In addition, the road infrastructure is becoming increasingly instrumented. All of this presents an opportunity to convert cars into intelligent environments.

In [33], an In-Vehicle Ambient Intelligent Transport System (I-VAITS) is proposed. It would use three sources of information: driver's health (attention, cognitive workload, psychological state and electronic health records), environment (crash data about road accidents, traffic information) and car (movement, repairs or fuel consumption). The information from the driver would be gathered by sensors such as bending sensors and hand pressure on the wheel. The middleware would analyze all of this information using context-awareness and data-mining techniques. It would then execute actions adapted to the context; helping the driver when dangerous or complex situations arise.

One of the main problems is assessing when information can be presented, as safety has to be the first concern. This is another example of the great importance of context-awareness in intelligent environments.

Some of these ideas are exemplified in [34], where a multimodal interface for an augmented driver simulation is presented. The simulator, using video data and biological signals, evaluates the attention level of the drivers and detects their fatigue. Analyzing video images of the driver's face and interpreting them enables the detection of gestures such as yawning, eye blinking or head rotation. The stress level of the driver is derived from an electrocardiogram and galvanic skin response. This information is used in the simulator to send signals to the driver (audio and visual messages, in addition to wheel vibrations) when fatigue is detected.

This example emphasizes the importance of multimodal interfaces and integration with other sciences, such as biology and psychology, in order to create more advanced systems.

13.6.7 Smart Laboratory

In natural science laboratories, experiments regarding the physical world are carried out, but the data obtained must be analyzed in the digital world. This means that there is an opportunity to use smart devices that transform laboratories into smart environments.

The most well-known project in this field is Labscape [35], which is a smart environment developed by the University of Washington and Intel to improve the experience of researchers in a cell biology laboratory.

Experiments in this type of laboratory consist of carrying out a series of processes with cells, annotating the observed results and analyzing them. The goal of the Labscape project is improve the efficiency of these mechanisms.

It is interesting to note that laboratories like these, before conversion into smart environments, do not demonstrate many traditional computing elements, as the work carried out is mostly physical and traditional computers are not suited to this environment, so traditional tools such as paper and pencil are used to make notes of the results of experiments.

One of the most interesting aspects of the project is that, in the beginning, it tried to use sensors and recognizers to obtain the data directly from the laboratory instruments in order to avoid wasting the biologist's valuable time. This approach failed because sensor technologies suited to the environment could not be identified and because the process of error correction in the recognition system was a new and difficult task.

The new approach consists of using directed acyclic graphs (DAG) to organize the activities carried out in the laboratory, and a combination of the biologists' personal computers on their desktops, combined with tablets-PC, USB numeric keypads and small wireless IR keyboards available in the shared work area; all of these are connected through wired and wireless networks.

The biologist completes data in the DAGs while carrying out the experiment. This offers the benefit of being able to change the plan, achieving better process tracking, and having more detailed information in the required format, which helps in the documentation phase.

In order to track location, an active badge system is used, but in combination with a traditional interface for correcting errors or for use in the absence of the active badge.

According to the researchers in the Labscape project, one of the important lessons that can be learned from their experience is that 'UI (User Interface) precedes AI,' that is, the good design of the user interface enables the gradual integration of more advanced AI technologies. In addition, user values are very important to the outcome of the system and in some cases, as in biology research where the recorded data are more important than the user experience, this favors using a more explicit interface.

13.6.8 Smart Library

The librarian robot system described in [36] presents several ubiquitous computing technologies. The goal of the system is to create an environment where the user can give books to a robot so that it arranges them in their correct place in the bookshelf.

In order to achieve this goal, the robot has to know where the book and the bookshelf are. To locate them, RFID tags are attached to the book, the bookshelf and the floor, so that the robot can navigate through the physical space. Additionally, a web application that uses information from the RFID tags in bookshelves shows the user the status of each slot in the shelves.

The librarian robot has several sensors so as to be able to take a book from a table, recognize its information and place it in the shelf. In addition to RFID readers, a vision system is used to recognize a book on a table and in the rack attached to the robot, where several books can be located, and a force sensor is attached to the gripper used for book manipulation.

The robot, the floor and the bookshelf are connected with a background infrastructure, where a database with information about the books is located. Communication is achieved by means of a middleware system based on web services.

This example demonstrates the use of several of the key technologies in pervasive systems, such as RFID tags, system oriented architectures and AI.

13.7   Conclusion

Smart devices and systems incorporate intelligence in environments, so that the vision of pervasive computing is realized. This chapter has analyzed the main components of smart devices (processing, memory, communication, sensing and actuating) and has proposed a taxonomy using three criteria: form factor, technology and purpose. The components of intelligent environments that are grouped into four layers (physical, communication, information and decision) have been described. Intelligent environments have been classified according to four criteria: where the instrumentation is placed, where the intelligence is localized, the location awareness that the system exhibits and its purpose.

The main trends that have been observed are, on the one hand, the march towards miniaturization with research projects such as Claytronics and, on the other, the success of the smart phone as a ubiquitous device. In addition, the nature of intelligent environments has inspired new interfaces, with multimodal interaction. These are gaining in importance and at the same time, new characteristics of the users, such as their psychological state, are taken into account.

The main limitations and challenges in intelligent environments have also been reviewed. These include handling the power consumption of smart devices, achieving more miniaturization, decreasing the cost, being able to track the location of several inhabitants in an environment, increasing the intelligence of the systems and integrating different intelligent environments, all of this taking into account the privacy concerns that the users can raise. In addition, metrics to compare different systems are required.

Finally, a view of what has been done and what the field can offer has been presented; describing several applications and case studies, which range from smart homes to smart laboratories, and include smart everyday objects, smart cars and smart offices.

References

[1] M. Weiser (1991) 'The computer for the twenty-first century,' *Scientific American* **265**(3): 94–104.
[2] F. Mattern (2001) 'The vision and technical foundations of ubiquitous computing,' *Upgrade* **2**: 2–6.
[3] M. Satyanarayanan (2001) 'Pervasive computing: vision and challenges,' *Personal Communications, IEEE* **8**: 10–17.
[4] F. Siegemund, C. Floerkemeier, and H. Vogt (2005) 'The value of handhelds in smart environments,' *Personal Ubiquitous Computing* **9**: 69–80.
[5] C. Thompson (2005) 'Smart devices and soft controllers,' *Internet Computing, IEEE* **9**: 82–5.
[6] P.S. Poslad (2009) *Ubiquitous Computing: Smart Devices, Environments and Interactions.* John Wiley & Sons.
[7] B. Warneke, M. Last, B. Liebowitz, and K. Pister (2001) 'Smart dust: communicating with a cubic-millimeter computer,' *Computer* **34**: 44–51.
[8] R. Bose (2009) 'Sensor networks – motes, smart Spaces, and beyond,' *IEEE Pervasive Computing* **8**: 84–90.
[9] S. Goldstein, J. Campbell, and T. Mowry (2005) 'Invisible computing: programmable matter,' *Computer* **38**: 99–101.
[10] C. Kintzig, G. Poulain, G. Privat, and P. Favennec, eds (2003) *Communicating with Smart Objects: Developing Technology for Usable Pervasive Computing Systems*, Kogan Page Science.
[11] F. Siegemund (2004) *Cooperating Smart Everyday Objects – Exploiting Heterogeneity and Pervasiveness in Smart Environments*, University of Rostock, Germany.

[12] M. Beigl, H. Gellersen, and A. Schmidt (2001) 'Mediacups: experience with design and use of computer-augmented everyday artefacts,' *Computer Networks* **35**: 401–9.
[13] S. Baeg, J. Park, J. Koh, K. Park, and M. Baeg (2007) 'Building a smart home environment for service robots based on RFID and sensor networks,' In *Proceedings of the International Conference on Control, Automation and Systems, 2007. ICCAS '07*, pp. 1078–82.
[14] S. Das and D. Cook (2006) 'Designing and modeling smart environments,' In *Proceedings of the International Symposium on a World of Wireless, Mobile and Multimedia Networks. WoWMoM 2006*, pp. 490–94.
[15] J. Hong, E. Suh, and S. Kim (2009) 'Context-aware systems: A literature review and classification,' *Expert Systems with Applications* **36**: 8509–22.
[16] M.H. Coen (1998) 'Design principles for intelligent environments,' In *Proceedings of the Fifteenth National/Tenth Conference on Artificial Intelligence/Innovative Applications of Artificial Intelligence*, Madison, Wisconsin, US, pp. 547–54.
[17] K. Ducatel, M. Bogdanowicz, F. Scapolo, J. Leijten, and J. Burgelman (2001) *Scenarios for Ambient Intelligence in 2010*, European Commission.
[18] C. Ramos, J.C. Augusto, and D. Shapiro (2008) 'Ambient intelligence – the next step for artificial intelligence,' *IEEE Intelligent Systems* **23**: 15–18.
[19] D.J. Cook, S.K. Das (2007) 'How smart are our environments? An updated look at the state of the art,' *Pervasive and Mobile Computing* **3**: 53–73.
[20] A. Brodt, S. Sathish (2009) 'Together we are strong – towards ad-hoc smart spaces,' In *Proceedings of the 2009 IEEE International Conference on Pervasive Computing and Communications, IEEE Computer Society*, pp. 1–4.
[21] K.M. Dombroviak and R. Ramnath (2007) 'A taxonomy of mobile and pervasive applications,' In *Proceedings of the 2007 ACM Symposium On Applied Computing*, Seoul, Korea, pp. 1609–15.
[22] M. Srivastava, R. Muntz, and M. Potkonjak (2001) 'Smart kindergarten: sensor-based wireless networks for smart developmental problem-solving environments,' In *Proceedings of the 7th Annual International Conference On Mobile Computing and Networking*, Rome, Italy, pp. 132–8.
[23] L. de Souza, P. Spiess, D. Guinard, M. Köhler, S. Karnouskos, and D. Savio (2008) 'SOCRADES: a web service based shop floor integration infrastructure,' *The Internet of Things*, pp. 50–67.
[24] K. Karpouzis, J. Soldatos, and D. Tzovaras (2009) 'Introduction to the special issue on emerging multimodal interfaces,' *Personal and Ubiquitous Computing* **13**: 1–2.
[25] B. Cook, S. Lanzisera, and K. Pister (2006) 'SoC issues for RF smart dust,' In *Proceedings of the IEEE* **94**: 1177–96.
[26] K. Connelly, K. Siek, I. Mulder, S. Neely, G. Stevenson, and C. Kray (2008) 'Evaluating pervasive and ubiquitous systems,' *Pervasive Computing, IEEE* **7**: 85–8.
[27] L.E. Holmquist, H. Gellersen, G. Kortuem, A. Schmidt, M. Strohbach, S. Antifakos, F. Michahelles, B. Schiele, M. Beigl, and R. Mazé (2004) 'Building intelligent environments with Smart-Its,' *IEEE Computer Graphics and Applications* **24**: 56–64.
[28] M. Chan, D. Estève, C. Escriba, and E. Campo (2008) 'A review of smart homes – present state and future challenges,' *Computer Methods and Programs in Biomedicine*, **91**: 55–81.
[29] G. Youngblood, L. Holder, and D. Cook (2005) 'Managing adaptive versatile environments,' In *Proceedings of the Third IEEE International Conference on Pervasive Computing and Communications, 2005. PerCom 2005*, pp. 351–60.
[30] F. Bagci, H. Schick, J. Petzold, W. Trumler, and T. Ungerer (2006) 'The reflective mobile agent paradigm implemented in a smart office environment,' *Personal and Ubiquitous Computing* **11**: 11–19.
[31] J. Neumann, J.R. Casas, D. Macho, and J.R. Hidalgo (2009) 'Integration of audiovisual sensors and technologies in a smart room,' *Personal and Ubiquitous Computing* **13**: 15–23.
[32] C. Wong, D. McFarlane, A. Ahmad Zaharudin, and V. Agarwal (2002) 'The intelligent product driven supply chain,' In *Proceedings of the 2002 IEEE International Conference on Systems, Man and Cybernetics* **4**: 393–8.
[33] A. Rakotonirainy, R. Tay (2004) 'In-vehicle ambient intelligent transport systems (I-VAITS): towards an integrated research,' In *Proceedings of the The 7th International IEEE Conference on Intelligent Transportation Systems*, pp. 648–51.
[34] A. Benoit, L. Bonnaud, A. Caplier, P. Ngo, L. Lawson, D. Trevisan, V. Levacic, C. Mancas, and G. Chanel (2009) 'Multimodal focus attention and stress detection and feedback in an augmented driver simulator,' *Personal and Ubiquitous Computing* **13**: 33–41.
[35] L. Arnstein, C. Hung, R. Franza, Q.H. Zhou, G. Borriello, S. Consolvo, and J. Su (2002) 'Labscape: a smart environment for the cell biology laboratory,' *IEEE Pervasive Computing* **1**: 13–21.
[36] B.K. Kim, T. Sugawara, K. Ohara, K. Kitagaki, and K. Ohba (2008) 'Design and control of the librarian robot system in the ubiquitous robot technology space,' In *Proceedings of the 17th IEEE International Symposium on Robot and Human Interactive Communication, 2008. RO-MAN 2008*, pp. 616–21.

Part Three

Pervasive Networking and Communications

14

Autonomic and Pervasive Networking[1]

Thabo K. R. Nkwe, Mieso K. Denko, and Jason B. Ernst
Department of Computing and Information Science, University of Guelph, Guelph, Ontario, N1G 2W1, Canada.

14.1 Introduction

The 21st century has seen the vast growth of the Internet and great technological advances. This development is intense and ongoing due to the increased demands of the services and benefits that the Internet has to offer both from a business and a social perspective (e.g. advertising, e-commerce, podcast, entertainment, networking, etc.). The nature of business and social relations is that they have no boundaries (in terms of time and place) and wireless communication structures (WPAs, WWANs, WMANs, etc.) have become an integral foundation designed to facilitate the needs of computing anywhere, anytime, also known as Pervasive computing and networking [1]. Pervasive computing is also known as ubiquitous computing which was defined by Marc Weiser at the Xerox PARC laboratory as 'the method of enhancing computer use by making many computers available throughout the physical environment, but making them invisible to the user' [16]. The goal of pervasive computing is to provide computing everywhere while remaining in the background from the perspective of end users. The paradigm of pervasive computing can also be applied to networking, since The Internet is available virtually everywhere with mobile and wireless communication networks. Rather than just having computing capabilities everywhere, it is now possible to have communication everywhere. Since the original definition of pervasive computing we have also seen the explosion in popularity of portable technological devices such as PDAs, cell phones, laptops, virtual input gadgets, iPods, pocket PCs and many more, together with sophisticated wireless technologies (ZigBee, Bluetooth, WI-Fi, Wi-MAX, IrDA, etc.) to enable their operation and utilize their capabilities. An example of Ubiquitous/pervasive architecture with some of these technologies can be seen in Figure 14.1. The addition of new devices and networking technologies enables a new level of pervasive computing that brings its own unique problems. For a more in-depth history on Autonomic Networking consult [18].

Although pervasive computing adds features to the network and has literally brought convenience and services to the users' finger tips, the technologies employed need to co-exist with the current infrastructure. This has led to increased network size, complex network structures, device heterogeneity, high maintenance costs and performance issues. These devices have to interoperate and the technologies must integrate. In a shared data network environment like this, security is also of great concern in order to maintain data integrity, safety and reliability. To further compound the increasing complexities, there is a significant shortage of skilled personnel qualified to handle this phenomenon. This state of affairs can render networks unmanageable and unusable [2]. Essentially this directs the industry to focus on maintenance and support rather than higher levels of technology.

[1] This research was supported in part by the Natural Sciences and Engineering Research Council of Canada (NSERC) Discovery Grant no: 400139 and the Botswana International University of Science and Technology (BIUST) in conjunction with the Botswana Ministry of Education (MOE).

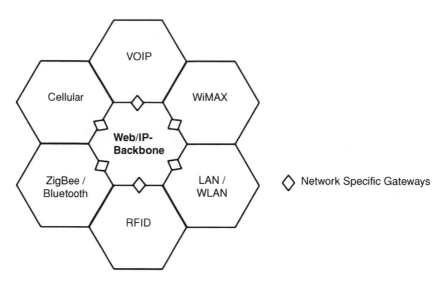

Figure 14.1 Ubiquitous/pervasive architecture (the symbol ◊ represents 'network specific gateways').

To address these issues, researchers have embarked on different research areas so as to facilitate pervasive communication(s) which have prompted the idea of autonomic networking [3, 4]. The fundamental idea behind autonomic networking is to create networks which can self manage, allowing them to cope with the rapid growth of the Internet (and other networks) and their complexities. Self-management must occur without compromising usability, efficiency, flexibility and manageability and at inexpensive maintenance costs. There should also be less or no physical human intervention. The core properties of the autonomic networking framework often referred as self-CHOP are: (a) self-configuration: the ability to adjust and adapt dynamically to changing environmental factors; (b) self-healing: the ability to resist or recover from failures, by anticipating and detecting any component or device malfunctions; (c) self-optimization: the capability to maximize the use and allocation of network resources dynamically and efficiently, for gradual performance improvement during its operation; and (d) self-protection: this enables the network to to defend against, prevent, anticipate, detect and recover from attacks.

To achieve these aims and the seamless integration of networks and technologies, the challenges involve potentially the redesigning of traditional strict layered Open System Interconnection (OSI) architecture [1], which is more suitable for wired than wireless networks.

Wireless networks device composition and mobility introduce reliability and resource availability concerns. For mobility the devices need to be smaller in size for portability and are thus limited in resources such as storage space, computational power, battery size and the like. This also requires lightweight software and applications in order to operate. Network connectivity cannot be guaranteed as mobile network devices can move at anytime, terminating the current link or connection. However, multimedia applications (e.g. gaming, podcasts, streaming, downloading, etc.) are part and parcel of network communications and require substantial power, storage, link stability and reliability. Other applications include pervasive health care systems, commercial systems, security systems, rescue and military systems and many more.

All of these applications, devices, technologies and application scenarios have evolved into different types of ubiquitous/pervasive networks, such as WMNs, WSNs, Opportunistic networks, WMSNs, MANETs, HWNs, etc. The objective of this chapter is to report on the current state of research that addresses autonomic pervasive networks with a focus on self-protection, self-organization and self-management in ubiquitous/pervasive networking and communications.

The chapter is organized as follows. Section 14.2 provides an overview of the architectures and applications of ubiquitous/pervasive networks. Section 14.3 applies autonomic computing principles to ubiquitous/pervasive networking. In this section the benefits of cross-layered protocols are also discussed with respect to autonomic properties. In Section 14.4 there is a detailed discussion on the main self-* properties as applied to ubiquitous/pervasive networking. Section 14.5 concludes the chapter and indicates possible future directions for research.

14.2 Ubiquitous/Pervasive Networks

14.2.1 Introduction

Ubiquitous/pervasive networks can generally be classified into two categories, which can be within the network (LAN) or across networks. A ubiquitous/pervasive LAN is when we have a LAN of dissimilar devices, for example: windows, Linux, Mac PCs. A ubiquitous/pervasive network can also be formed across networks (LANs), by using different technologies, for example; wireless/Ethernet LAN connecting to a cellular network. These networks connect to form a ubiquitous/pervasive WAN. There are many disparities in these types of networks and their properties, which raises performance and QoS concerns. That is, delays might occur during switching of networks, as well as demand for more storage space for buffering during handoffs and packet losses, due to mobility or link quality. Table 14.1 shows a comparison of many wireless access technologies that may be used for pervasive networking. In the following subsections, we will look at ubiquitous/pervasive networks architecture and applications.

14.2.2 Ubiquitous/Pervasive Networks Architecture

Ubiquitous/pervasive network architecture typically has three hierarchies consisting of LANs, WAPs (Wireless Access Point) or RBS (Relay Base Stations) and the Internet/IP backbone. The LANs, which can also be ubiquitous/pervasive, are connected to each other, internally and to the Internet by the WAPs or RBS serving as a gateway.

LAN: a composite LAN is characterized by a variety of network clients, fixed or mobile, interconnected wirelessly or by a physical connection, using a variety of interface mediums. A LAN usually spans a smaller geographical area, for example home, office and small networks. Types include WMNs (Wireless Mesh Networks), PANs (Personal Area Networks), MANETs (Mobile Ad hoc Networks) and many more. Within the LANs makeup, there is a composition of incompatible physical and network layer protocols, network interfaces, network operating systems, cables types and wireless connections.

BS: base stations exist in both computer and radio communication networks. They are often fixed in position and act as a central hub for wireless networks and gateways between wired and wireless networks. There are three types of BS usually categorized in terms of their placement, namely remote BS, relay BS and home BS connecting to an IP backhaul. These include Towers, satellites dish, antennae and many more.

IP Backbone: all of these networks connect to a TCP/IP backhaul, also known as the backbone infrastructure. Over the years the TCP/IP has become a networking standard protocol that has completely replaced protocols such as AppleTalk, IPX and the like. This network backbone is often burdened with management tasks/systems including QoS, security, billing, and network management to connect to the worldwide Internet backbone. Each of the interconnected LANs has their own LAN managers, to address QoS, performance, security and other issues.

14.2.3 Ubiquitous/Pervasive Networks Applications

The reason why pervasive networks should be supported is because of the large range of possible applications. Rather than restricting networks to being homogenous, we standardize the features which must be supported

Table 14.1 Comparison of IEEE access technologies

Technology	Coverage	Data Rate	Mobility	Multi-Hop	Wireless
IEEE 802.3 Ethernet	Single PC	1 Gb/s +	No	Yes	No
IEEE 802.3 Optical	Single PC	10 Gb/s +	No	Yes	No
IEEE 802.15 WPAN	< 10 m	< 55 Mb/s	Limited	No	Yes
IEEE 802.15 WMN	< 10 m/ device	< 55 Mb/s	Limited	Between devices	Yes
802.11n WLAN*	< 200 m	< 600 Mb/s	Limited	No	Yes
802.11s WMN*	< 200 m/ device	< 600 Mb/s	Yes	Between MRs	Yes
802.16 WiMAX	10 s of kms	< 100 Mb/s < 1 Gb/s (F)	Yes	Between BSs	Yes
Cellular (LTE,HSPA)	10 s of kms	< 100 Mb/s	Yes	Between BSs	Yes

*Note: Data Rates are based on proposed PHY layer theoretical maximums for 802.11n.

across the entire pervasive network. This allows networks made up of different technologies to communicate using the basic supported feature set while the individual devices bring their own unique abilities to the network. For example, consider a personal area network equipped with body sensors which provides feedback to a hospital on the condition of patients. This type of application could save the hospital and patients money and be more convenient to the patient. In the military, GPS units could be provided to soldiers and equipment and combined with other wireless transmission technology, allowing for location tracking and planning. This could result in more efficient operations and execution of plans. This could also be extended to include devices with multiple sensors and processing units such as drone devices and robotic soldiers to enable communication and planning between the devices in sophisticated digital armies. The biggest and most time consuming problems with many of these applications are (a) getting the existing standards to work together; (b) configuring the devices properly; and (c) reconfiguration of the devices under changing conditions. This is where autonomic networking principles may be applied to pervasive networking to help keep the technology transparent for the users.

14.3   Applying Autonomic Techniques to Ubiquitous/Pervasive Networks

A considerable fraction of pervasive systems rely on an underlying network of sensing devices for information about their operating environment.

14.3.1   Introduction

Since pervasive networking makes use of many component networks with various technologies, the pervasive network is very complex. It has become difficult to manage and maintain even homogenous networks, so the only solution to make pervasive networking practical is to automate the process of setup, administration and maintenance wherever possible. The topic of autonomic networking has been studied extensively in recent years; however, there are some unique challenges to pervasive computing, in particular around the configuration and interoperation of multiple technologies in a ubiquitous/pervasive network. Table 14.2 provides a comparison of some recent automation applications in pervasive networks in terms of network technology and mechanisms which are automated.

14.3.2   Autonomic Networking and Computing Paradigms

According to IBM, the autonomic computing paradigm should be analogous to the human nervous system [15]. In [24], the authors propose an environmental nervous system for autonomic wireless sensor applications. The human nervous system is a large complex system made up of many smaller systems, similar to how the pervasive network is a large network consisting of many smaller networks. An important observation made by IBM is that it is vital to automate the entire system as a whole and not just the individual parts. In the case of the pervasive network, it is often the case that the many smaller systems are not able to maintain themselves and provide feedback to each other in a way that cooperation between the systems is possible. For example, QoS in a WPAN may be quite different than in a WiMAX network. There is often no mechanism to provide feedback from one network to the other and what may be good QoS in one network may cause poor QoS in another.

Table 14.2   Summary of a Autonomic applications in pervasive networks

Project / Reference	What is Automated	Network Technology
CONMan [20]	Configuration	Technology Independent
ANA Project [21]	Completely reconfigurable network based on conditions	Technology Independent
Co-operative caching for AWMNs [4]	Caching and Clustering	WMNs (802.11s, 802.15 mesh, 802.16 etc.)
Autonomic QoS [34]	Quality of Service Requirements	802.11 Networks
Mobility and QoS [35]	QoS, Vertical Handoff	802.11 and 802.16

14.3.2.1 Autonomic Applications

Applying autonomic principles to networking has many applications. The configuration of network addresses using DCHP is one example of autonomic computing applied to networking. Before DHCP was created it would have been the task of a network administrator to assign the IP addresses of the computers statically. If this had continued, the Internet would have required many IT professionals to manage these addresses and it is likely that it would not have grown to the size it is today. Other applications of autonomic networking include the discovery of devices in the network, management of mobility and distributed caching. CONMan [20] is an autonomous configuration system which tries to minimize the amount of configuration required by human administrators. The ANA Project [21] aims to be a completely autonomous network which can reconfigure itself based on changing conditions in the network. In [4] the authors propose a cooperative caching solution for WMNs which allows for data replication at MRs around the network to provide rapid access to frequently used data. The authors of [34] and [35] propose QoS applications. Each makes use of cross-layer information. In [34] a video application which is able to adjust its encoding based on the conditions within the network is proposed. In [35] the authors also automate vertical handoff between 802.11 and 802.16 networks while maintaining QoS requirements.

14.3.2.2 Autonomic and Pervasive Hardware

Autonomic and pervasive hardware is essentially the same hardware we use for communication now. The differences often lie in the software on the device. Many devices that we already have are pervasive, for example mobile phones, video game consoles and media devices which are linked to a network. However, these devices are often not autonomic. Still, we are required to set up some aspect of the device to connect to networks, share files and discover other devices. This technology is not yet invisible to the user. So by making these devices more autonomic, the pervasiveness of the device increases since the technology becomes more transparent to everyday users.

14.3.2.3 Autonomic and Pervasive Networks

Autonomic networking allows for easy management and configuration of the network. As networks grow larger they become increasingly difficult to administer and set up requiring skilled IT professionals to manage them. Autonomic networking principles may be combined with pervasive computing in order to create a practical, easy to manage ubiquitous/pervasive network. Autonomic principles would allow for more complex new applications to be built on top of a pervasive network that have not yet been imagined.

14.3.2.4 Autonomic Networking Architecture (ANA) and Cross-Layer Design Optimizations

The Internet owes much of its success to the strict separation of layers because separation allowed for easily maintainable and extendible code which simplified software development. However, there are many cases where strict OSI layering is inadequate. One example is in wireless networks where channel conditions may vary and cross-layered design may provide feedback from physical or link layers to the network layer in order to make smarter routing decisions.

There are several ways to classify cross-layer autonomic network architectures. In [18] the architectures are classified into global view and local view cross-layer solutions. Global layer solutions are more centralized but try to optimize the network as a whole. Conversely, local view deals with individual nodes or small subsets of nodes within the network. There are pros and cons to each architecture type. Since global layer solutions are often centralized and require information from each node in the network, they often do not scale well. Local view solutions may optimize within only a limited area and while decisions made may be optimal for an individual node, the solution may not apply well to the entire network. Perhaps the best architecture is a hybrid which makes use of both local and global aspects.

14.3.3 Cross-Layer Interactions

In recent years, there has been a great deal of research into cross-layered design and optimizations, especially in wireless networks. [33] identifies the traditional layered design as one of the major challenges in next-generation

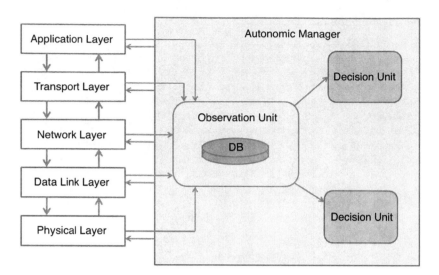

Figure 14.2   Example architecture for cross-layer autonomic networks [34].

networking. The authors argue that in order to apply autonomic and adaptive solutions to networking, a more specialized design should be used that abandons some aspects of the original design. The following sections will provide an overview on how the layers can be used in cross-layering along with examples of existing solutions involving each layer.

Figure 14.2 [34] illustrates a proposed autonomic cross-layer architecture which was designed for improved QoS in a wireless network. Feedback is supplied to the autonomic manager which is responsible for tuning network parameters automatically towards some goal providing feedback from various layers. Within the autonomic manager there are separate components such a databases for storing observations within the network and units in order to make decisions based on these observations. This general framework can also be applied to form a cross-layer autonomic solution for other problems. In [4], a similar architecture is proposed which makes use of fewer layers. This particular example is for a caching scheme in autonomic wireless mesh networks. It still has an abstracted autonomic manager which handles the tuning of the network parameters.

14.3.3.1   Application Layer

Since applications can be very unique, it is difficult to design a mechanism which is compatible with all current and future applications within a cross-layer framework. However, it is possible that application information could be provided to the cross-layer autonomous network and used to yield performance improvement. For example, the type of traffic could be provided and priorities could be assigned to the traffic. Video and other delay-intolerant data could be given higher priority than delay-tolerant traffic such as web pages and emails. The presentation layer and session layers do not provide much opportunity for performance improvement; however, there may be some forms of autonomic network which may make use of them to some extent.

In [34] the authors propose a video application which changes its encoding rate depending on network conditions. Additionally, their solution makes use of GPS data for locations of the nodes which also may be taken from an application layer program and stored in a database for use by other layers. The application layer can also provide QoS requirements to the transport and network layers which affect how routes are selected.

14.3.3.2   Transport and Network Layer

Transport and Network layer optimizations are usually made by incorporating information from the other layers into making routing or congestion avoidance decisions. Additionally, these layers can provide information on queue sizes to layers such as the data link layer which can be used when scheduling which device should send next. In the normal layered approach this is not possible, so in some schemes time or other resources are given to devices which do not necessarily require any at a given moment in time.

The cooperative caching application proposed in [4] makes use of feedback from the network layer and data-link layer to help determine a clustering strategy and when to cache data at the cluster heads.

14.3.3.3   Data Link Layer

Link layer optimizations often make use of information from other layers. In a wireless network, the link layer is responsible for packet and device scheduling. Based on parameters such as queue length, channel quality and others, the link layer can better decide when to schedule each device. In [35], the authors propose an autonomic cross-layer solution which supports mobility across technologies. The client is able to achieve vertical handoff between 802.11 and 802.16 networks with this solution. Information is exchanged between the network, data link and physical layers so that predictions can be made about when a node is associated with a particular network and when communications can resume after moving between networks.

14.3.3.4   Physical Layer

One of the most popular layers to include in cross-layer optimizations is the physical layer. Much of the information available at this layer is useful in many other layers so as to enable more intelligent decision making. For example, the physical layer could provide information on the quality of a channel at a particular node. This information can assist the network in using the best possible channels to ensure packet loss is minimized. The physical layer can also give information on the data rates at which the channel is capable of transmitting, so links which have poor capabilities can also be avoided.

14.3.3.5   Summary of Cross-Layer Interactions

In Table 14.3 below, the interactions between various layers are summarized to present a clear comparison between different autonomic applications to pervasive networking and how cross-layering has been applied.

14.3.4   Limitations of Existing Networks

There are many limitations of traditional networks when compared to the concept of an autonomic network. Often, the limitations have to do with the amount of human intervention and administration required in the existing networks. By automating the network, there is less dependence on humans which is important with constantly growing, complex systems such as the Internet. It is rapidly becoming too difficult and requires far too much human effort to keep these systems functioning. Costs for maintenance and administration are becoming restrictive.

Networks attached to the Internet require constant attention from network administrators who are required to monitor and update software to prevent against attacks and viruses. There is no mechanism for trust management or intrusion detection in many systems. Furthermore, the software must be updated manually by the administrator, often on many machines, since this system is also dependent on human intervention.

In [33], the authors identify several limitations of existing networks. First they address the inadequacy of traditional models. This includes message-passing, client-server and shared memory. They argue that traditional models do not support any context-awareness or meaningful interactions. Additionally, they rely on assumptions

Table 14.3   Summary of cross-layer interactions

Reference	Application	Layers
[34] Jooma *et al.*	QoS Improvement in Wireless	Application, Transport, Network, Data Link, Physical
[4] Nkwe and Denko	Co-operative caching in AWMN	Network, Data Link
[35] Ali- Yahiya *et al.*	QoS and Mobility in Wireless 4G	Network, Data Link, Physical

that components must live in the same network and at the same period of time, which is not always the case in many modern applications. Lastly, the authors argue that the traditional layered design is flawed, especially from an adaptive point of view. Many solutions have shown that cross-layering can improve performance since more intelligent decisions may be made at various layers with additional information.

In general, many of the limitations related to the self-* properties may be provided by autonomous networks. Each of these properties will be covered in detail in Section 14.4.

14.4   Self-* (star) In Autonomic and Pervasive Networks

Autonomic networking concepts are guided firmly by the four main self-CHOP characteristics. However, they are not only limited to those and can be extended to many others, hence referred to as self-* characteristics. In this section we will focus more on self-protection, self-organization, and self-management. To conclude this section we provide an example network, the autonomic wireless mesh network (AWMN), which supports all of the self-CHOP characteristics.

14.4.1   Self-Protection

The vast proliferation of pervasive wireless ubiquitous/pervasive networks not only increased network complexity but also made security provisioning in terms of detecting and protecting against attack more difficult [6]. Even the 'best' of networks are useless if not secure. Static secure techniques and dependence on human intervention are not sufficient to contend with advancing attacks. Networks need to be proactive and reactive in order to self-protect. The key idea behind this feature is to enable the network to be able to defend/prevent, anticipate, detect and recover from environment attacks. The varying forms of attacks can be malicious (hackers, viruses, worms, etc.), accidental or trickled failures that evade the self-heal feature [5]. This feature is constituted of self-defense, self-awareness and self-reconfiguration. Since IBM's autonomic vision of security, the self-protection intelligence capability has received attention from the industry and research community at large.

In 2004 CISCO collaborated with its network associates Symantec and Trend Micro to incorporate self-defending capabilities in their routers [7]. The software system is based on network admission control (NAC) technology. The way it works is that it only grants access to network complaint devices based on device software information such as antivirus state and OS patch levels. This system will be implemented on a small scale and rolled out progressively into large scale networks. In this way the network can detect and defend itself against malicious attack. Enterasys is another company that has incorporated the same security features in its products. These security products are good and will attract many network users or owners. It is important to diversify network security products for autonomic security for enhanced protection. Although single vendor security products may be cost effective, by diversifying it will be even more worthwhile in the future to prevent catastrophic results in a case where a particular system is compromised.

One of the leading American companies in defense systems, Raytheon Company, released a prototype system for aircraft self-protection security (ASPSS) in 2006 [8]. The system provides electronic perimeter security for aircraft and uses near object detection technology (NDDS). This system is for parked aircraft and to complement it they also developed the vigilant eagle airport protection system that protects airplanes from danger during take off, landing and other close proximity threats to parked planes.

Academic research is focused largely on the design and framework of self-protection systems. Rvan He et al. [9], define a component-based end-to-end framework for ubiquitous systems. The framework is made up of a three level architecture consisting of two control loops at the network and node level. At the network level, context awareness and self organization are important in order for a network to adapt itself to its environment. At the node level, a node should be able to reconfigure to the changing environment (e.g. security updates) and the component level deals with the architectural deformities that may result from incorporating third party applications. Samer Fayssal et al. [10] propose a novel framework designed for wireless anomaly based intrusion detection and response systems. They also present a wireless self-protection system (WSPS). The framework is structured on multi channel online monitoring and analysis of wireless network features using data mining techniques. The WSPS consists of wireless networks probes, network flow generator, behavioral analysis and action modules. By using online monitoring and data mining techniques to analyze the misuse and anomalies in network features, the WSPS is able to self-protect.

Another framework and component based self-protection system was proposed in [11]. In [12], a software agent is proposed based on a self-protecting framework for varying network attacks based on predetermined matrices; the software agents use them to deploy the appropriate self-protection mechanism. In [13], the self-protection is designed particularly for servers as they host most of an organization network's sensible data and provide services. Although the system is not sufficient to protect against all types of attack, it has the capability of being integrated with more application specific mechanisms so as to strengthen security.

Just a year after IBM's launch of the autonomic computing paradigm [14], ZHU *et al.* proposed self-protection agents designed to work with existing distributed intrusion detection techniques such as firewalls. The self-protection agent was implemented using an OS kernel packet filter and was redesigned in a loosely coupled manner to work with the IDS for more secure self-protection capabilities.

Security should be the number one priority in every network or any data shared environment, as they are prone to attack; hence the self-protection property in the autonomic networking paradigm should be promoted to the top of the list of the self-* properties as an unsecured network is a useless network.

14.4.1.1   Network Security

Network security ensures information security. Information security encompasses every aspect of protecting information resources as well as network security and data loss prevention. Some of the main security threats are denial of service and unauthorized access, which always develops or leads to a variety of malicious effects. An even bigger concern is that usually most of the attacks are often from one's own network.

The more specific seven classifications of wireless networks as described in [10] are identity spoofing, network analysis, eavesdropping, vulnerability attacks, DoS, replay attacks, rogue access points, man in the middle, session hijacking and ARP poisoning. These types of security attack can be classified further according to the layers targeted, such as MAC layer and physical layer attacks. The current security controls or measures have been predefined and are often reactive and unable to deal with the more sophisticated attacks or adapt to changing environments.

14.4.1.2   Security Controls

Some self-sense security controls have been proposed in [11], not necessarily to replace the existing controls but to enable coordination between them for self-protection and to deal with issues in a timely manner, rather than reporting to administrators to take action which could potentially allow more time for destruction. In some cases security controls are duplicated in an attempt to strengthen security. These controls include:

- *Authentication*: This is often done in the form of username and passwords, security tokens and biometrics.
- *Network virus scanners*: Network administration tools perform regularly to keep the network clean from different threats. This can be done by use of active antivirus protection on client computers.
- *Quarantine*: This is the isolation of threats to files, curing or deleting them.
- *Client assessment*: This is achieved by defining client access level controls by the use of firewalls to enforce access policies. Activities such as audit logging and assessment.
- *Access control mechanisms*: These can be defined according to IP address, MAC address, IP and MAC addresses, by using VLANs, DHCP, control ARP poisoning, etc.
- *Intrusion Detection Systems (IDS) and Intrusion Prevention Systems (IPS)*: These can be hardware or software and operate in a DMZ or network boarders. IPSs are viewed as an IDS extension. They both exist in different types including host base, network based, protocol based, application based, etc.
- *Vulnerability assessment*: This includes identifying and guarding against potential threats.

Automating and coordinating these network controls by use of information context awareness and other autonomic concepts will help secure networks without physical human intervention.

14.4.1.3   Usability vs. Security

Over the years, security and usability have always conflicted. In fact, that is when we first experienced the bio inspiration and had it integrated into technology in order to provide strong security and yet maintain systems' usability. In an effort to strengthen authentication security, passwords are required to be complex leading to the problem of being forgotten or miss-kept by writing them down as a reminder. These are dangerous practices and expose the systems' vulnerabilities to attacks or hackers. To rectify this, security tokens (e.g. access cards) and biometric (e.g. finger prints, retina) techniques were implemented in order to authenticate users. This form of security is expensive and is usually only applied to highly sensitive data.

Other ways in which security and usability conflicted are the locking out of a computer when not in use in order to safeguard against inappropriate access and making sure that the user is the valid user. This could be achieved by a face recognition system; to lock the system when the user is not the right one. Although these systems are expensive, in the name of autonomic networking and achieving self-protection they are worth every penny and will go a long way in protecting the systems without compromising usability.

Another scenario where security and usability seemsto get in each other's way is the question of which wireless networks are protected by which software security agent (e.g. Cisco clean access agent). Every time a client self-reconfigures or self-organizes to establish a connection because of mobility (moving from one university building to the next), the user has to re-authenticate (re-log on) to resume or restore network services which are always terminated in the name of security.

14.4.2   Self-Organization

Self-organization is the ability for the network to support the addition and removal of nodes, while retaining the ability to update routing and path information. This also includes the ability to self-heal when certain paths become congested or, in the case of a wireless network, where interference occurs. Autonomic Wireless Mesh Networks (AWMNs) are a good example of wireless self-organizing networks (WSON), since they have self-organizing capabilities. More detail on this type of network will be presented in Section 14.4.4. Self-organization induces self-configuration which results in some sort of self-healing. That is to say when a network re-organizes due to mobility and application needs, the current links are broken; thus clients need to reconfigure to restore or establish new connections, which is a self-healing way to maintain network connectivity. In many cases, the Internet is also a self-organizing network to some extent. When a wired computer is connected to the network, they are often assigned an IP address automatically. However, this solution often makes the assumption that the device is non-mobile. In recent years, we have seen devices that are wireless and mobile. The devices' connection point to the Internet or network is no longer static and thus unique solutions are required in order for the network to continue to service the device.

14.4.2.1   Self-Organizing Networks

There are many networks which in some way are self-organizing. As mentioned previously, to an extent the Internet is self organizing in that nodes can be added and removed with little or no human intervention. DHCP servers allow nodes to obtain an IP address when joining a network such as a LAN or WLAN. DNS servers abstract the name resolution function within the network which allows IP addresses to be translated in to readable names. Updates to DNS tables are propagated (for the most part) autonomously throughout the Internet.

Wireless sensor networks (WSNs) are another good example of this. Sensors are scattered in the area where data is to be collected. The nodes discover their neighbors, routing paths and sometimes elect cluster heads, all automatically. Some solutions [23] manage data collection and routing based on the power levels of devices with the objective of prolonging the life of the network since this is a common problem in WSNs. In [24], it is noted that a wireless sensor network should be able to continue to function despite the loss of one or several nodes. This is not limited solely to the functioning of the network, but also includes the ability to draw conclusions from the data from the sensors. This can be achieved by various mechanisms such as increasing the remaining sensors' sampling rates and reconfiguring the network topology to route around the broken sensors. With recent advances in robotics, it is also possible to have the sensors adjust their position physically to repair the network, avoid network partitioning and retain good coverage [25].

Another good example of a self-organizing network is the Autonomic wireless mesh network (AWMN). More detail on this type of network is covered in Section 14.4.4. The ANA Project [21] aims to create a network of self-organizing nodes that are able to re-adapt to changing network conditions. The project makes use of test-bed equipment, simulations and some interesting novel techniques such as distributed open collaboration (similar to SETI@Home and Folding@Home) to explore how the network performs.

The reasons for re-organization of the network may vary. Network conditions may require a re-organization such as poor link quality or failing links or hardware in particular areas. In a WSN, the sensors may be placed in extremely harsh conditions where nodes are prone to failure such as near volcanoes, in outer space or in the arctic. In commercial networks more or fewer resources may become available at a given time. The network may also need to be reorganized dynamically and frequently owing to the economic models of the providers or the social needs of the users. Often these rapid changes in organization of the network may be too much for a human administrator so the process must be automated.

14.4.2.2 Self-Configuration on Organization

After the network has organized itself, there should also be a self-configuration process. One example of the self-configuration process is [17] which proposes self-configuring routers in an autonomic network. [20] also provides an auto-configuration solution which can be used to configure a variety of tools which are used by network administrators such as SNMP, DNS and routing table configuration. Certain re-organizations of the network may require changes to settings in network services which would normally be accomplished by a human operator.

14.4.2.3 Mobility vs. Organization

As mentioned previously, unique solutions are required in order for mobility to be supported in networking. When mobility is supported the network needs a mechanism to keep track of where the devices are currently attached to the network. As a device moves from one point of attachment to another, the network must forward packets to the new location. This is often solved through the use of mobility agents [29]. In traditional wired networks, mobility was often not a problem since nodes were always fixed to a particular point of attachment in the network. In [26] a self-configuring/self-organizing WSN is proposed where a subset of the nodes is assumed to be static. The data discovery nodes (sensors) are assumed to be mobile, while the data dissemination is static. This arrangement is similar to a WMN with the mesh routers as the dissemination nodes and the sensors as the data discovery nodes. Other solutions [27, 28] treat all nodes as equal; however, the authors of [26] note that it may be inappropriate for certain nodes such as acoustic sensors or specialized cameras to perform data dissemination tasks.

Often, as mobility increases, the ability to organize the network becomes more difficult. The difficulty increases as the number of mobile nodes increases in the network. Problems such as network partitioning can occur when the nodes are so mobile that splits occur in the network and communication cannot occur.

Mobility in the network can also introduce problems of congestion when many nodes migrate to the same region of a network, overwhelming the fixed resources provided. In the case of wireless networks, interference and channel limitations can occur with many nodes migrating to the same region. One solution to this is [26] where clustering is introduced so that data can be aggregated at cluster-heads which decreases the overall overhead in the network.

14.4.3   Self-Management

As opposed to self-organization, which deals with the ability of the network to add, remove and organize nodes, self-management deals more with provisioning, resource allocation, scheduling and other operations which affect QoS. Included in the category of self-management are operation, administration, maintenance and provisioning of the network, each of which will be covered in detail in the following sections.

14.4.3.1   Operation

Network operation is another area which would be benefit significantly from automation. Consider the case of a user with a laptop computer in a wireless network. The user is mobile in the network which is made up of various wireless access points (WAPs). Each time the user moves from one access point to another they are required to re-authenticate to the network since the network does not support mobility. This is an extreme inconvenience to the user and could have been avoided if a more automated approach was taken in the design of the wireless network. This problem may be solved if the network supports Mobile IP [30], or makes use of mobile agents [29] to support node mobility. However, many network deployments do not support this feature.

Another way in which operation can benefit from autonomy is the detection of poor or good operating conditions. For example, if a user has access to several types of network for its connection to the Internet, it would be beneficial for the client system to choose the best connection based on some characteristics. These could be lowest delay, highest bandwidth, most reliable and so forth. In these cases we consider the transition between networks to be a vertical handoff if the client is changing from one technology to another [31]. However, the client may also be in an area where multiple networks using the same technology exist. There should also be some mechanism which prevents the client from switching back and forth between these competing networks as well as ensuring good performance.

14.4.3.2 Administration

The administration of the network should also be automated as much as possible. One example of this would be rather than having an administrator admit new users to the particular network on a case-by-case basis, trust management systems could be implemented in the network to decide which users should be able to gain access to network resources and which users may be malicious. The CONMan [20] project is an autonomous networking framework which tries to minimize the amount of configuration and administration of the network by humans.

Applying autonomic principles to administration is often related to security issues such as admission control or access to resources since this is by and large the greatest burden placed on network administrators, who must maintain these themselves in current systems. Normally, this is managed through some type of policy control system. The challenge here is to keep the autonomic system synchronized with constantly evolving policies at the institutional level while maintaining a high enough level of autonomy.

One example of an administrative policy being enforced using automated techniques is network packet inspection [32]. This also overlaps with provisioning and self-protection, whereby users which are determined to be greedy or detrimental to the network by making an unfair use of resources are penalized. Packet inspection could occur at MRs or GWs and traffic patterns which are deemed harmful to the network may be penalized so that performance for other users does not suffer. In a situation where autonomic control is not applied, network administrators would have to identify troublesome users manually and either suspend access or send out warning notices to the parties involved. This technique is somewhat controversial since net-neutrality supporters claim that network operators should not interfere with traffic and should instead invest in more capacity. At the same time, in wireless networks especially, it is not always possible to add capacity so there must be some mechanism to balance the traffic so that good performance can be experienced by all users.

14.4.3.3 Maintenance

Maintenance of the network can be automated through the use of automated monitoring and protection tools. The network should be able to detect conditions when devices need to be reset, software should be updated and whether there are threats to the system. The network should be able to make decisions based on these conditions and act accordingly.

14.4.3.4 Provisioning

Provisioning in a network can mean a variety of things. It can be provisioning of resources such as time, frequency or bandwidth. It can also be provisioning of resources such as CPU or memory in the devices of the network. In any case, as networks become very large this task becomes increasingly difficult for humans to undertake efficiently. Much research is being done in the areas of scheduling and resource allocation [22] in networks; however, these solutions are often specific to particular types of network. An autonomic network which is ubiquitous/pervasive may not function well with existing solutions. A solution that is ideal in one network may cause problems in another. For example, a scheduling scheme that is designed for 802.11 (WIFI) may not work well when the autonomic network also contains an 802.16 (WiMAX) network owing to the differences in range, architecture and other unique characteristics of each network. The autonomic network must support some mechanism for coordinating provisioning between many different types of network.

14.4.4   *Autonomic Wireless Mesh Networks*

Autonomic Wireless Mesh Networks (AWMNs) are a special type of network which exhibits many of the self-CHOP characteristics by definition and because of this more detail will be provided on this type of network in this section. AWMNs are made up of mesh routers (MRs) which form the backbone of the network, mesh clients (MCs) which are user devices such as laptops, PDAs, etc. and gateways (GWs) which connect the network with the Internet. Figure 14.3 illustrates an AWMN which is made up of GWs, MRs and various MCs. The AMWN should not be dependent on a particular technology, but rather it should be inter-operable. This section will describe an ideal AWMN which would support all of the self-CHOP features.

Self-configuration will occur when MRs, MCs or GWs are added or removed from the network. This means that it is possible to set up the network with minimal human intervention. Also, adding or removing capacity to specific regions of the wireless coverage area becomes much easier than before. This is usually accomplished by implementing autonomic techniques at the GWs and MRs depending on whether the solution is centralized or distributed. A more centralized solution would depend more on the gateways while a distributed solution would make more use of the MRs. It is also possible to have some form of self-configuration at the MCs in order to make the network easier to use for the end-users.

When paths in the network are compromised by congestion, interference or mobility the network should adapt. The mesh property of the network allows for multiple paths between MRs and the network should be able to self-heal itself in this case. Congestion control, interference avoidance and mobility management is again handled at the GWs or MRs. In some solutions, such as in mobile telecommunications, additional units are added to the network in order to manage mobility and to translate addresses between different network types (for example, between IP and GSM). Mobility management can also be assisted with information provided from the mobile devices as well such as velocity, signal strength of potential hand-off targets and so forth. In some solutions the addition to hand-off is made at the client's side and in some it is made on the network side (in MRs or GWs).

Being a wireless network, security should also be kept in mind since these networks are especially susceptible to attacks from malicious users. In this case the network should be self-protecting. Often, trust management schemes are implemented at the MRs. One example of this is a currency scheme where routers exchange packets as if it were currency. In this way greedy MRs run out of tokens and are eventually excluded from the network.

Lastly, the network should try to balance and minimize loads on nodes and paths in the network. This can be done through a variety of techniques including scheduling, load balancing, cooperative caching [4] and context-aware

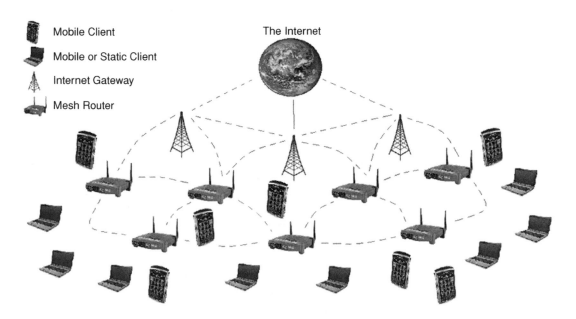

Figure 14.3   Example of an AWMN.

routing. In this way the network is self-optimizing. Load balancing and scheduling can be implemented not only in the MRs, but also in the GWs if the network supports multiple GWs.

Unfortunately, many WMNs do not currently implement all of these characteristics as they often focus on only one aspect of the self-CHOP properties. The goal in the future should be to bring more autonomy to the WMN so that administrative costs of running the network are reduced, security is increased and performance is improved.

14.5   Autonomic and Pervasive Networking Challenges

There are several challenges still remaining in pervasive networking from an autonomic point of view. The following is a list of each of the challenges which will be touched upon briefly in the following.

1. Integration of Existing Solutions
2. Scalable Solutions
3. Security
4. Inter-operability

One of the greatest challenges in autonomic and pervasive networking at present is designing an autonomic pervasive network which integrates many of the existing systems into one system. Many systems focus on a very small subset of the larger problem by automating only a very small portion of the tasks which are left to administrators and technicians. Finding a good way to link what has already been done is a major challenge. There are no clear guidelines on implementation in this area and there may be incompatibility between systems that have been deployed already. While one system may perform auto-configuration, another may be a good intrusion detection and security system; however, it is rare to find many of the self-CHOP or self-* components implemented together.

As in any large scale system, scalability is a major concern. Great care must be taken to ensure this property when designing an autonomic system since the main goal of these systems is to support and minimize human support and interaction with large-scale complex systems. Given the scale of the Internet, one can assume that wireless and mobile networks will eventually grow similarly so it is extremely important to plan for this from the start.

As devices evolve in open and dynamic networks, some stringent security mechanisms should also be designed in order to protect software running on the device from architectural conflicts or from attacks by malicious nodes. A strict security model is, however, usually considered as incompatible with ease of reconfiguration. Furthermore, in a large-scale distributed system, the cost of security administration is generally very heavy.

Lastly, inter-operability is also a major concern. With many different types of network available it is important for autonomic techniques to be transparent in terms of the underlying network. Standards for inter-operability should be designed and followed so that all networks are able to take advantage of autonomic properties. With the scale cost and complexities of configuration, administration and maintenance of communication networks this is impossible to ignore.

14.6   Conclusions and Future Directions

This chapter has provided a brief overview of autonomic and pervasive networking. We have identified the importance of ubiquitous/pervasive networks and the many applications of these types of networks. The importance of the self-* characteristics of the autonomic network have been covered, and in particular self-CHOP properties were covered in detail. The challenges and problems in existing networks have been outlined and may be used as a starting point for future research in autonomic and pervasive networking.

References

[1] U. Hansmann, L. Merk, M.S. Nicklous, and T. Stober (2003) *Pervasive Computing: The Mobile World* 2nd edn, Springer-Verlag, Berlin, Germany.
[2] X. Gu, X. Fu, H. Tschofenig, and L. Wolf (2006) 'Towards self-optimizing protocol stack for autonomic communication: initial experience,' *Autonomic Communication* 186–201.

[3] M.K. Denko, L.T Yang, and Y. Zhang (2009) *Autonomic Computing and Networking* 1st edn, Springer Publishing, USA.

[4] T.K.R. Nkwe, M.K. Denko (2009) 'Self-optimizing cooperative caching in autonomic wireless mesh networks,' *14th IEEE Symposium on Computers and Communications (ISCC 09) – Sousse*, Tunisia pp. 411–16.

[5] M. Salehie and L. Tahvildari (2005) 'Autonomic computing: emerging trends and open problems,' In *Proceedings of the 2005 Workshop on Design and Evolution of Autonomic Application Software*, pp. 1–7.

[6] http://en.wikipedia.org/wiki/Wireless_Self-Protection_System.

[7] http://news.cnet.com/From-Cisco,-self-defense-weapons-for-networks/2100-7355_3-5239359.html.

[8] http://www.spacewar.com/reports/Raytheon_Delivers_First_Aircraft_Self_Protection_Security_System.html

[9] R. He, M. Lacoste (2008) 'Applying component-based design to self-protection of ubiquitous systems,' *SEPS'08*, July 6, Sorrento, Italy.

[10] S. Fayssal, Y. Al-Nashif, B. Uk Kim, and S. Hariri (2008) 'A proactive wireless self-protection system,' *ICPS'08*, July 6–10, Sorrento, Italy.

[11] B. Claudel, N. De Palma, R. Lachaize, and D. Hagimont (2006) 'Self-protection for Distributed Component-Based Applications,' In *Stabilization, Safety, and Security of Distributed Systems, 8th International Symposium, SSS 2006*, no. 4280 in *Lecture Notes in Computer Science*, Springer.

[12] G. Qu, S. Hariri, S. Jangiti, G. Zhang, M. Parashar (2004) 'Online monitoring and analysis for self-protection against network attacks,' In *Proceedings of the International Conference on Autonomic Computing (ICAC'04)*.

[13] S.R. Zhu, W.Q. Li (2003) 'Design and implementation of self-protection agent for network-based intrusion detection system' *Journal of Central South University of Technology* – http://www.csa.com.

[14] http://www.research.ibm.com/autonomic.

[15] IBM Corporation (2001) *Autonomic Computing: IBM's Perspective on the State of Information Technology* IBM Corporation, New Orchard Road, Armonk, NY 10504.

[16] M. Weiser (1993) 'Some computer science problems in ubiquitous computing,' *Communications of the ACM*, July (Reprinted as *Ubiquitous Computing*. Nikkei Electronics, pp. 137–43).

[17] T. Bullot, D. Gaiti (2004) 'Towards autonomic networking and self-configuring routers: The integration of autonomic agents,' *IFIP International Federation for Information Processing* **213**, *Networking Control and Engineering for QoS, Security, and Mobility*, **V**, ed. D. Gaiti, pp. 127–42.

[18] M.A. Razzaque, S. Dobson, and P. Nixon (2007) 'Cross-layer architectures for autonomic communication' *Journal of Network and Systems Management* **15**(1): 13–27.

[19] M. Huebscher, J.A. McCann (2008) 'A survey of autonomic computing – degrees, models and applications' *ACM Computing Surveys* **40**(3), Article 7.

[20] H. Ballani, P. Francis (2007) 'CONMan: a step towards manageability,' In *Proceedings of Applications, Technologies, Architectures and Protocols for Computer Communication*, Kyoto, Japan, pp. 205–16.

[21] ANA – Autonomic Network Architecture Project http://www.ana-project.org/web/. Accessed 23 August 2009.

[22] J.B. Ernst, M.K. Denko (2009) 'Fair scheduling with multiple gateways in wireless mesh networks,' In *Proceedings of 23rd IEEE Conference on Advanced Information Networking and Applications (AINA 2009)*, Bradford, UK, pp. 106–12.

[23] R. Doss, L. Gang, V. Mak, Y. Shui, and M. Chowdry (2009) 'Improving the QoS for information discovery in autonomic wireless sensor networks' *Journal of Pervasive and Mobile Computing* **5**(4): 334–49.

[24] D. Marsh, R. Tynan, D. O'Kane, and G.M.P. O'Hare (2004) 'Autonomic wireless sensor networks' *Journal of Engineering Applications of Artificial Intelligence* **17**(7): 741–8,.

[25] A. Osmani, M. Dehghan, H. Pourakbar, and P. Emdadi (2009) 'Fuzzy-based movement-assisted sensor deployment method in wireless sensor networks,' *1st International Conference on Computational Intelligence, Communication Systems and Network (CICSYN 2009)*, pp. 90–5.

[26] L. Subramanian, R.H. Katz (2000) 'An architecture for building self-configurable systems,' In *Proceedings of 1st ACM International Symposium on Mobile Ad Hoc Networking & Computing*, pp. 63–73. Boston, Mass.

[27] D. Estrin, R. Govindan, J. Hiedeman, and S. Kumar (1999) 'Next century challenges: scalable coordination in sensor networks,' In *Proceedings of the 5th Annual ACM/IEEE Conference on Mobile Computing and Networking*, pp. 263–70. Seattle, Wash.

[28] W.R. Heinzelman, J. Kulik, and H. Balakrishnan (1999) 'Adaptive protocols for information dissemination in wireless sensor networks,' In *Proceedings of the 5th Annual ACM/IEEE Conference on Mobile Computing and Networking*, pp. 174–85. Seattle, Wash.

[29] C.T. Chou, K.G. Shin (2007) 'Smooth handoff with enhanced packet buffering-and-forwarding in wireless/mobile networks' *Wireless Networks* **13**(3): 285–97.

[30] C. Perkins (1998) 'Mobile networking through mobile IP' *IEEE Internet Computing* **2**(1): 58–69.

[31] F. Jin, C. Ayeong-Ah, K. Jae-Hoon, O. Se-Hyun, I. Jong-Tae, S. JungKyo, and C.I. Hyeong (2007) 'Cost-based approach to access selection and vertical handover decision in multi-access networks' *Lecture Notes in Computer Science, Networking*, pp. 1245–8.

[32] T. Choi, S. Yoon, S. Kim, D. Kang, and J. Lee (2009) 'Policy-based monitoring and high precision control for converged multi-gigabit IP networks' *Lecture Notes in Computer Science, Management Enabling the Future for Changing Business and New Computing Services* **5787**, pp. 112–21.

[33] S. Dobson, S. Denazis, A. Fernandez, D. Gaiti, E. Gelenbe, F. Massacci, P. Nixon, F. Safre, N. Schmidt, and F. Zambonelli (2006) 'A survey of autonomic communications' *ACM Transactions on Autonomous and Adaptive Systems* **1**(2): 223–59.

[34] W.B. Jooma, H. Yousseff, S. Lohier, and G Pujolle, (2009) 'A cross-layer architecture for QoS support in wireless networks,' In *Proceedings of the 1st Wireless Days Conference*, Paris, France, pp. 1–6.

[35] T. Ali-Yahiya, T. Bullot, A. L. Beylot, and G. Pujolle (2008) 'A cross-layer based autonomic architecture mobility and QoS supports in 4G networks,' In *Proceedings of the 5th Consumer Communications Conference*, pp. 79–83.

15

An Adaptive Architecture of Service Component for Pervasive Computing[1]

Fei Li,[1] Y. He,[1] Athanasios V. Vasilakos,[2] and Naixue Xiong[3]
[1]School of Computer Science, Wuhan University, Wuhan, P.R. China.
[2]Department of Computer and Telecom. Engineering, University of Western Macedonia, Greece.
[3]Department of Computer Science, Georgia State University, Atlanta, GA, USA.

15.1 Introduction

Nowadays, CBSE (Component-based Software Engineering) is an approach to application development in which ready-made pieces of software are assembled to enable the rapid construction of applications for a Pervasive system [2]. To be successful in CBSE, a large number of reusable components should be developed and reused by developers. However, practical experience shows that 'as-is' reuse seldom occurs [19] and reusable components generally need to be adapted to match the application requirements [3].

Generally speaking, component adaptation is a technique which enables component composition when the components are functionally compatible, though their visible interfaces are not compatible [19]. During the process of adaptation, originally independent components become logically dependent on each other. That is, the process of adaptation introduces 'use relation' between components [15].

When there is 'use relation' between component A and component B, assuming A uses B, there are two typical ways in which A can invoke B's function. One is that A invokes B's interface directly, the other is that A uses some naming mechanism, such as JNDI [16], to locate one instance of B and invoke its function. Neither of these methods is satisfactory because there remains invocation dependency and name dependency among components.

Dependency results in coupling which brings potential risks to applications for losing flexibility for future changes. Modifying a component interface is dangerous because it would have a ripple effect on other components which depend on it. Such a dependency would be scattered and duplicated and the process of modification is tedious and fallible. How to enable interaction amongst components without invocation and name dependency is a crucial requirement for component adaptation technique. Another important requirement is 'Delay Adaptation'. It enables the development of a component which requires other components' services without specifying accurate providers and interfaces. It is useful for third-party developers, since they will always want to develop components without considering the specific requirements. And the different composers may make the final decisions.

This chapter proposes a framework which focuses on a thorough resolution of functional dependency among components and enabling 'Delay Adaptation'. This framework introduces an 'Adaptation Table' to describe components' interfaces and their interaction. Since functional dependency comes from interaction among components,

[1] This research was supported by a grant from the National High Technology Research and Development Program of China (863 Program) No. 2007AA01Z138.

this 'Adaptation Table' brings all dependency under control. A toolkit called DAT is developed and improved based on this framework.

Differing from other invocation mechanisms, the component which requires service sends its own identification to DAT. By matching the identification of components in the 'Adaptation Table', DAT identifies component which provides a corresponding service and enables those components' interaction with a reflection mechanism [16]. With this mechanism, a component acquires the service from DAT but not the provider component. When some components' interfaces need to be changed, developers only need to modify the 'Adaptation Table' and the related adaptation code specified by this table. Also, a component developer need not make any assumption about the interface contract which provides a service to the products. Practical experience shows that such a centralized mechanism improves the maintainability and flexibility of applications.

This chapter is organized as follows: Section 15.2 describes the motivation for this work. In Section 15.3, the model of DAT is represented. A case study is depicted in Section 15.4. Section 15.5 introduces the related work. Section 15.6 is the conclusion and discusses future work.

15.2 Motivation

In this section, we focus on a scenario of CBSE and the requirements of Component Adaptation.

15.2.1 Scenario

Assume that A is a COTS developer and is developing a component which is able to report run-time warning information. A decides to adopt some third-party component that could provide a reporting service with a flexible configuration mechanism.

There are many third-party components which could satisfy this requirement. But like other customers, A would like to choose its own preferred functional components, such as Company B.

B is an independent component integration company; it provides a professional service for integrating third-party components into applications. B's customers are various and most have their own specific requirements. It is a nightmare for engineers to replace components that satisfy such customized requirements.

Both A and B want to resolve dependency among components and decide which components are to be composed. This latter requirement is described as 'Delay Adaptation' in this chapter.

Generally, components have two types of interface: a 'require' interface which asks for a service from other components; and a 'provide' interface (called 'provider') which provides a service to other components. The mechanism of Delay Adaptation enables a 'require' interface to invoke some 'provide' interfaces which are 'unknown'. When the 'provide' interface is specified, the 'require' interface and 'provide' interface are connected. To enable component interaction, the component composer could add adaptation code without modifying both the 'require' and 'provide' interfaces.

15.2.2 Requirements For Component Adaptation

Researchers have proposed many requirements in order to evaluate different component adaptation techniques. These requirements focus on different aspects. To evaluate the techniques at the same standard, this chapter extracts eight requirements from Bosch [3], Heineman [8, 7], Kim [12] and Martin [5]. These eight requirements, which cover the most important factors in component adaptation, are widely accepted by most researchers. Since there is some variance in terminology, the most commonly used and explicit phrases are chosen. The detailed descriptive information goes as follows:

1. *Transparent.* Both the user of the adapted component and the component itself are unaware of the adaptation between them [3, 8].
2. *Black-box.* The adaptation techniques have no knowledge of the internal implementation [3, 7].
3. *Identity.* The component which is adapted should retain its identity [7, 10].
4. *Configurable.* The adaptation technique should be able to parameterize and apply a particular adaptation (the generic part) to many different components (the specific part) [3].
5. *Reusable.* One should be able to reuse the code written to adapt a component [18].

6. *Framework independent*. The adaptation technique must not be dependent upon the component framework to which it belongs [7, 12].
7. *Language independent*. The adaptation mechanism must not be dependent upon the language used to implement the component [3].
8. *Architectural focus*. The specification and/or implementation of the adaptation should be visible at the architectural level [7].

Many component adaptation techniques have been proposed and meet the requirements above: Copy and Paste, Inheritance, Adaptor [6], Superimposition [3], Generic Wrappers [5], Binary Component Adaptation [10] and Type-Safe Delegation [13]. These techniques do not provide a solution to thoroughly resolve the dependency among components and take the requirement of 'Delay Adaptation' into consideration. Furthermore, some techniques, though they provide very strong mechanisms, cannot be implemented with the support of current mainstream programming languages. That is, there is a need to strengthen the compiling and running environment in order to support them.

The motivation of this chapter is to propose a framework which could resolve dependency among components and support 'Delay Adaptation' within the framework of the current mainstream of programming languages.

15.3   An Overview of the Delaying Adaptation Tool

We implement a tool in order to support our idea. The tool is called the Delaying Adaptation Tool (DAT). The two main implementation technologies of DAT are XML and 'reflection'. XML is a universal standard and many programming languages have provided API for parsing XML files. Reflection is a mechanism which allows programmatic access to information about the fields, methods and constructors of loaded classes, as well as the use reflected fields, methods, and constructors to operate on their underlying counterparts in objects, within security restrictions [16]. Reflection is supported by many programming languages.

With these two technologies, DAT resolves component dependency and supports 'Delay Adaptation'. DAT records all of the interfaces' signature information in a well-defined XML file (adaptation table). Each interface in this table has a unique identification. The adaptation table also glues the require interface and the provide interface. When the request interface asks for service, it sends its own unique identification to DAT and DAT identifies and invokes the corresponding provide interface. In this way, DAT decouples components by removing the straight function invocation.

Because XML and reflection are widespread technologies, it is very easy for users to understand the working process of DAT. Additionally, if users want to implement DAT in some different programming language, the popularity of these technologies would make it easier.

The following sections introduce the framework, working process and definition of the adaptation table in DAT. These represent the static structure, dynamic behavior and key implementation mechanism of DAT. Some additional features are also presented.

15.3.1   Framework

DAT is an extension to the adapter design pattern which is limited by direct function invocation. [6].

In a component system, the customer and provider are tightly coupled [13]. If the interface's name has been changed, the adapter should be changed and complied again.

DAT extends the adaptor pattern into a general adaptor which gathers all of the interaction information and routing messages among components via an adaptation table. Figure 15.1 presents the framework of DAT and the description of its elements goes as follows.

1. *Requester and Provider*. The Requester is the component that requests a service from other components and Provider is the component that provides a service to others. Such a definition makes the model much clearer for tracing interaction messages in the model.
2. *Abstract Organizer*. The Abstract Organizer is the core of this model. The main task of Abstract Organizer is receiving messages sent from Requester, forwarding the messages to the appropriate Provider and returning the result. Differing from the normal function invocation and naming mechanism, DAT fulfils this task by message

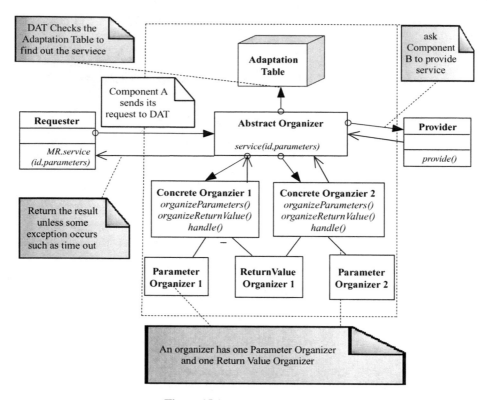

Figure 15.1   Framework of DAT.

forwarding. The requester does not know who the exact provider is or the interface definition of the provider. This design resolves invocation and name dependency among components.

3. *Parameter Organizer and ReturnValue Organizer.* The Parameter Organizer reorganizes the requester's parameters into the format required by the provider. The ReturnValue Organizer reorganizes the return value into the format required by requester.

4. *Concrete Organizer.* The Concrete Organizer binds the Parameter Organizer and ReturnValue Organizer together and this information is also recorded in the adaptation table.

5. *Adaptation Table.* The Adaptation Table records all information about components, interfaces and routers. The detailed mechanism and definition is described in next subsection.

15.3.2   Adaptation Table

Figure 15.2 presents a definition of the adaptation table in XML-schema format. This definition is designed to be consistent with Garlan [1] and Bracciali [4].

The root node of the adaptation table is a 'RoutingTable'. Components and Routers are the first level child nodes. Because the definitions of components and routers are very important, they are discussed individually.

1. *Component*: There are many interfaces in components, each one of those interfaces has a unique identification, which is defined as an attribute of the interface node. There are two types of interface. One requests service, i.e. the 'request Service' nodes in Figure 15.2; the other provides services, i.e. the 'provide Service' nodes.

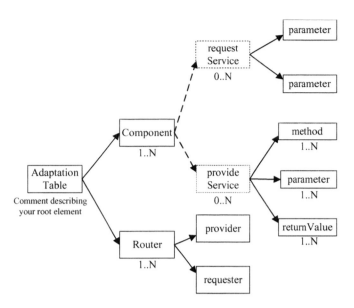

Figure 15.2   Adaptation table in XML-Schema format.

The interface that requests a service has a list of child nodes. These child nodes represent the parameters. Each parameter node has an implicit serial number, which keeps the sequence of the parameter nodes unchanged.

The interface that provides a service has to determine which class and method with which provide the service. There is also a list of parameter nodes which are required to invoke it. These interfaces also have child nodes which define the type of return value.

2. *Router*. Each router node defines a specific 'Router' in Figure 15.1. This node binds together the interfaces which request services and interfaces which provide service. In Figure 15.2, the corresponding two nodes are child nodes of Router. Each of these two nodes has a signature attribute whose value corresponds to the interface signature that is defined under interface nodes. DAT can identify the required interface through this signature attribute.

15.3.3   Working Process

The working process of DAT is depicted in Figure 15.3 as a sequence diagram. The setails are as follows.

In step 1, the Requester sends its request to the Service Organizer. The request includes the Requester's own identification as defined in adaptation table and its parameters.

```
DAT. service (String requesterID, Object[] parameters);
```

The Service Organizer then checks the adaptation table as to whether there is a corresponding service available. If not, the Service Organizer throws exceptions and terminates this requesting process. In reorganizing the parameters step, the Service Organizer transforms the parameters into the format required by the Provider. If necessary, the Service Organizer reports the parameters incompatible error information to the Requester and terminates the current handling process. When the prerequisite work is finished, the Service Organizer sends the request to the Provider with packaged parameters. If the Service Organizer fails to receive response from the Provider within the appointed time interval, it sends an error message back.

As the results are received, the Concrete organizer reorganizes the return value to the format required. There may be some exceptions and the handling process is similar to the step of 'Reorganizing parameters'. It then returns the appropriate result to the Requester.

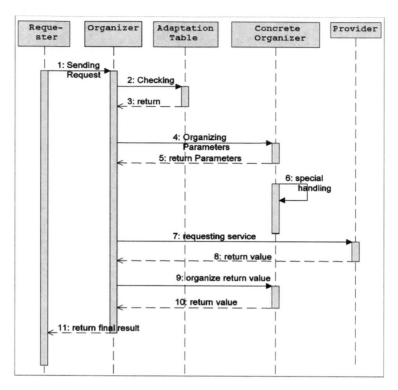

Figure 15.3   The working process.

15.3.4   Supplementary Mechanism

In order to enable DAT to achieve a practical balance between flexibility, robustness, low overhead and ease of use we add some additional features. These mechanisms are Dynamic Updating, Fault Tolerance and Performance Enhancement. The detailed implementing mechanism is introduced in the following.

1. *Dynamic Updating*. Dynamic Updating enables the application to replace a component even when it is running. Since the components are loosely coupled and messages are forwarded by Service Organizer, the application can update components dynamically and transparently. DAT enables dynamic updating by using a message queue mechanism. When developers change a component's status to 'updating' in the adaptation table and begin to update that component, DAT can capture this change and caches the requests in a predefined message queue. When the status is changed back to 'running', the requests are sent to it from the message queue.
2. *Fault Tolerance*. DAT provides fault tolerance mechanisms to handle exceptions during working process. Three categories of exceptions are listed below.

   Unavailable Exception. This exception declares that the service required has no corresponding component recorded in the Adaptation Table or that the required component is not ready for use.

   Updating Exception. DAT starts a time counter when the system needs dynamic updating. When the component providing a service is not available after a certain time interval, DAT records an Updating Exception in the log file and reminds users to retry after a comparatively longer time.

   Mismatching Exception. This exception may occur when the parameters or return value are not consistent with the contract.

3. *Performance Enhancement*. In DAT, the Adaptation Table is correlated to many kernel operations. This means that the performance of an application system would be debased for vast I/P operations. Taking this into

consideration, file buffering mechanisms are adopted. As an application starts to run, DAT loads the Adaptation File in a Hash Table. When access to the Adaptation File is requested, DAT checks whether the file has been modified since the last loading. This mechanism keeps the Adaptation File consistent with the data in the memory and enhances performance.

15.4   Case Study

15.4.1   Background

In our earlier research, we rebuilt a toolkit which is called the Performance Analysis and Characterization Environment (PACE) [20]. PACE can analyze and evaluate the performance of a component before running. In the current project, the Research of the Key Problems on Dynamic Web Service Composition and Performance Analysis (863 program), a toolkit is needed in order to carry out preprocess and estimation of services before choosing a suitable composition scenario. Thus, PACE is utilized in this project with a necessary modification.

15.4.2   Case Process

To achieve stable performance, the events and exceptions must be recorded by PACE. The recording modules in PACE are independent of the recording component. DAT is introduced into PACE and the first version of DAT's adaptation table is defined as Figure 15.4. There is only one component's definition and we can see that it is irrelevant to the definition of the recording component.

During the final stage of PACE's design, we choose Log4j, an open-source software, to log the important information. In the Feedback module (one of the basic modules in PACE), Log4j has been used many times. At the start, version 1.0.4 of Log4j was utilized by developers. As an example, in Log4j's manual, interfaces of Log4j are used as:

```
Category cat = Category.getInstance
(''whu.cs.pace.FeedBackClient'');
cat.warn (''feedback is not permitted'');
```

The typical means of reusing is direct function invocation, since the component that requests a service is coupled with the component that provides the service (Log4j). If Log4j was used in this way in PACE, the system would depend on Log4j's function and any modification of Log4j would be a nightmare. With the support of DAT, the adaptation table is defined as in Figure 15.5.

As Log4j evolves into version 1.3.0, the interface class is changed from Category to Logger:

```
Logger logger = Logger.getLogger
(''whu.cs.pace.FeedBackClient'');
logger.warn (''Low fuel level.'');
```

```
<AdaptationTable>
   <Component name="PACE.Feedback" status="running">
   <requestService signature="PACE.Feedback.logging" >
      <parameter name="class Name" type="java.lang.String" />
      <parameter name="information" type="java.lang.String" />
   </requestService>
   </Component>
</AdaptationTable>
```

Figure 15.4   The early version of the adaptation table.

```
<AdaptationTable>
  <Component name="PACE.Feedback" >
    <requestService signature="PACE.Feedback.logging" >
      <parameter name="class Name" type="java.lang.String" />
      <parameter name="information" type="java.lang.String" />
    </requestService>
  </Component>

  <Component name="Log4j"  status="running">
    <provideService signature="Category.Instance" >
      <method class="org.apache.log4j.Category" method="getInstance"/>
      <parameter name="class Name" type="java.lang.String" />
      <returnValue type="org.apache.log4j.Category" />
    </provideService>

    <provideService signature="Category.Info">
      <method class="org.apache.log4j.Category" method="info"/>
      <parameter name="class Name" type="java.lang.String" />
      <returnValue type="void" />
    </provideService>
  </Component>

  <Router class="whu.cs.router.LoggingRouter.InstanceRouter">
    <role component = "PACE.Feedback" requester="PACE.Feedback.logging" />
    <role component = "Log4j" provider="Category.Instance" />
  </Router>

  <Router class="whu.cs.router.LoggingRouter.InfoRouter">
    <role component = "PACE.Feedback" requester="PACE.Feedback.logging" />
    <role component = "Log4j" provider="Category.Info" />
  </Router>
</AdaptationTable>
```

Figure 15.5   The first version when Log4j is introduced.

As the new version of Log4j has provided some useful features, the system has been updated so as to adapt to the new interfaces. Normally, it is a nightmare to change all the code for there are more than one hundred instances referencing Log4j's interface. And the maintainers of the system will also wish to make this modification during running time without interrupting users' visits.

Fortunately, we have introduced DAT so as to adapt components at the first stage. With DAT, the modification is very convenient and time saving. The only task is to make some modifications in the adaptation table. During the process of updating, the system does not need to be stopped.

Also, the exception handling framework identifies the 'Unavailable Exception' and 'Updating Exception' during the course of the process.

The entire updating process entails only three steps: (i) Change the component's status; change the component Log4j from the status of 'running' to that of 'updating', as shown in Figure 15.6; (ii) Deploy the new version of the component; (iii) Modify the adaptation table for the new adaptation. Figure 15.7 depicts the new adaptation table for the higher version of Log4j.

```
<Component name="Log4j" status="updating">
    <provideService signature="Category.Instance" >
        <method class="org.apache.log4j.Category" method="getInstance"/>
        <parameter name="class Name" type="java.lang.String" />
        <returnValue type="org.apache.log4j.Category" />
    </provideService>
    <provideService signature="Category.Info">
        <method class="org.apache.log4j.Category" method="info"/>
        <parameter name="class Name" type="java.lang.String" />
        <returnValue type="void" />
    </provideService>
</Component>
```

Figure 15.6   Modifying the status of Log4j.

15.4.3   Discussion

1. *Adaptation Delaying and Resolving Dependency.* Noting the difference between Figure 15.4 and Figure 15.5, it is clear that the development of the component 'Feed Back' is not dependent on the interface's definition of the printing tool. Whenever the composer decides which specific component should be introduced and composed, it simply has to deploy it and modify the content of the adaptation table. That means:
   (i) the dependency between the require interface and provide interface is resolved (not only invocation dependency but also name dependency);
   (ii) the adaptation is delayed and later modification of the adaptation will not cause internal code modification.
2. *Meeting Requirements.* To be comparable to other techniques, this chapter evaluates DAT with the eight different aspects presented in Section 15.2:
   (i) *Transparent.* Both the component requesting service and the component providing service communicate through the message routing mechanism provided by DAT; this means that they are unaware of each other.

```
<Component name="Log4j">
    <!--interface provides service-->
    <provideService signature="Category Instance" >
        <method class="org.apache.log4j.Category" method="getInstance"/>
        <!--parameters for calling-->
        <parameter pname="class Name" ptype="java.lang.String" />
        <!--return value-->
        <returnValue type="org.apache.log4j.Category" />
    </provideService>
    <!--interface provides service-->
    <provideService signature="Category.Info">
        <method class="org.apache.log4j.Category" method="info"/>
        <!--parameters for calling-->
        <parameter pname="class Name" ptype="java.lang.String" />
        <!--return value-->
        <returnValue type="void" />
    </provideService>
</Component>
```

Figure 15.7   The new version of the adaptation table.

(ii) *Black-box*. DAT does not access the internals of the component. The changing of a component's internal process will not effect the behavior of DAT.

(iii) *Identity*. During the whole working process, DAT does not change the identity of original component.

(iv) *Configurable*. DAT provides a highly flexible configuration mechanism with the adaptation table. Components and interfaces can be appointed dynamically by editing the configuration freely. DAT can perceive the modification of the adaptation table and adjust behavior automatically. This parameterized mechanism enables a user to apply a particular adaptation to any component.

(v) *Reusable*. Since components request service from DAT in the same way, the general part of DAT is definitely reusable. The specific part, the specific router, may not always be reusable because it contains the process of reorganizing parameters. But the parameter organizer and the return value organizer in Figure 15.1, which are integrated into a specific router, can be reused by many different specific routers. So, as a whole, DAT is reusable.

(vi) *Framework independent*. DAT does not depend on a specific component framework. Log4j is a pre-packaged class library written in Java language and does not measure up to any specific component standard such as EJB and CORBA. DAT also provides good support for such components.

(vii) *Language independence*. DAT requires language to support the mechanism of finding the interface reference by name. We argue that this function is quite common in current mainstream languages though the implementing mechanism may be different. For example, the Java and C; programming languages provide reflection to support it; and c++ provides a function pointer to do so. From such examples, we might come to the conclusion that DAT is at least partly language-independent.

(viii) *Architectural focus*. As stated above, the adaptation table defines the components and routers at an architectural level. We also referred to some architecture specifications designed to achieve such a definition. It would be very easy to transfer an Adaptation Table into a visual architecture view.

In [8], Heineman has provided detailed comparisons for existing component adaptation techniques. We merge his results with ours for DAT in Table 15.1. Table 15.1 is divided into three parts. Techniques which could be implemented with current languages are in the first part (the upper one). DAT is put in the middle. Techniques which could not be implemented with current languages are listed in the bottom. From Table 15.1, it is clear that DAT meets most of the requirements.

The methods in the table are listed by increasing order of complexity. In this sequence, DAT is at the middle. Though it seems to be more complex than some simpler methods, it is much simpler than the methods in third part which cannot be supported directly by current mainstream programming languages.

3. *Additional Features*. In addition, the case study shows that DAT has some superior features. The most extraordinary functionality of DAT is that it supports dynamic updating. DAT helps customers avoid stopping the service while updating and clearly improves a component system's availability. Since the process of dynamic updating is quite complex, exception handling is also regarded as important. The exception handling framework defined in DAT helps to improve PACE's reliability, especially during the process of dynamic updating.

Table 15.1 Comparison of component adaptation techniques

RM	Transparent	Black-box	Identity	Config-urable	Reusable	Framework independent	Language independence	Architectural Focus
Copy-Paste	Yes	No	No	No	No	Yes	Yes	No
Inheritance	Yes	Yes	Yes	No	No	Yes	Yes	No
Wrapping	No	Yes	Yes	No	No	Yes	Yes	No
Binary Component Adaptation	Yes	Yes	Yes	No	No	Yes	No	No
Superimposition	Yes	Yes	Yes	Yes	Yes	Yes	No	No
Generic Wrappers	Yes	Yes	Yes	No	No	Yes	Yes	No
DAT	Yes	Yes	Yes	Yes	Yes	Yes	Yes	Yes

15.4.4   Summary

From the analysis above and from practical experience, the conclusion may be drawn that: (i) DAT enables the mechanism of 'Delay Adaptation'. It helps the designers of PACE to delay making decisions as to which third-party functional components will be chosen; (ii) DAT can meet all of the key requirements for component adaptation; (iii) DAT supports dynamic updating and a provides fault tolerance mechanism. This makes it more practical.

15.5   Related Work

This chapter is based on previous researchers' hard work and contributions. Those researchers have summarized the requirements for component adaptation technique [3, 5, 8, 7, 11, 12]. These requirements are valuable for evaluating component adaptation techniques at a unified standard.

In this chapter, a new requirement, 'Delay Adaptation' is proposed. This requirement does not conflict with others' requirements and, in some sense, overlaps with the requirements of 'Transparent' and 'Black-box'. Though 'Transparent' and 'Black-box' have emphasized the independence of adaptor and adapter [7], 'Delay Adaptation' motivates a stronger mechanism.

These mechanisms enable the required interface of components that are declared explicitly and are independent with the interface provided. Many component adaptation techniques are proposed. Common techniques for component adaptation are Copy-Paste, Inheritance and Adapter Design Pattern [6].

Many researchers have identified the drawbacks of these techniques [8, 9, 12]. There are also many comparatively new ideas for component adaptation, such as Superimposition [3], Binary Component Adaptation [10] and Generic Wrapper [5].

Most of these techniques have been proposed with a requirement framework and each of them has made an outstanding contribution at one or two points. A summary and evaluation is provided in Heineman's technical report [8]. However, they did not take the requirement of 'Delay Adaptation' into consideration which is the main contribution of this chapter. Additionally, some of the mechanisms and techniques above, such as Generic Wrapper and BCA, cannot be implemented with current mainstream programming languages; that is, some new programming language must be devised in order to support them. The toolkit developed in this chapter's idea, DAT, is implemented with the core Java language. This makes DAT more practical and understandable.

Implicit invocation means that a component can announce (or broadcast) one or more events and when the event is announced the system itself invokes the procedures that have registered an interest in the event [14]. Implicit invocation is an event-based mechanism that depends on a specific programming language. One limitation of implicit invocation is that a component registers events in its source code explicitly. The consequence is that the usage of implicit invocation is limited. Implicit invocation is used mainly in UI applications. The framework is somewhat similar to Chiron-2(C2). C2 is an architecture style directed at supporting larger grain reuse and flexible system composition. A key aspect of C2 is that components are not constructed with any dependencies [17]. Differing from C2, this chapter focuses on providing more of an implementation mechanism and creating a practical support toolkit.

15.6   Conclusions

This chapter focuses on resolving functional dependency among components and enabling a new mechanism for a pervasive computing system, 'Delay Adaptation', which helps COTS providers to develop a component's required interface without knowing the specific definition of the provider. Our solution is presented and a toolkit based on it (called DAT) is developed. With practical experience of using DAT in PACE, which is a part of a key science and technology project supported by the government in China, DAT achieves our goal. We also extract many requirements for component adaptation techniques from other researchers' contributions and evaluate DAT under these requirements. The case study shows that DAT meets all of the requirements. DAT is also practical because it is implemented in current mainstream programming languages.

There are three limitations to this framework. First, though it decouples components, it requires developers to declare the components' required interface in a specific form. Secondly, the DAT toolkit clusters all of the communication among components. This may be a potential risk to the application's performance. Thirdly, DAT cannot enable dynamic semantic interaction among components.

These limitations are under consideration and will be resolved in our future work. Furthermore, the authors will apply this framework and toolkit in further engineering projects to ascertain any other drawbacks. Our long-term goal is to merge this toolkit into an application server so as to support components' loose coupling communication.

References

[1] R. Allen, D. Garlan (1994) 'Formalizing architectural connection,' In *ICSE '94: Proceedings of the 16th International Conference on Software Engineering*, pp. 71–80, Los Alamitos, CA, USA. IEEE Computer Society Press.

[2] B. Barn, A.W. Brown, and J. Cheesman (1998) 'Methods and tools for component based development' *Technology of Object-Oriented Languages, 1998. TOOLS 26. Proceedings* pp. 385–395 Santa Barbara, CA, USA.

[3] D. Niebuhr, A. Rausch (2007) 'A concept for dynamic wiring of components,' *Sixth International Workshop on Specification and Verification of Component-Based Systems (SAVCBS 2007)*, Cavtat near Dubrovnik, Croatia.

[4] A. Bracciali, A. Brogi, and C. Canal (2005) 'A formal approach to component adaptation,' *the 28th Australasian Computer Science Conference* 74 (1): 45–54, Newcastle, Australia.

[5] M. Buchi, W. Weck (2000) 'Generic wrappers,' In *Proceedings of the 14th European Conference on Object-Oriented Programming*, pp. 201–25, Sophia Antipolis and Cannes, France.

[6] E. Gamma, R. Helm, R. Johnson, and J. Vlissides (1995) *Design Patterns: Elements of Reusable Object-Oriented Software* Addison-Wesley, Reading, Massachusetts.

[7] G.T. Heineman (1998) 'A model for designing adaptable software components,' In *Proceedings of the 22nd Annual International Computer Software and Applications Conference*, pp. 121–7, Washington, DC, USA, IEEE Computer Society.

[8] G.T. Heineman, H. Ohlenbusch (1999) 'An evaluation of component adaptation techniques,' In *ICSE'99 Workshop on CBSE*, pp. 126–38, Washington, DC, USA. IEEE Computer Society.

[9] U. Holzle (1993) 'Integrating independently-developed components in object oriented languages,' In *ECOOP '93: Proceedings of the 7th European Conference on Object-Oriented Programming*, pp. 36–56, London, UK, Springer-Verlag.

[10] R. Keller, U. Holzle (1998) 'Binary component adaptation,' In *ECOOP '98: Proceedings of the 12th European Conference on Object-Oriented Programming*, pp. 307–29, London, UK, Springer-Verlag.

[11] I. Mouakher, A. Lanoix, and J. Souqui'eres (2006) 'Component adaptation: specification and verification,' *http://research.microsoft.com/~cszypers/events/WCOP2006/wcop06-Lanoix.pdf*.

[12] J.A. Kim, O.-C. Kwon, J. Lee, and G.-S. Shin (2001) 'Component adaptation using adaptation pattern components,' *The 2001 IEEE International Conference on Systems, Man, and Cybernetics.*, pp. 1025–9 **2** Tucson, AZ, USA.

[13] G. Kniesel (1999) 'Type-safe delegation for run-time component adaptation,' In *ECOOP '99: Proceedings of the 13th European Conference on Object-Oriented Programming*, pp. 351–66, London, UK, Springer-Verlag.

[14] D. Notkin, D. Garlan, W.G. Griswold, and K.J. Sullivan (1993) 'Adding implicit invocation to languages: three approaches,' In *Proceedings of the First JSSST International Symposium on Object Technologies for Advanced Software*, pp. 489–510, London, UK, Springer-Verlag.

[15] D.L. Parnas (1978) 'Designing software for ease of extension and contraction,' In *Proceedings of the 3rd International Conference on Software Engineering, IEEE*, pp. 264–77.

[16] Sun. Java API Specification. *http://www.java.sun.com*.

[17] R.N. Taylor, N. Medvidovic (1996) 'A component- and message-based architectural style for gui software,' *IEEE Transactions on Software Engineering* **22**(6): 390–406.

[18] G.T. Heineman (1999) 'Adaptation of software components,' *Technical report, WPI*.

[19] D.M. Yellin, R.E. Strom (1997) 'Protocol specifications and component adaptors' *ACM Trans. Program. Lang. Syst.* **19**(2): 292–333.

[20] L. Fei, H. Yanxiang, and Z. JingJing (2005) 'Implementing grid load balance based on performance prediction and agent' *Computer Engineering* **31**(15): 104–6.

16

On Probabilistic k-Coverage in Pervasive Wireless Sensor Networks[1]

Habib M. Ammari

WiSeMAN Research Lab., Department of Computer Science, Hofstra University, Hempstead, NY 11549, USA.

16.1 Introduction

Advances in wireless communications and sensor technology led to the design of *pervasive wireless sensor networks* (WSN), which have rapidly become a reality. A WSN consists of tiny, low-powered sensors that communicate with each other in a multi-hop fashion and report their sensed data to a central gathering point, called the *sink*, for further analysis and processing. Typical civilian, environmental, natural and military applications of WSNs include health monitoring, environmental monitoring, seismic monitoring and battlefield surveillance. WSNs suffer from highly limited resources, such as battery power (or *energy*), bandwidth, CPU, and storage, with energy being the most crucial.

One of the important research issues in WSNs is *sensor scheduling* with the goal being to achieve (*redundant*) *coverage* of a field. Indeed, there are several applications, such as intruder detection and tracking, which require that each location in a field be sensed by at least one sensor. In these types of application, the accuracy of the decision-making process depends greatly on the availability of data gathered during monitoring. In particular, k-coverage is a solution which demands that each point in a field be sensed by k sensors at least. For energy saving, it is important to deploy as few a number of sensors as possible in order to achieve k-coverage for a given application. Thus, it is necessary that redundant sensors be turned *off* so as to save energy and extend the network's lifetime. One related problem is the *art gallery*, which is a visibility problem that has been much studied in computational geometry. The problem is to find the minimum number of observers (or security cameras) that is required to guard a polygon field. A solution to this problem exists if for any point in the polygon field there is at least one observer such that the line segment connecting the point and the observer lies entirely within the polygon field.

The problem of coverage has been much studied in the literature [3, 16, 17, 19]. The issue of coverage-preserving scheduling (or duty-cycling) has also received considerable attention [12, 13, 32, 37]. Several works on k-coverage in WSNs assumed a perfect sensing model (also known as a *deterministic sensing model*), where a point in a field is guaranteed to be covered by a sensor provided that this point is within the sensor's sensing range [37]. A few works considered a more realistic sensing model, called a *stochastic sensing model*, in the design of sensor scheduling protocols while preserving either coverage [20, 23, 41] (respectively, k-coverage of a field [30, 32, 41]), where a point is covered (or *sensed*) by a sensor (respectively, k sensors) with some probability. While some approaches focused on coverage only [1], others considered coverage and connectivity in an integrated

[1] This work is partially supported by the US National Science Foundation (NSF) grant 0917089 and a New Faculty Start-Up Research Grant from Hofstra College of Liberal Arts and Sciences Dean's Office.

framework [32, 37]. Indeed, coverage deals with all locations in a field, and thus informs on how well a phenomenon is monitored, whereas connectivity is related to the locations of the sensors, and thus quantifies how well the active sensors communicate with each other and forward sensed data on behalf of each other to the sink. Accordingly, sensing coverage and network connectivity should be considered jointly so as to ensure the correct operation of the network.

In this chapter, we focus on the sensor scheduling problem in order to guarantee *sensing coverage*, where each point in a field is *covered* by at least one sensor, while maintaining the *network connectivity* which is needed for data routing. More precisely, we focus on the problem of probabilistic coverage, where each point in a field is covered probabilistically (or covered with some probability). In particular, we focus on the problem of probabilistic *k*-coverage in pervasive wireless sensor networks, where each point in a field is *k*-covered probabilistically, i.e. covered by at least *k* sensors. To the best of our knowledge, only the coverage configuration protocol (CCP) [32] and the stochastic *k*-coverage protocol (SCP$_k$) have considered redundant coverage [2]. First, we present the existing work on how to compute the sensor density in order to achieve coverage or *k*-coverage of a field of interest. Also, we provide a brief overview of protocols in the literature on coverage in pervasive wireless sensor networks. Then, we provide a detailed description of CCP and SCP$_k$ protocols.

The remainder of this chapter is organized as follows. Section 16.2 reviews sample approaches to computing sensor spatial density for coverage in pervasive wireless sensor networks and describes a few coverage protocols. Sections 16.3 and 16.4 focus on the problem of probabilistic sensing *k*-coverage. While Section 16.3 discusses the coverage configuration problem (CCP) [32], Section 16.4 presents the stochastic *k*-coverage protocol (SCP$_k$) [2]. Section 16.5 concludes the chapter.

16.2   The Coverage Problem

The issue of coverage was thoroughly investigated by Koushanfar *et al.,* [12] using computational geometry and graph theoretic techniques. In particular, they addressed the coverage problem from different perspectives, including deterministic, statistical, worst and best case and illustrative examples for each domain [12]. In this section, we present some approaches for computing sensor spatial density in order to cover a field with a minimum number of sensors. We also provide an overview of coverage protocol for pervasive wireless sensor networks.

16.2.1   Computing Density for Coverage

Adlakha and Srivastava [1] proposed an exposure-based model in order to find the sensor density required to achieve full coverage of a desired region based on the physical characteristics of the sensors and the properties of the target. They showed that the number of sensors required to cover an area of size A is in the order of $O(A/\hat{r}_2^2)$, where $\hat{r}_2$ is a good estimate of the radius r of the sensing disk of the sensors. Specifically, r lies between $\hat{r}_1$ and $\hat{r}_2$, where $\hat{r}_1$ overestimates the total number of sensors required to cover an area of size A, while $\hat{r}_2$ underestimates it. Our approach, however, provides an exact value of the minimum sensor spatial density required for k-coverage based on the exact value of the sensing range of the sensors. Franceschetti *et al.* [11] investigated the number of disks of given radius r, centered at the vertices of an infinite square grid, which are required to cover entirely an arbitrary disk of radius r placed on the plane. Their result depends on the ratio of r to the grid spacing.

Kumar *et al.* [17] proved that for random deployment with uniform distribution, if there exists a slowly growing function $\phi(np)$ such that $np\pi r^2 \geq \log(np) + k\log\log(np) + \phi(np)$, then a square unit area is k-covered with high probability when n sensors are deployed in it, where p is the probability that a sensor is active. It is worth noting that n also represents the sensor spatial density given that the area of the square region is equal to 1. Hence, the inequality above can be written as $np \geq \frac{\log(np)+k\log\log(np)+\phi(np)}{\pi r^2}$, which means that the minimum sensor density required for k-coverage of a unit square region is equal to $\frac{\log(np)+k\log\log(np)+\phi(np)}{\pi r^2}$. If we set $p = 1$ (i.e., every sensor is active), we obtain $\frac{\log(n)+k\log\log(n)+\phi(n)}{\pi r^2}$. In [17], it was showed that the minimum number of sensors needed to achieve k-coverage with high probability is approximately the same regardless of whether the sensors are deployed deterministically or randomly, if the sensors fail or sleep independently with equal probability. Recently, Balister *et al.* [4] computed the sensor density necessary to achieve both sensing coverage and network connectivity in finite region, such as thin strips (or annuli) whose lengths are finite. Balister *et al.* [4] applied this result to achieve barrier coverage [16] and connectivity in thin strips, where sensors act as a barrier that ensures that any moving object that crosses the barrier of sensors will be detected.

Zhang and Hou [36, 38] proved that the required density for *k*-coverage of a square field, where sensors are distributed according to a Poisson point process and are always active, depends on both the side length of the field and *k*. Put precisely, Zhang and Hou [36, 38] found that a necessary and sufficient condition of complete *k*-coverage of a square field with side length *l* is that the sensor density is equal to $\lambda = \log l^2 + (k + 1) \log \log l^2 + c(l)$, where $c(l) \to +\infty$ as $l \to \infty$. In another paper, given a wireless sensor network deployed as a Poisson point process with density λ and every sensor is active, Zhang and Hou [35] provided a sufficient condition for *k*-coverage of a square region with area *A*. They proved that assuming $\lambda = \log A + 2k \log \log A + c(A)$, if $c(A) \to \infty$ as $A \to \infty$, then the probability of *k*-coverage of the square region approaches 1. Zhang and Hou [35] provided the same result in the case where sensors are deployed according to a uniformly random distribution. Both results are based on the following statement: the square region is divided into square grids with side length $s = \frac{\sqrt{2}r}{\log A}$, where *r* stands for the radius of the sensing range of the sensors. For a grid *i* to be completely *k*-covered, it is sufficient that there are at least *k* sensors within a disk centered at the center of the grid and with radius $(1 - u)r$, denoted by $B_i((1 - u)r)$, where $u = 1/\log A$.

Wan and Yi [29] showed that with boundary effect, the asymptotic $(k + 1)$-coverage of a square with area *s* by Poisson point process with unit-area coverage range requires that the sensor density be equal to $\log s + 2(k + 1) \log \log s + \xi(s)$ with $\lim_{s \to \infty} \xi(s) = \infty$. Without the boundary effect, however, the asymptotic $(k + 1)$-coverage requires that the sensor density be computed as $\log s + (k + 2) \log \log s + \xi(s)$ with $\lim_{s \to \infty} \xi(s) = \infty$.

16.2.2 Overview of Coverage Protocols

We will now discuss a sample of deterministic and probabilistic coverage protocols for pervasive wireless sensor networks.

16.2.2.1 Deterministic Approaches

An optimal deployment strategy was proposed in [3] for full coverage and 2-connectivity regardless of the relationship between *R* and *r*. Centralized and distributed algorithms for connected sensor cover were proposed in [12] so the network can self-organize its topology in response to a query and activate the necessary sensors to process it. Distributed protocols to guarantee both coverage and connectivity of WSNs were proposed in [13]. Efficient distributed algorithms were proposed in [19] to solve optimally the best-coverage problem with the least energy consumption. Optimal polynomial time worst and average case algorithms were proposed in [22] for coverage calculation based on the Voronoi diagram and graph search algorithms. A distributed algorithm was proposed in [37] to keep a small number of active sensors regardless of the relationship between *R* and *r*.

A *k*-barrier coverage protocol for intrusion detection and an optimal deployment pattern for achieving it were proposed in [16]. The work in [32, 37] was improved by proving that if the original network is connected and the identified active nodes can cover the same region as all of the original nodes, then the network formed by the active nodes is connected when the communication range is at least twice the sensing range [27]. The problem of selecting a minimum size connected *k*-cover was discussed in [41] and a greedy algorithm to achieve *k*-coverage with a minimum set of connected sensors was proposed in [41].

16.2.2.2 Probabilistic Approaches

Lazos and Poovendran [18] formulated the coverage problem in heterogeneous WSNs as a set intersection problem. They also derived analytical expressions, which quantify the coverage achieved by stochastic coverage. Meguerdichian *et al.*, [23] studied the exposure in WSNs, which is related to the quality of coverage provided by WSNs, based on a general sensing model, where the sensing signal of a sensor at an arbitrary point is by a function that is inversely proportional to the distance between the sensor and point. Liu and Towsley [20] studied three coverage measures, area coverage, node coverage and detectability using the general sensing model defined in [23].

Liu *et al.* [21] presented a joint scheduling scheme based on a randomized algorithm for providing statistical sensing coverage and guaranteed network connectivity. This scheme works without the availability of per-node location information. Zou and Chakrabarty [41] proposed a distributed approach for the selection of active sensors to cover a field fully based on the concept of a connected dominating set. This approach is based on a probabilistic sensing model, where the probability of the existence of a target is defined by an exponential function that represents the confidence level of the received sensing signal.

16.3 Coverage Configuration Problem

The problem of integrated coverage and connectivity was addressed originally by Xing *et al.* [32]. In particular, the concept of connected k-coverage in pervasive wireless sensor networks was also first introduced by the same research team [31]. They analyzed the relationship between coverage and connectivity in a network and showed that this relationship depends on the ratio of the communication range to the sensing range of the sensors. Before discussing their coverage protocol, we present some key assumptions and definitions.

16.3.1   Key Assumptions

All of the sensors are static and homogeneous with regard to all of their capabilities. In particular, they have the same sensing and communication ranges. Also, the sensing and communication ranges of the sensors are represented by disks of radii R and r, respectively. Moreover, every sensor is aware of its precise location in the field though some location system, such as a global positioning system (GPS), or using some localization technique [6].

16.3.2   Definitions

A point p is said to be *covered* by a sensor s_i if p is located within the sensing disk of s_i. In this case, we say that the field is *covered* if any point in the field is covered by at least one sensor. A more general concept of coverage is k-coverage. A point p is said to be *k-covered* if p is covered by at least k sensors, i.e. p is located within the sensing disks of at least k sensors simultaneously.

A sensor s_i can *directly communicate* with a sensor s_j if the Euclidean distance between s_i and s_j does not exceed R, where R stands for the radius of their communication range. The *communication graph* of a network, denoted by $G = (S, E)$, consists of a set of sensors S and a set of communication links E such that for any pair of sensors s_i and s_j, we have $(s_i, s_j) \in E$ if s_i can communicate directly with a sensor s_j. The *connectivity* of a communication graph (i.e., network) is the number of nodes whose removal disconnects the graph into at least two sub-graphs.

16.3.3   Fundamental Results

Given a communication graph of a network, Xing *et al.* [32] provided the following result, which establishes a relationship between coverage and connectivity:

- **Theorem 1:** *For a set of nodes that at least 1-cover a convex region* A, *the communication graph is connected if* R $\geq$ 2 r, *where* R *and* r *stand for the radii of the communication and sensing ranges of the sensors.*

Furthermore, Xing *et al.* [32] extended their result to account for k-coverage. Specifically, they proposed a fundamental result that relates k-coverage to network connectivity. This result is expressed as follows:

- **Theorem 2:** *A set of nodes that* k-*cover a convex region* A *forms a* k-*connected communication graph if* R $\geq$ 2 r.

As can be seen, the theorem above shows that if a network is configured to provide k-coverage, the network is connected and its connectivity is equal to k provided that the radius of the communication range of the sensors is at least double the radius of their sensing range.

16.3.4   Protocol Description

For the sensors to k-cover a field, each runs an *eligibility algorithm* to decide whether to become active and participate in the k-coverage process. More precisely, each sensor checks whether its sensing range is already *k*-covered by other active sensors. If not, the sensor will become active. In order to characterize *k*-coverage of a region of interest, let us give the following definitions:

- A point *p* is said to be an *intersection point* between sensors s_i and s_j if *p* is an intersection point of the sensing circles of s_i and s_j. That is, *p* is located on the boundary of the sensing range of s_i and s_j.
- A point *p* is said to be an *intersection point* between sensors s_i and an area to be covered *A* if *p* is a point within *A* and the Euclidean distance between *p* and s_i is equal to *r*.

Based on these two definitions, Xing *et al.* [32] gave the following result, which is the basis for the eligibility algorithm:

- **Theorem 3:** *A convex area A is k-covered by a set of nodes if all the intersection points between the sensing circles of the sensors are k-covered, and all the intersection points between the sensors and A are k-covered.*

Based on this theorem, Xing *et al.* [32] proposed a coverage configuration protocol (CCP) to k-cover a field while maintaining connectivity provided that the condition $R \geq 2\,r$ holds. Indeed, when $R \geq 2\,r$ is not true, network connectivity cannot be ensured. Therefore, in order to deal with a more general case, and particularly when $R < 2\,r$, the CCP protocol is integrated with a connectivity maintenance protocol (SPAN) [8]. The latter is an energy conservation protocol that turns unnecessary nodes off while maintaining a communication backbone that consists of active nodes, which are connected to each other. Moreover, all inactive nodes are directed to at least one active node in the communication backbone.

The state of a node (active or inactive) when $R < 2r$ depends on the eligibility rules of CCP and SPAN can be stated as follows:

- An inactive node is eligible to become active based on its eligibility according to CCP or SPAN.
- An active node becomes inactive based on its eligibility according to CCP or SPAN.

While *k*-coverage is guaranteed by CCP, active nodes may not be connected to each other when $R < 2r$. In this case, SPAN will activate more nodes, each of which can reach an active node located in its transmission range.

An updated version of SPAN is used by CCP in which a HELLO message includes the coordinates of each neighboring coordinator. This allows each node to maintain in its neighborhood table the locations of all two-hop neighboring coordinators. The motivation behind this table structure is to reduce the number of active nodes.

Now, we focus on the probabilistic coverage model which relaxes the assumption made about the sensing capability of the sensors. In this case, the sensors may have non-uniform and irregular communication and sensing ranges. Moreover, each point within the sensing range of a sensor is sensed with a certain probability P $(0 < P < 1)$.

Under this probabilistic model, the coverage configuration problem can be stated as follows:

Given a convex region *A*, and application-specific parameters *k* $(k \geq 1)$ and β $(0 < \beta \leq 1)$, select a minimum number of sensors to remain active while the probability that each point in *A* is sensed by at least *k* sensors, is no lower than β. This probabilistic coverage model is denoted by (k, β)-coverage.

The use of CCP to provide probabilistic coverage requires mapping the (k, β)-coverage requirement to a pseudo-coverage degree k'. The latter, which will be an input to the CCP protocol, is computed as follows:

$$1 - \sum_{i=0}^{k-1} \binom{k'}{i} P^i (1 - P)^{k'-i} \geq \beta$$

where *P*, *k*, and β are known. The value of the lower bound of the pseudo-coverage degree will be used by CCP to achieve probabilistic coverage of a convex deployment region. Figure 16.1 shows that the lower bound of the pseudo-coverage degree k' increases roughly linearly with *k*. It is worth noting that the (k, β)-coverage model can be applied to several real-world sensing applications, such as distributed target detection based on the constant false alarm rate [28].

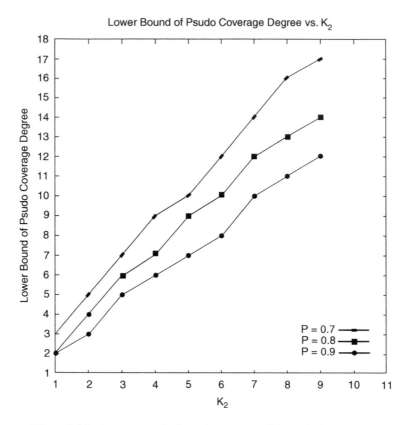

Figure 16.1   Lower bound of pseudo coverage k' depends linearly on k.

16.4   Stochastic k-Coverage Protocol

In this section, we consider the problem of *stochastic k-coverage* in WSNs, where the degree of coverage $k \geq 3$. More specifically, we divide this problem into two sub-problems, namely *stochastic k-coverage characterization* and *stochastic k-coverage-preserving scheduling*, which can be formulated as follows:

- *Stochastic k-coverage characterization sub-problem:* Find a tight sufficient condition so that every point in the field is covered probabilistically by at least k sensors with a probability no less than p_{th}, called *threshold probability*, under a stochastic sensing model and compute the required minimum number of sensors.
- *Stochastic k-coverage preserving scheduling sub-problem:* Show how to select and schedule the sensors while providing *stochastic k*-coverage of the field as well as connectivity between the selected sensors.

First, we analyze the k-coverage problem in WSNs using the deterministic sensor model. Then, we adapt this analysis to our stochastic sensing model.

This choice of $k \geq 3$ is motivated by the existence of at least the following three applications. First, in order to cope with the problem of sensor failures due their fragility, the design of sensor networks for planet exploration [26] should be as reliable as possible since failed sensors in space cannot be easily diagnosed and replaced. In [26], a Confidence Weighted Voting technique is simulated on top of k-cover deployment strategy, which guarantees data redundancy when $k \geq 3$. Second, multiple-sensor data fusion was found to be useful for at least a three-sensor system, i.e. a system whose degree of coverage is $k \geq 3$ [14]. This type of k-covered WSN helps detect, classify and track the target objects. As at least all three sensors participate in the decision, it is unlikely that a false target would be detected as a true target [14]. And third, the design of triangulation-based positioning systems requires that each point in a field be covered by at least $k \geq 3$ sensors to increase the position accuracy [24].

16.4.1   Stochastic Sensing Model

In the *deterministic sensing model*, a point ξ in a field is covered by a sensor s_i based on the Euclidean distance $\delta(\xi, s_i)$ between ξ and s_i. In this chapter, we use 'coverage of a point' and 'detection of an event' interchangeably. Formally, the coverage $Cov(\xi, s_i)$ of a point ξ by a sensor s_i is equal to 1, if $\delta(\xi, s_i) \leq r$, and 0, else. As can be seen, this sensing model considers the sensing range of a sensor as a *disk*, and thus all sensor readings are precise and have no uncertainty. However, given the signal attenuation and the presence of noise associated with sensor readings, it is necessary to consider a more realistic sensing model by defining $Cov(\xi, s_i)$ using some probability function. That is, the sensing capability of a sensor must be modeled as the probability of successful detection of an event, and thus should depend on the distance between it and the event as well as the type of propagation model being used (free-space vs. multi-path). Indeed, it has been showed that the probability that an event in a distributed detection application can be detected by an acoustic sensor depends on the distance between the event and the sensor [9]. A realistic sensing model for passive infrared (PIR) sensors that reflects their non-isotropic range was presented in [7]. This sensing irregularity of PIR sensors was verified by simulations [7]. Thus, in our *stochastic sensing model*, the coverage $Cov(\xi, s_i)$ is defined as the *probability of detection* $p(\xi, s_i)$ of an event at point ξ by sensor s_i as follows:

$$p(\xi, s_i) = \begin{cases} e^{-\beta\delta(\xi, s_i)^\alpha} & \text{if } \delta(\xi, s_i) \leq r \\ 0 & \text{otherwise} \end{cases} \tag{16.1}$$

where β represents the physical characteristic of the sensors' sensing units and $2 \leq \alpha \leq 4$ is the path-loss exponent. To be specific, we have $\alpha = 2$ for the free-space model, and $2 < \alpha \leq 4$ for the multi-path model. Our stochastic sensing model is motivated by that introduced by Elfes [10], where the sensing capability of a sonar sensor is modeled by a Gaussian probability density function. A probabilistic sensing model for coverage and target localization in WSNs was proposed in [42]. This sensing model considers $\delta(\xi, s_i) - (r - r_e)$ instead of $\delta(\xi, s_i)$, where r is the detection range of the sensors and $r_e < r$ is a measure of detection uncertainty.

16.4.2   *k*-Coverage Characterization

In order to characterize *k*-coverage, we need to compute the maximum size of an area (*not* a field) that is surely *k*-covered with exactly *k* sensors. According to the deterministic sensing model defined above, a point is covered if the distance between it and at least one sensor is at most equal to the radius of the sensing range of the sensors. Put precisely, we want to determine the shape of this *k*-covered area. To this end, we present a fundamental result of convexity theory, known as *Helly's Theorem* [5], which characterizes the intersection of convex sets. We will exploit this theorem to characterize *k*-coverage of a field with the help of a nice geometric structure, called *Reuleaux triangle* [43]. Then, we compute the minimum number of sensors required for *k*-coverage.

- **Helly's Theorem [5]:** *Let* E *be a set of convex sets in* R^n *such that for* m $\geq$ n+1, *any* m *members of* E *have a non-empty intersection. Then, the intersection of all members of* E *is non-empty.*

Lemma 1 is an instance of Helly's Theorem [5] in a two-dimensional space that characterizes the intersection of *k* sensing disks.

- **Lemma 1 [2]:** *Let* k $\geq$ 3. *The intersection of* k *sensing disks is not empty if and only if the intersection of any three of those* k *sensing disks is not empty.*

Based on Lemma 1, Lemma 2 gives a *sufficient condition* for *k*-coverage of a field [2]. For the sake of completeness, we also give the proof of Lemma 2.

- **Lemma 2 [2]:** *Let* k $\geq$ 3. *A field is* k*-covered if any* Reuleaux triangle *region of width* r *(or simply* slice*) in the field contains at least* k *active sensors.*

- *Proof.* Let *A* be the intersection area of the sensing disks of *k* sensors. From Lemma 1, it is clear that the width of *A* should be upper-bounded by *r*, the radius of the sensing disks of the sensors, so that any location in *A* is *k*-covered by these *k* sensors. First, we consider the case of three sensors (i.e. *k* = 3). Then, we generalize

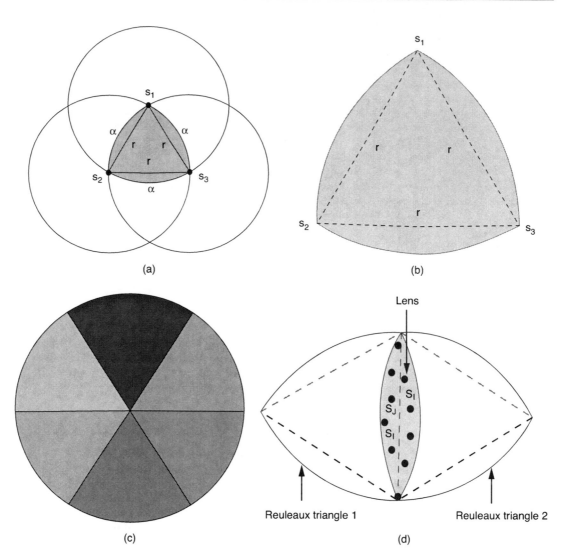

Figure 16.2   (a)–(b) Reuleaux triangle (c) Slicing of sensing disk (d) Adjacent Reuleaux triangles.

our analysis for any value of $k > 3$. Using the Venn diagram given in Figure 16.2(a), the maximum size of the intersection of the sensing disks of sensors s_1, s_2, and s_3, called Reuleaux triangle [43] and denoted by $RT(r)$, is obtained when s_1, s_2, and s_3 are located symmetrically from each other so that the distance between any pair of sensors is equal to r as shown in Figure 16.2(a)–(b). Notice that $RT(r)$ has a constant width equal to r. as can be seen, each sensor is able to sense the location of the other two sensors. Hence, the center of the sensing disk of each sensor is located at the intersection of the other two. By definition of the intersection set operator, the intersection of k sensing disks is always less than or equal to that of any three among these k sensing disks such that the distance between any pair of sensors it at most equal to r. Thus, the maximum size of A that is surely k-covered with exactly k sensors is equal to the area of the Reuleaux triangle of width r.   ∎

It is worth noting that *tiling* a field by Reuleaux triangles is impossible unless we allow overlapping between them. Figure 16.2(c) shows that the sensing disk of a sensor can be covered by six Reuleaux triangles of width r with a minimum overlap when two sides of the triangles associated with two slices are coinciding totally with each other. Thus, the minimum overlap area of two adjacent slices forms a *lens* (Figure 16.2(d)). Lemma 3 characterizes k-coverage based on this notion of a lens.

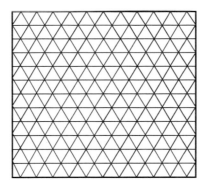

Figure 16.3   Slicing grid of a square sensor field.

- **Lemma 3:** *k active sensors located in the lens of two adjacent slices in a field k-cover both slices.*

- *Proof.* The proof is verbatim: the distance between any of these k sensors located in the lens and any point in the area of the union of both slices is at most equal to r.   ■

  Based on Lemma 3, Theorem 4 states a *tighter sufficient condition* for *k*-coverage of a field.

- **Theorem 4:** *Let* k ≥ 3. *A field is* k-*covered if for any slice of width* r *in the field there is an adjacent slice of width* r *such that their* lens *contains at least* k *active sensors, where* r *is the radius of the sensing disks of the sensors.*

- *Proof.* Assume that a field is decomposed into slices, thus forming a *slicing grid* (Figure 16.3), where for any slice in the field there exists an adjacent slice such that their *lens* contains at least *k* active sensors. Given that the width of a slice is constant and equal to *r*, it is always true that the maximum distance between any of these *k* sensors and any point in both adjacent slices cannot exceed *r*. Thus, any point in the union of these two adjacent slices is *k*-covered. Since this applies to any slice in the field, the field is also *k*-covered.   ■

16.4.3   Protocol Description

In this section, we exploit the results of Section 16.4.2 to characterize probabilistic *k*-coverage in WSNs based on our stochastic sensing model. Theorem 5 computes the *minimum k-coverage probability* $p_{k,\min}$ such that every point in a field is probabilistically *k*-covered.

- **Theorem 5:** *Let* r *be the radius of the nominal sensing range of the sensors and* k ≥ 3. *The minimum* k-*coverage probability so that each point in a field is probabilistically* k-*covered by at least* k *sensors under our stochastic sensing model defined in (1) is given by*

$$p_{k,\min} = 1 - \left(1 - e^{-\beta r^{\alpha}}\right)^{k} \tag{16.2}$$

- *Proof.* First, we identify the least *k*-covered point in a field so we can compute $p_{k,\min}$. By Theorem 4, *k* sensors should be deployed in the lens of two adjacent slices of a field to achieve *k*-coverage with a minimum number of sensors. By looking at the lens in Figure 16.4 and using the result of Theorem 4, it is easy to check that the point ξ is the least *k*-covered point when all deployed *k* sensors are located on the bottom boundary (or arc) of the lens. Indeed, ξ is the farthest point from the bottom arc of the lens. Moreover, ξ is equidistant from the *k* sensors $(s_1, .., s_k)$ located on the bottom arc of the lens as shown in Figure 16.4. Hence, the distance between ξ and each of these *k* sensors is equal to *r*. Thus, the *minimum k-coverage probability* for the least *k*-covered point ξ by *k* sensors under the stochastic sensing model in equation (16.1) is given by

$$p_{k,\min} = 1 - \prod_{i=1}^{k}(1 - p(\xi, s_i)) = 1 - \left(1 - e^{-\beta r^{\alpha}}\right)^{k}$$

                                                                                                          ■

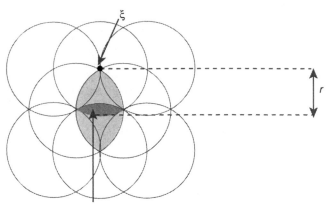

k sensors located on the boundary of a lens

Figure 16.4   Least k-covered point.

The stochastic k-coverage problem is to select a minimum subset $S_{min} \subseteq S$ of sensors such that each point in a field is k-covered by at least k sensors and that the minimum k-coverage probability of each point is at least equal to some given threshold probability p_{th}, where $0 < p_{th} < 1$. This helps us compute the stochastic sensing range r_s, which provides probabilistic k-coverage of a field with a probability no less than p_{th}. Lemma 4 computes the value of r_s.

- **Lemma 4:** *Let* k ≥ 3 *and* $2 \leq \alpha \leq 4$. *The stochastic sensing range* r_s *of the sensors that is necessary to probabilistically* k*-cover a field with a minimum number of sensors and with a probability no lower than* $0 < p_{th} < 1$ *is given by*

$$r_s = \left(-\frac{1}{\beta} \log\left(1 - (1 - p_{th})^{1/k}\right) \right)^{1/\alpha} \tag{16.3}$$

   where β represents the physical characteristic of the sensors' sensing units.

- *Proof.* $p_{k,min} \geq p_{th} \Rightarrow r_s \leq \left(-\frac{1}{\beta} \log\left(1 - (1 - p_{th})^{1/k}\right) \right)^{1/\alpha}$ ∎

The upper bound on the stochastic sensing range r_s of the sensors computed in equation (16.3) will be used as one of the input parameters to the *k-coverage candidacy* algorithm, which will be described below. Figure 16.5 shows r_s for different values of p_{th} and k while considering the free-space model ($\alpha = 2$) (Figure 16.5(a)) and the multi-path model ($\alpha = 4$) (Figure 16.5(b)). As can be seen, rs decreases as a function of p_{th}, k, and α. This is due to the fact that $p_{k,min}$ of the same location by multiple sensors decreases as p_{th}, k, and α increase.

Similarly to Lemma 2 and Theorem 4, while Lemma 5 states a sufficient condition for probabilistic k-coverage, Theorem 6 states a tighter sufficient condition for probabilistic k-coverage based on our stochastic sensing model, p_{th}, and k, respectively.

- **Lemma 5:** *Let* k ≥ 3. *A field is probabilistically* k*-covered with a probability no lower than* $0 < p_{th} < 1$ *if any slice of width* r_s *in the field contains at least* k *active sensors.*

- **Theorem 6:** *Let* k ≥ 3. *A field is* k*-covered probabilistically with a probability no lower than* $0 < p_{th} < 1$ *if for any slice of width* r_s *in the field there exists an adjacent slice of width* r_s *such that their* lens *contains at least* k *active sensors, where r_s is the stochastic sensing range of the sensors.*

Lemma 6 states a sufficient condition for connectivity between sensors under our stochastic sensing model.

- **Lemma 6:** *Let* k ≥ 3. *The sensors that are selected to* k*-cover a field with a probability no less than* $0 \leq p_{th} \leq 1$ *under the stochastic sensing model defined in equation (16.1) are connected if the radius of their communication range is at least equal to* $\sqrt{3}r_s$, *where* r_s *is the stochastic sensing range of the sensors.*

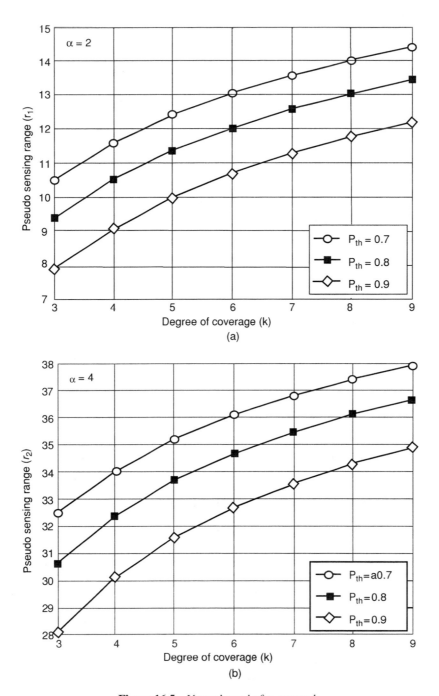

Figure 16.5　Upper bound of r_s versus k.

16.4.3.1　*k*-Coverage Candidacy Algorithm

A sensor turns active if its sensing disk is not *k*-covered. Based on Theorem 5, a sensor randomly decomposes its sensing range into six slices of width r_s (Lemma 4) and checks whether the lens of each of the three pairs of adjacent slices contains at least *k* active sensing neighbors. If any of the three lenses does not have *k* active sensors, a sensor checks whether each slice of width r_s contains at least *k* active sensors based on Lemma 5. Otherwise, it is a candidate to become active.

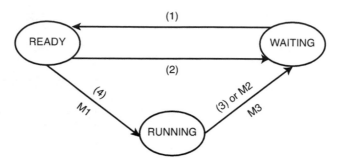

(M1) Sending out NOTIFICATION-RUNNING message
(M2) Receive READY-to-RUNNING message
(M3) Sending out NOTIFICATION-WAITING message

(1) t_{waist} expires (2) t_{ready} expires (3) $t_{transit}$ expires
(4) ($t_{inactive}$ and $t_{transit}$ expire) or k-coverage candidate

Figure 16.6   State transition diagram.

16.4.3.2   The State Transition of SCP$_k$

Figure 16.6 shows a state transition diagram associated with our *stochastic k-coverage protocol* (SCP$_k$) and that has three states: READY, WAITING, and RUNNING.

- READY state: A sensor listens to messages and checks its k-coverage candidacy to switch to RUNNING state.
- RUNNING state: A sensor is active and can communicate with other sensors and sense the field.
- WAITING state: A sensor is neither communicating with other sensors nor sensing a field, and thus its radio is turned *off*. However, after some fixed time interval, it switches to the READY state to check its candidacy for k-coverage and receives messages.
- At the beginning of their monitoring task, all sensors are in the RUNNING state. Moreover, each sensor chooses randomly and independently of all other sensors a value t_{check} between 0 and t_{check}_max after which it runs the k-coverage candidacy algorithm to check whether it stays active or switch to the WAITING state. Our intuition behind this random selection of t_{check} is to avoid as much as possible higher or lower coverage of any region in a field.
- When a sensor runs the k-coverage candidacy algorithm and finds out that it is a candidate, it sends out a NOTIFICATION-RUNNING message to inform all its neighbors. While in the READY state, a sensor keeps track of its sensing neighbors that are in the RUNNING state. If it finds out that its sensing area is k-covered, it will switch to the WAITING state.
- For energy efficiency purposes, a sensor may wish to switch from RUNNING state to WAITING state. For this purpose, a sensor broadcasts a RUNNING-to-WAITING message and waits for some transit time $t_{transit}$. If $t_{transit}$ expires and it has not received any RUNNING-to-WAITING message, it switches to the WAITING state and sends a NOTIFICATION-WAITING message, where it can stay there for t_{wait} time. When t_{wait} expires, it switches to the READY state, where it can stay for t_{ready} time. When a sensor in the READY state receives a RUNNING-to-WAITING message from its sensing neighbor, it runs the k-coverage candidacy algorithm.

If a sensor in the READY state finds out that it has not been active for some $t_{inactive}$ time, it will broadcast a READY-to-RUNNING message and wait for some $t_{transit}$ time. If $t_{transit}$ expires without receiving any other READY-to-RUNNING , it will send out a NOTIFICATION-RUNNING message and switch to the RUNNING state. Otherwise, it stays in the READY state. a sensor in the READY state would also apply the same process if it finds out that it has not heard from one of its sensing neighbors within some t_{alive} time. This means that this sensing neighbor has entirely depleted its energy and died. To this end, each sensor in the RUNNING state should broadcast an ALIVE message after each t_{active} time. We assume each sensor stays active for at least t_{active} time.

16.5 Conclusion

In this chapter, we surveyed existing protocols on stochastic *k*-coverage in pervasive wireless sensor networks. Several approaches were proposed to compute the sensor spatial density to achieve coverage or *k*-coverage in this type of pervasive network, where $k > 1$. We have discussed a sample of these approaches. As can be noticed, in order gain more insight into the problem of stochastic *k*-coverage in pervasive wireless sensor networks, the approaches that we have reviewed considered a deterministic sensing model and then extended their analysis to a stochastic sensing model.

The performance of all of these protocols using either a deterministic or probabilistic model for sensing are based on their characterization of coverage or *k*-coverage of a field. As of now, there is no exact bound on the sensor spatial density for *k*-coverage. It is still an open problem and this is due mainly to the morphology of the sensing capability of the sensors. Indeed, it is well known that it is impossible to tile a planar field with disks without any overlap. Furthermore, a random deployment strategy, which is a widely used assumption in the design of protocols for pervasive wireless sensor networks, is another handicap that makes it impossible to compute an exact lower bound on sensor density, and hence *k*-cover a field using a minimum number of sensors. Another research of interest is the design of *k*-coverage protocols for heterogeneous pervasive wireless sensor networks using stochastic sensing model. In fact, real-world sensing applications may use sensors with various capabilities.

References

[1] S. Adlakha and M. Srivastava (2003) 'Critical density threshold for coverage in wireless sensor networks,' In *Proceedings of IEEE WCNC*, 2003, pp. 1615–20.
[2] H.M. Ammari (2009) 'Stochastic *k*-coverage and scheduling in wireless sensor networks,' *Invited paper*, *Proceedings of the International Conference on Wireless Algorithms, Systems and Applications (WASA)*, Springer LNCS, Boston, Massachusetts, USA, August 16–18.
[3] X. Bai, S. Kumar, D. Xuan, Z. Yun, and T.H. Lai (2006) 'Deploying wireless sensors to achieve both coverage and connectivity,' In *Proceedings of ACM MobiHoc*, pp. 131–42.
[4] P. Balister, B. Bollobas, A. Sarkar, and S. Kumar (2007) 'Reliable density estimates for achieving coverage and connectivity in thin strips of finite length,' In *Proceedings of ACM MobiCom*, pp. 75–86.
[5] B. Bollobás (2006) *The Art of Mathematics: Coffee Time in Memphis*, Cambridge University Press.
[6] N. Bulusu, J. Heidemann, and D. Estrin (2000) 'GPS-less low cost outdoor localization for very small devices,' *IEEE Personal Communications Magazine* 7(5): 28–34.
[7] Q. Cao, T. Yan, J. Stankovic, and T. Abdelzaher (2005) 'Analysis of target detection performance for wireless sensor network,' In *Proceedings of DCOSS*, LNCS 3560, pp. 276–92.
[8] B. Chen, K. Jamieson, H. Balakrishnan, and R. Morris (2002) 'Span: An energy-efficient coordination algorithm for topology maintenance in ad hoc wireless networks,' *ACM Wireless Networks* 8(5): 481–94.
[9] M. Duarte, Y. Hu (2003) 'Distance based decision fusion in a distributed wireless sensor network,' In *Proceedings of IPSN*, pp. 392–404.
[10] A. Elfes (1989) 'Using occupancy grids for mobile robot perception and navigation,' *IEEE Computer* 22(6): 46–57.
[11] M. Franceschetti, M. Cook, and J. Bruck (2002) 'A geometric theorem for wireless network design optimization,' *Tech. Report*, http://paradise.caltech.edu/papers/.
[12] H. Gupta, Z. Zhou, S. Das, and Q. Gu (2006) 'Connected sensor cover: Self-organization of sensor networks for efficient query execution,' *IEEE/ACM TON* 14(1): 55–67.
[13] C. Huang, Y. Tseng, and H. Wu (2007) 'Distributed protocols for ensuring both coverage and connectivity of a wireless sensor network,' *ACM TOSN* 3(1): 1–24.
[14] L.A. Klein (1993) 'A boolean algebra approach to multiple sensor voting fusion,' *IEEE Trans. Aerospace and Electronic Systems* 29(2): 317–27.
[15] F. Koushanfar, S. Meguerdichian, M. Potkonjak, and M. Srivastava (2001) 'Coverage problems in wireless ad-hoc sensor networks,' In *IEEE Proceedings Infocom*, pp. 1380–7.
[16] S. Kumar, T. Lai, and A. Arora (2005) 'Barrier coverage with wireless sensors,' In *Proceedings of ACM MobiCom*, pp. 284–98.
[17] S. Kumar, T. Lai, and J. Balogh (2004) 'On *k*-coverage in a mostly sleeping sensor network,' In *Proceedings of ACM MobiCom*, pp. 144–58.
[18] L. Lazos, R. Poovendran (2006) 'Stochastic coverage in heterogeneous sensor networks,' *ACM TOSN* 2(3), pp. 325–58.
[19] X. Li, P. Wan, O. Frieder (2003) 'Coverage in wireless ad-hoc sensor networks,' *IEEE TC* 52, 753–63.
[20] B. Liu, D. Towsley (2004) 'Study of the coverage of large-scale sensor networks,' In *Proceedings of IEEE MASS*, pp. 475–83.

[21] C. Liu, K. Wu, Y. Xiao, and B. Sun (2006) 'Random coverage with guaranteed connectivity: Joint scheduling for wireless sensor networks,' *IEEE TPDS* **17**(6): 562–75.

[22] S. Meguerdichian, F. Koushanfar, M. Potkonjak, and M. Srivastava (2005) 'Worst and best-case coverage in sensor networks,' *IEEE TMC* **4**(1): 84–92.

[23] S. Meguerdichian, F. Koushanfar, M. Potkonjak, and M. Srivastava (2001) 'Exposure in wireless ad-hoc sensor networks,' In *ACM MobiCom*, pp. 139–50.

[24] D. Nicules, B. Nath (2003) 'Ad-hoc positioning system (APS) using AoA,' In *Proceedings og IEEE INFOCOM*, pp. 1734–43.

[25] S. Shakkottai, R., Srikant, and N. Shroff (2003) 'Unreliable sensor grids: Coverage, connectivity and diameter,' In *Proceedings of IEEE INFOCOM*, pp. 1073–83.

[26] T. Sun, L. J. Chen, C. C. Han, and M. Gerla (2005) 'Reliable sensor networks for planet exploration,' In *Proceedings of IEEE Int'l Conf. Networking, Sensing and Control*, pp. 816–21.

[27] D. Tian, N. Georganas (2005) 'Connectivity maintenance and coverage preservation in wireless sensor networks,' *Ad Hoc Networks* **3**: 744–61.

[28] P. Varshney (1996) *Distributed Detection and Data Fusion*. Spinger-Verlag, New York, NY.

[29] P.-J. Wan, C.-W. Yi (2006) 'Coverage by randomly deployed wireless sensor networks,' *IEEE Transactions on Information Theory* **52**(6): 2658–69.

[30] Y.-C. Wang and Y.-C. Tseng (2008) 'Distributed deployment schemes for mobile wireless sensor networks to ensure multi-level coverage,' *IEEE TPDS* **19**(9): 1280–94.

[31] X. Wang, G. Xing, Y. Zhang, C. Lu, R. Pless, and C. Gill (2003) 'Integrated coverage and connectivity configuration in wireless sensor networks,' In *Proceedings of ACM SenSys*, pp. 28–39.

[32] G. Xing, X. Wang, Y. Zhang, C. Lu, R. Pless, and C. Gill (2005) 'Integrated coverage and connectivity configuration for energy conservation in sensor networks,' *ACM TOSN*, **1**(1): 36–72.

[33] S. Yang, F. Dai, M. Cardei, and J. Wu (2006) 'On connected multiple point coverage in wireless sensor networks,' *Int. Journal of Wireless Information Networks* **13**(4): 289–301, 2006.

[34] F. Ye, G. Zhong, J. Cheng, S. Lu, and L. Zhang (2003) 'PEAS: A robust energy conserving protocol for long-lived sensor networks,' In *Proceedings of ICDCS*, pp. 1–10.

[35] H. Zhang, J. Hou (2006) 'Is deterministic deployment worse than random deployment for wireless sensor networks?,' In *Proceedings of IEEE Infocom*, pp. 1–13.

[36] H. Zhang, J. Hou (2005) 'On the upper bound of α-lifetime for large sensor networks,' *ACM Transactions on Sensor Networks* **1**(2): 272–300.

[37] H. Zhang, J. Hou (2005) 'Maintaining sensing coverage and connectivity in large sensor networks,' *Ad Hoc & Sensor Wireless Networks* **1**(1–2): 89–124.

[38] H. Zhang, J. Hou (2004) 'On deriving the upper bound of α-lifetime for large sensor networks,' In *Proceedings of ACM MobiHoc*, pp. 121–32.

[39] Z. Zhou, S. Das, and H. Gupta (2004) 'Connected k-coverage problem in sensor networks,' In *Proceedings of ICCCN*, pp. 373–8.

[40] G. Zhou, T. He, S. Krishnamurthy, and J. Stankovic (2004) 'Impact of radio irregularity on wireless sensor networks,' In *Proceedings of MobiSys*, pp. 125–38.

[41] Y. Zou, K. Chakrabarty (2005) 'A distributed coverage- and connectivity-centric technique for selecting active nodes in wireless sensor networks,' *IEEE Transaction on Computers* **54**(8): 978–91.

[42] Y. Zou, K. Chakrabarty (2004) 'Sensor deployment and target localization in distributed sensor networks,' *ACM TOSN* **3**(2): 61–91.

[43] mathworld.wolfram.com/ReuleauxTriangle.html.

17

On the Usage of Overlays to Provide QoS Over IEEE 802.11b/g/e Pervasive and Mobile Networks

Luca Caviglione,[1] Franco Davoli,[2] and Piergiulio Maryni[3]
[1]*Institute of Intelligent Systems for Automation (ISSIA) – Genoa Branch, of the National Research Council of Italy (CNR), Genova, Italy.*
[2]*Department of Communications Computer and Systems Science (DIST), University of Genova, Genova, Italy.*
[3]*Datasiel S.p.A., Genova, Italy.*

17.1 Introduction

Pervasive computing is characterized by collaborative work performed by 'thinking' devices that are distributed at all stages throughout a network. It is also known as the 'everyware' paradigm [1]. Within this paradigm, applications might require some form of guarantee from the network, in terms of bandwidth or access and transfer delay. This demand for Quality of Service (QoS) can be addressed by different mechanisms, depending on the type of network being traversed.

In this context, one interesting way to provide QoS can be through pervasive techniques such as those adopted in peer-to-peer (P2P) communities, in which every P2P component participates locally in the provisioning of global QoS within a self-organizing framework. Providing QoS in a self-organizing framework (automatically and configuration-free) could be very important in pervasive environments where mobility and heterogeneity of connected devices and networking elements can be very high.

Whereas end-to-end QoS provision is still an open problem, several possibilities now exist in local environments and this chapter attempts to make a contribution in this direction, by focusing on the Wireless Local Area Network (WLAN) framework, which represents a good example of a 'local environment' where peer-to-peer techniques might have an interesting impact.

The remainder of the chapter is organized as follows: Section 17.2 investigates the capabilities offered by a P2P overlay for distributed bandwidth management in a wireless pervasive network scenario. In addition, it also showcases briefly other techniques adopted to support QoS requirements over WLANs. Section 17.3 discusses in detail both the design and the techniques, based upon P2P principles, to manage resources in WLANs, by describing specific use cases, and introduces a mechanism that enables: (i) to complement the functionalities of the IEEE 802.11e protocol and to enhance them to handle local mobility and dynamic requests for bandwidth; and (ii) to create an application level resource management to recover the lack of QoS support in lower layers, e.g. when employing IEEE 802.11b/g, without 802.11e functionalities. Section 17.4 analyzes the effectiveness of the

Pervasive Computing and Networking, First Edition. Edited by Mohammad S. Obaidat, Mieso Denko, and Isaac Woungang.
© 2011 John Wiley & Sons, Ltd. Published 2011 by John Wiley & Sons, Ltd.

proposed overlay-based approach, by means of simulations and prototypes deployed in an emulated environment, and Section 17.5 concludes the chapter. The detailed formalization of the distributed algorithm carrying out bandwidth requests is described in the Appendix.

17.2 A Glance at P2P Overlay Networks and QoS Mechanisms

In a P2P environment nodes have no distinguished role and algorithms that make the network run properly are – in principle –distributed completely.

Thus, in a pure P2P network, there is no single point of bottleneck or failure. Examples of such algorithms are: service discovery (name, address, route, metric, etc.), neighbor status tracking, application layer routing (based possibly on content, interest, etc.), resilience and handling link or node failure.

A completely P2P application must implement only peering protocols that do not recognize the concepts of 'server' and 'client'. However, such pure peer applications and networks are rare: many P2P systems use stronger peers (super-peers, super-nodes) as servers and client-peers are connected in a star-like fashion to a single super-peer. Nonetheless, despite the lower level of algorithm distribution, redundancy in the number of servers still makes such networks highly robust, and P2P technology is often used to create overlays: it works at the application layer, often directly over the TCP-UDP/IP level. IP itself could be perceived as an overlay over a heterogeneous network infrastructure, e.g. the reconfigurable core network at the physical layer.

The many different types of P2P existing nowadays can be divided roughly into two different groups: unstructured and structured. Unstructured systems, such as Gnutella, WinMx and KaZaA, are characterized by a general lack of structuring: protocols are simple, but it might take a little longer to locate resources. Conversely, P2P structured systems are built for optimizing the lookup phase, by maintaining and organizing data structures in order to 'rationalize' the overlay. The cost for locating resources is smaller than that of unstructured networks, but an overhead is paid in terms of join/leave messaging and local table maintenance. Popular examples of structured systems are Chord [2] and Kademlia [3], both based on Distributed Hash Tables (DHTs) and on virtual ring topology.

The P2P philosophy became popular along with file-sharing phenomena and it is still one of the most widely adopted technologies on the Internet [4]. Owing to its intrinsic scalability and resistance against failures, it could be adopted as a basic building block for a number of telecommunication-related issues. For instance, we have used it successfully for enhancing satellite access [5] or to organize the QoS management operations of wireless sensor networks [6, 7]. In [8] innovative solutions and guidelines for middleware development tweaked specifically for such technology are presented with regard to ad-hoc solutions for wireless sensor networks. [6] deals in more detail with a P2P architecture implementing a distributed bandwidth shaper, based on schedulers, limiting the data flows produced by the application layers and implementing an example of what we defined as 'loose QoS'. In the forthcoming sections we will emphasize the use of P2P mechanisms for providing QoS in a WLAN framework.

Within the WLAN context, the IEEE 802.11 standard was developed as a simple and cost-effective wireless technology, providing connectivity on the fly and in the most disparate places. Notwithstanding the high popularity of this standard, only the IEEE 802.11e revision introduces effective solutions in the Medium Access Control (MAC) sub-layer to cope with QoS requirements. In [9] a detailed analysis of the different behaviors of IEEE 802.11e is described, emphasizing the effort required for creating a suitable simulation environment.

However, despite the availability of IEEE 802.11e standardized and built-in QoS support, a huge amount of algorithms has been proposed in order to enhance the QoS capability of the plain IEEE 802.11 a/b/g family of standards, and much work is still ongoing.

Reference [10] deals with adaptive channel allocation for supporting specific QoS requirements in wireless networks, according to the emerging widespread use of real-time multimedia applications over wireless systems. Also, in [11] a system to enhance IEEE 802.11 MAC behavior in congested networks is proposed; even though it is not strictly related to QoS provisioning, it introduces some countermeasures to cope with a highly chaotic transmission environment. In [12] (and the references therein) an exhaustive analysis of the most advanced techniques and algorithms to guarantee QoS requirements over wireless systems is presented. Other recent work deals with the use and the performance analysis of IEEE 802.11e, as well as the coexistence of mixed environments ([13–15], among others). Additionally, in [16] the algorithms explained in Section 17.3 to build an overlay for QoS management have been tested in a 'rural' scenario, so as to emphasize the versatility of the proposed approach, while in [17] we thoroughly tested the approach discussed in this chapter designed to handle mobile environments.

In the following, we describe an operating environment in the Bandwidth on Demand (BoD) P2P architectural framework, created at the application layer, to convey the required signals among nodes participating in an overlay network. Its most relevant features may be summarized as follows: (i) distributed bandwidth management through a traffic shaping discipline acting at the application layer; (ii) bidirectional cross-layer communications between applications and the MAC layer in order to achieve full control over the channel and its actual occupation; and (iii) distributed capability of reacting dynamically to malfunction and *churn* (i.e. the process of continuous node arrivals and departures).

17.3   Design of Overlays to Support QoS

Overlay technologies are normally adopted to juxtapose another layer over a physical deployment, in order to create a sort of 'unified' playground in which to develop algorithms and applications. As presented in Section 17.2, this design paradigm has already been utilized to accomplish several networking tasks and functionalities. Nevertheless, classical 'server-centric' protocol architectures, such as the Session Initiator Protocol (SIP), are also extending their capabilities to support P2P and overlay-based scenarios [18, 19]. From this perspective, overlays can be also deployed to perform operations devoted to supporting QoS.

However, the increasing availability of different IEEE 802.11-based standards increases the range of possible scenarios. In fact, IEEE 802.11e introduces mechanisms that manage packet queues at the MAC layer, to support services for QoS management. Thus, deploying an overlay on top of a WLAN to provide support for pervasive and mobile networks can rely on these functionalities and provide enhanced capabilities.

From this perspective, we can identify two main scenarios:

1. Overlays for QoS management are deployed over WLANs without algorithms or mechanisms to support QoS requirements. This is where networks rely upon plain IEEE 802.11a/b/g standards. As a consequence, the functionalities to be exploited by the overlay may range from signaling to 'specific' techniques to enforce flows to be compliant with the proper service level agreements.
2. Overlays for QoS management are deployed over WLANs supporting QoS functionalities within the MAC layer. This is in the case of IEEE 802.11e WLANs. In this particular case, the overlay is relieved of those strategies devoted to managing and taming traffic flows, thus reducing it to a coordination/signaling mechanism.

It is therefore possible to develop QoS policies and mechanisms in both cases, but with different levels of complexity and, of course, with different precision regarding traffic control and partitioning of the bandwidth. As a paradigmatic example, if the MAC protocols already support some form of QoS guarantee, it will be possible to use more fine-grained policies. Conversely, enforcing QoS at the application layer, or – broadly speaking – 'at some distance' from the lower layers, will be reflected intuitively in a more coarse-grained result. In this vein we define as '**loose-QoS**', the QoS provided by enforcing mechanisms acting at the application layer. Additionally, we define as '**strict-QoS**', the QoS provided by exploiting mechanisms available at the MAC layer. We point out that the two schemes can be mixed, in order to achieve additional enforcement to partition resources managed in a best-effort fashion, in the presence of QoS support from the MAC layer.

The following sub-sections will describe both methodologies. We will perform a direct comparison in terms of achieved performance in Section 17.4.

17.3.1   Overlays for Guaranteeing Loose-QoS

We have identified our overlay mechanism for providing loose QoS under the name PRIMUS (*PRoviding Independent-of-MAC User QoS*).

The PRIMUS system organizes nodes in an unstructured overlay; the latter is only employed for the purpose of managing routing and bandwidth exploitation. Consequently, the actual traffic flows produced by the applications (e.g. file transfer operations) are performed outside the overlay. This approach resembles the Gnutella file-sharing system, but here, instead of files, bandwidth partitions are shared, transferred from a node to another, and are therefore managed. From this perspective, the nodes composing the overlay are called bandwidth peers, so as to emphasize their relationship in the overlay.

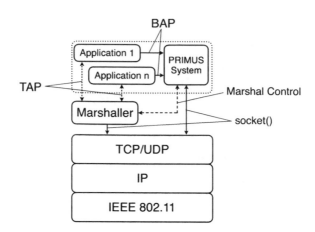

Figure 17.1 General architecture and interactions among the different system components.

PRIMUS has been designed to be a lightweight P2P-based signaling MAC-independent method oriented to providing QoS, and it does not assume any support from the lower layers in carrying out this task. This independence will be exploited in Section 17.3.2 to map application requirements (that could be assimilated roughly to a TSPEC – i.e. a traffic specification) to layers supporting QoS natively. This is one of the key benefits of developing the QoS scheme over an overlay, assuring a kind of network-independent design. Therefore, when needed, cross-layer techniques and signaling could be deployed to reach particular features made available by the protocol hierarchy, but without re-designing the overall scheme. The architecture is depicted in Figure 17.1.

The system provides applications with a transport layer and controls connections, through the Transport Access Point (TAP) and the Bandwidth Access Point (BAP), respectively. TAP lies between the application and transport layer and carries the information flow. It is no different from a standard *socket()* system call; however, it is responsible for wrapping the dataflow and forcing it to be policed by a *marshal*. The marshal works by taking into account the amount of bandwidth assigned via the PRIMUS overlay. BAP makes available a simple connection to the application, with the QoS control element (the PRIMUS core system) that implements the Bandwidth on Demand (BoD) distributed algorithm. It is used to send/receive the communications carrying bandwidth requests and releases and it thus can be perceived as the 'wire' connecting the PRIMUS middleware through the remaining protocol stack.

Control traffic is exchanged among the PRIMUS core systems that are present in each node: overlay P2P maintenance operations, as well as resource provisioning and management, are handled transparently from the application viewpoint. The PRIMUS overlay is perceived by the application as a simple bandwidth allocation service accessible through the BAP. It is worth stressing that the application does not bother with P2P issues. Indeed, the fact that the bandwidth management takes place within a P2P framework is completely transparent to applications.

Bandwidth constraints are enforced by adopting a self-limiting approach: each node implements a bandwidth marshaling function that guarantees a correct bandwidth usage. The overall rate to be enforced by the marshal is set by the PRIMUS core system.

In order to assure the management of the shared resource (i.e. the bandwidth), each node updates a number of local state variables. The update algorithm is performed by the PRIMUS core system, considering both local and external bandwidth requests, that is, coming from the application layer and from other peers. The state variables are updated locally according to released, used and received bandwidth. For example, during a node's lifetime, the residual bandwidth could increase or decrease, depending upon applications running or in response to external bandwidth requests. A detailed description of such mechanisms is provided in the Appendix.

Summing up, the overall architecture can be viewed as a distributed bandwidth shaper, without any central point of coordination. The algorithm acts at the application layer independently of the actual MAC, resulting in a loose QoS support without strong real-time guarantee (as, instead, provided by other distributed algorithms, such as Virtual Clock [20]). Despite this, the PRIMUS architecture, as will be seen in Section 17.5, is able to sustain real-time traffic in a satisfactory manner.

17.3.2   Overlays for Guaranteeing Strict QoS

The aim of the system is again to satisfy the bandwidth requests of each host and, when necessary, to offer proper traffic control over the physical medium. Conversely, we do assume to have mechanisms to enforce QoS at the MAC layer. Therefore, we can use the overlay again both to maintain the 'status' of the bandwidth shares and to convey signaling through nodes, but we can map requests over the MAC and relieve the supporting middleware of the services needed to tame the data flows. The resulting software mechanism will not present functionalities devoted to bandwidth marshaling.

Concerning the MAC layer, we rely on IEEE 802.11e, which manages several priorities in order to transmit data with different constraints. To provide a service to applications, we have to 'present' such priorities to the higher layers. Besides, to prevent breaking the 'information hiding' properties provided by the protocol hierarchy, as well as to avoid increasing the complexity when accessing the bandwidth reservation service, we decided to partition all of the priority range into three classes of possible application profiles.

From the QoS perspective, applications can then be subdivided into three different groups:

- Applications with *quantitative* specifications of QoS: these require granted QoS services through explicit mechanisms of resource allocation and admission control. Nodes running such applications will be categorized as 'QoS nodes'.
- Applications with *qualitative* specifications of QoS: these require higher than best-effort performance in terms of QoS, but do not have specific requirements to describe the desired treatment.
- *Best effort* applications: these do not require specific QoS enforcement. Nodes running only such applications will be called 'Best-Effort nodes'.

Nodes running various application categories simultaneously are called 'Mix-Mode'. Table 17.1 presents the mapping between priorities defined at the application level and those used by the IEEE 802.11e standard.

Each node can receive bandwidth only from a node with the same or lower priority. Thus, it is possible that the request cannot be satisfied. In this case, the algorithm will proceed by incrementing the priority of that node locally and starting a new bandwidth request. In this way, at every cycle of request the probability of obtaining the needed resource increases. In order to avoid greedy applications hogging the system completely, we set an upper bound to the priority class that can be achieved by a lower priority application, and we force it to return to the original class as soon as some amount of bandwidth has been obtained.

The IEEE 802.11e standard has only eight static User Priorities (UP) available at the MAC layer. Our proposed scheme acts at the application layer with 12 dynamic priorities, which are mapped, wherever possible, to the MAC layer. The underlying MAC priorities are masked by the overlay, and the bandwidth assignment algorithm takes care of modifying the UPs at run-time (within the limits defined by the node type, i.e. Best-Effort, QoS, or Mix-Mode). The higher the UP, the higher is the probability of the host to obtain the needed bandwidth. Consequently, the data flow will benefit from the corresponding QoS services offered by the IEEE 802.11e layer.

Obviously, priorities must be engineered and assigned according to the specific type of service to be deployed over the wireless infrastructure. For instance, higher priorities should be employed for VoIP (Voice over IP) communications and video-conferencing, medium-range for Web 2.0 and interactive applications, while lower values should be representative of delay insensitive services, such as file-sharing and bulk data transfers. In this vein, the overlay could be perceived as another layer in the 'cloud' of services to be accessed through the wireless loop.

17.3.3   Bootstrapping and Overlay Maintenance

As with any P2P algorithm, the 'bootstrap' mechanism gathers all participating nodes in a P2P community (with a unique identifier) and, optionally, provides them with some initial amount of bandwidth. The way such an overlay

Table 17.1   Mapping between priorities defined at the application level and IEEE 802.11e standard priorities

IEEE 802.11e Reference Zone	Best Effort Zone								QoS Zone			
User Priorities (UP) IEEE 802.11e	1	2	0			3			4	5	6	7
Overlay Priority	0	1	2	3	4	5	6	7	8	9	10	11

network is created is not within the scope of this work (a detailed study of how to kickstart a P2P system can be found, e.g. in [21]). Moreover, starting from whatever bandwidth assignment point, the distributed algorithm proposed quickly adjusts bandwidth properly among the peers. Therefore, the resulting overlay will be established without any external intervention. Some useful parameters must be identified before starting the algorithm, as the total amount of bandwidth manageable must be known *a priori*. This information is critical in order to assure that the algorithm operates correctly, but it could be disseminated in the architecture in a simple way.

For example, if the proposed overlay-based frameworks are adopted for managing public hot-spots, the required parameters could be hard-coded inside a small software client in order to download. Concerning the effective amount of bandwidth manageable with the proposed architecture, there are three important considerations:

- A portion of the total bandwidth must remain allocated to carry the signaling flows needed to maintain the overlay. Even if this quantity is very low, it is not negligible.
- IEEE 802.11 is usually unable to sustain the full rate declared by the standard [22]. In addition, the environment could influence the effective speed of the network. From this perspective, a margin must be considered in order to avoid congestion and malfunctioning.
- Owing to mobility, fading channels and performance anomalies, the actual transmission rate could be lower than that granted by the current bandwidth share. For instance, this is the case when a node has 2 Mbit/s of assigned bandwidth, but noise and interference in the channel force the wireless interface to a lower rate, e.g. 1 Mbit/s. In this vein, proper feedback, exploited by a cross-signaling mechanism, must be in place, to avoid resource trashing and misbehavior.

In order to kick start the system, there is a need to distinguish each host in an unambiguous way. Each host must be identified with a unique ID, generated randomly. Owing to the low population of a WLAN (usually, in the order of tens to hundreds), a technique such as that used in eMule [23] to generate user hash could be adopted (e.g. by applying a SHA1-based hashing algorithm [24] to the IP address). After the IDs' generation phase, a Bandwidth Mediation Point (BMP) must be established. Each host broadcasts its ID within the local link: the host with the lowest ID is elected as BMP. The notion of 'lowest' should be taken literally; hence, the user hash must be assimilated to a positive integer.

To minimize the traffic and the latency introduced by the election phase of the BMP, a snoop-based protocol is proposed. A host with an ID greater than the one that has just been transmitted will not send its host ID. If a new host joins the network and has the lowest available ID, the BMP will bow out. Obviously, the BMP has to be re-elected if it leaves the overlay.

The BMP unbalances the network architecture: it resembles a client-server architecture rather than a P2P one. At this stage, the BMP 'owns' all the available bandwidth except the fraction allocated for the signaling traffic. Each node asks the BMP for bandwidth. Note that the BMP is only employed to kick start the network.

In order to cope with changes within the overlay (i.e. peers coming in or going out) each peer participating in the overlay for bandwidth management sends a keep-alive packet periodically, embedded in a UDP datagram, and maintains a table, called 'neighborhood table', which retains the status of the overlay. Each entry in the table has the following items:

- *IP address*: the IP address of the peer. PRIMUS can handle IPv4 and IPv6 addresses.
- *Last Seen*: the most recent time the remote peer has notified its existence (via a keep-alive packet).
- *Last Talked*: the most recent time the remote peer has been notified (via a keep-alive packet).

New entries are added on the neighborhood table each time a new keep-alive is received from a peer that has never been seen before. On the other hand, when a peer fails to receive 10 subsequent keep-alives from a corresponding one in its list, the remote peer is removed from the list and declared dead.

Concerning the implementation details, these depend strictly on the overlay structure and are shared for the two proposed schemes (i.e., loose and fine QoS). The core functionality of the proposed framework relies on an unstructured overlay, employed for routing bandwidth requests.

In more detail, each node has to manage three kinds of overlay communication:

1. **Keep-alive mechanism**: each peer participating in the overlay will send a keep-alive packet periodically, embedded in a UDP datagram, each second. After 10 subsequent keep-alives are lost, the recipient peer is

declared dead and its allocated bandwidth (if any) is declared as available. If the dead node was the BMP, a new BMP election has to be made.

2. **Bandwidth request issued**: bandwidth is requested through UDP packets sent in turn to all the known hosts. If a host cannot obtain the required resources from the others it will increase its UP (when possible).

3. **Bandwidth request received**: when a host with bandwidth in use receives a bandwidth request, it will compare its priority with that of the requesting peer's. Only if the requesting peer has a higher priority, the host will release the necessary bandwidth. Unused bandwidth is released in any case.

A comment about the keep-alive mechanism must be made. Keep-alives are not only employed to detect nodes leaving the overlay, but also to handle mobility. Even if mobility is perceived and handled by the system as a leaving node, this mechanism is the de-facto standard to handle churn, which is the continuous process of nodes' arrival and departure in P2P system. Summing up, we use a unique keep-alive mechanism both to maintain the consistency of the overlay and to update the user population.

17.4 Performance Evaluation

In order to test the effectiveness of the proposed approaches designed to provide QoS via overlays over IEEE 802.11 WLANs, we discuss the results collected in two different trials and evaluation settings. The first one concerns testing a prototype in an IEEE 802.11b/g WLAN, where we rely only on the application layer scheduling capabilities introduced by the middleware. The second one relies on a prototype acting in an emulated environment that supports the IEEE 802.11e QoS capabilities; as a consequence, the middleware will be able to 'map' the QoS parameters into the traffic services delivered by the MAC layer.

17.4.1 Performance Evaluation in the Absence of QoS Support From the MAC Layer

In this section we present the main results collected when using overlays to guarantee QoS when the MAC layer does not provide QoS support. Lab trials have been performed on an IEEE 802.11 wireless network with heterogeneous computing devices (e.g. PCs, laptops, handheld devices), access points and operating systems (i.e., Windows XP SP2, Linux Slackware 10.2 and 2.6 kernel branch and MacOSX Tiger), by adopting middleware 'specialized' to provide QoS in plain WLANs by means of application layer schedulers, which has been named PRIMUS. The required client interfaces comprising the overall test bench have been coded in Java 5 [25] Standard Edition (SE). To manage priorities, three different levels have been introduced, with level 3 considered to be the most important.

In all of the trials, bandwidth requests are propagated for one hop and a single UDP packet carries both requested bandwidth and associated priority. Lost control packets are detected by acknowledgement mechanisms along with a retransmission timeout. In order to keep the propagation traffic low, request packets are not sent to all neighbors: each node sends (pipelines) its requests up to a maximum number of neighbors. The system parameters have been set to the following values:

- retransmission timeout: 10 seconds;
- keep-alive packet frequency: 1 second;
- maximum number of neighbors (request pipeline, parallel requests): 5.

Concerning the bootstrap phase, this consists simply of assigning the total amount of bandwidth to one node only and letting the distributed algorithm described in Section 17.2 and in the Appendix to work out bandwidth partitioning among the peers.

The first trial has the objective of showing a normal session ruled by PRIMUS and verifying that the distributed control algorithm behaves as expected. The test bench is composed of five wireless devices producing TCP traffic within an IEEE 802.11b wireless network. The total amount of bandwidth has been set to 3 Mbit/s.

Figure 17.2 depicts the 'traffic dump' of such devices. Node 1 produces, for about one minute, a flow requiring 1.5 Mbit/s with priority 1. Node 2 behaves exactly as Node 1, but with priority 3. Slightly before time 24:30 Node 1 flow ends and Node 3 asks for 0.6 Mbit/s for a flow with priority 2. Node 4 does the same with priority 3. The

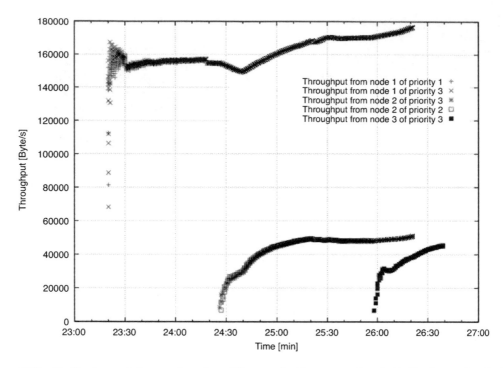

Figure 17.2 Traffic dump of three nodes when different priorities are employed. Traffic flows of node 1 are overlapped, as well as traffic flows of node 2.

total bandwidth required, right now, is of about 2.7 Mbit/s: the net is not stressed. At time 26:00, Node 5 asks for 0.6 Mbit/s with priority 3. Such flow obtains the required bandwidth because, under the effect of PRIMUS, priority Node 3 releases it.

Summing up, the PRIMUS distributed control mechanism behaves as expected. Note that since PRIMUS works at a very high level, it is not able to partition exactly the bandwidth in high load conditions: Node 1 and Node 2 require 1.5 Mbit/s but they get a little less: only when Node 1 leaves Node 2 can fully obtain the required bandwidth.

Figure 17.3 shows a more complex traffic pattern with a mix of TCP and UDP traffic. Within the same network setup Nodes 1, 2 and 3 transmit non-real-time (TCP) flows at priority 1, while Nodes 4 and 5 transmit real time (UDP) streams at priority 3. Note that, after a while, the UDP stream of Node 1 requires more bandwidth and, because of PRIMUS, the first TCP stream of Node 2 is forced to stop. In this experiment we also simulated a node (i.e. Node 1) that exits the wireless coverage gradually and thus experiences higher latencies, retransmissions and undelivered acknowledgments. Even in such extreme situations, as long as keep-alive packets are exchanged, the overlay mechanism is able to 'hold' that node and an overlay re-configuration is not performed. As a general consideration, when in the presence of UDP traffic, the system appears to be less sensitive to fluctuations.

Figure 17.4 shows yet another setup, still composed of five streams, but within an IEEE 802.11g network. This trial attempts to model a heterogeneous scenario in terms of applications with different behavior. It includes one UDP traffic flow with priority 3 and four TCP flows with priority 1. As expected, the collected trace highlights that the UDP stream is sufficiently regular, hence receiving by PRIMUS a rough QoS support, while the TCP flows tend to have higher oscillations. However, on average, all nodes have the bandwidth (requested and) obtained via the PRIMUS system (almost) guaranteed. The high spread factor of traffic flows is due to the loosely coupled scheduling discipline and to the dynamics of each node's MAC protocol mechanism.

Figure 17.5 depicts a trace concerning Round Trip Time (RTT): in the same plot it represents both RTT and throughput of a TCP session, in which bandwidth allocation changes over time, demanding and releasing bandwidth according to both the session's need and other nodes' requests. As expected, when bandwidth assignment is below the required one (see the central part of the plot, where PRIMUS takes bandwidth out from the session), the RTT increases rapidly.

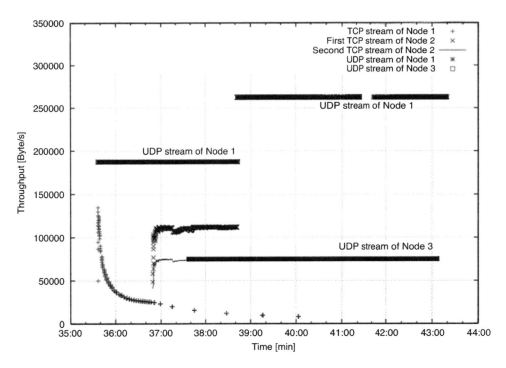

Figure 17.3 Traffic dump of three nodes when different priorities are employed. The traffic is composed by a mixture of TCP and UDP. The UDP traffic is set at the highest priority. Node 1 is moving, exiting the network.

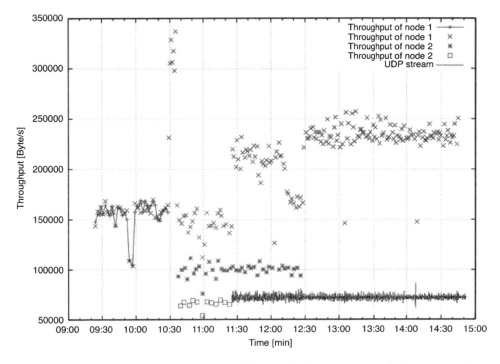

Figure 17.4 Traffic dump of three nodes when four different priorities are employed. The network is composed of eight stations.

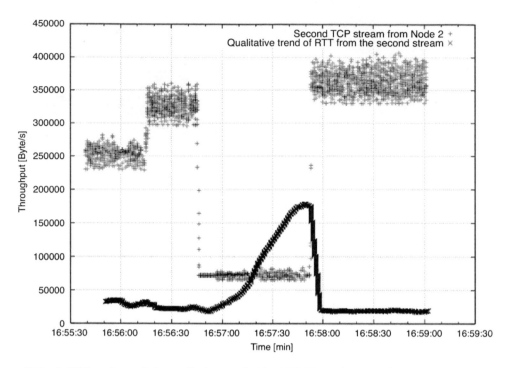

Figure 17.5 A TCP session and the qualitative trend of its RTT. The scale and axis of the graph are consistent with the throughput, while the RTT plot is overlapped to correlate its form factor.

An interesting aspect, revealed by repeating the trials several times and collecting traces from the scheduler inside end nodes, is that the chaotic nature of the MAC might sometimes become prominent. If a node must transmit with a smooth and regular delay statistic, it is 'good practice' to require a little more bandwidth than needed in order to balance the burden in the channel. This well-known fact, if exploited properly, can yield very good results without requiring sophisticated hardware, device drivers or software codecs.

So as to prove the effectiveness of a seamless integration of the PRIMUS system with pre-existing commercial software and standard devices, we conducted a number of tests with a popular software for personal entertainment (iTunes, a software for playing and organizing audio files, [26]) and an Airport Express base station, an IEEE 802.11g base station that enables it to be connected directly – via a standard lineout – to standard loudspeakers [27].

We quantified experimentally in less than 0.6 Mbit/s the amount of bandwidth required for streaming high-quality music without delays: thus, we requested 0.6 Mbit/s for streaming it via PRIMUS. Then, we prepared the PRIMUS test bench with nine hosts producing FTP transfers. Note that iTunes uses TCP for streaming data. Figure 17.6 depicts the result of this trial. iTunes is turned on and starts streaming a song. While the PRIMUS system is operating, the RTT value is kept low, and the music is streamed correctly, without delays and any loss of quality. From 39:42 to 39:53 PRIMUS has been turned off, bringing the WLAN in a state of 'anarchy'. The traffic produced by other stations quickly saturated all the available bandwidth and the music stopped. We then restarted the PRIMUS system and, after re-configuring the policies on the WLAN and restoring the enforcements, the music playback started again without any glitches.

A final consideration is PRIMUS signaling bandwidth. In all scenarios, we measured a very low signaling traffic: for instance, in a test bench with eight nodes, the traffic generated by each node is around 73 bytes/s for signaling and 12 bytes/s for bandwidth management.

17.4.2   Performance Evaluation When the MAC Supports QoS

In this section we showcase the results collected when the MAC supports QoS; we relied in this case on a simulated environment, with heterogeneous wireless interfaces, so as to emphasize aspects concerning mobility issues. The results presented in this section have been borrowed from [28], where other scenarios are also presented.

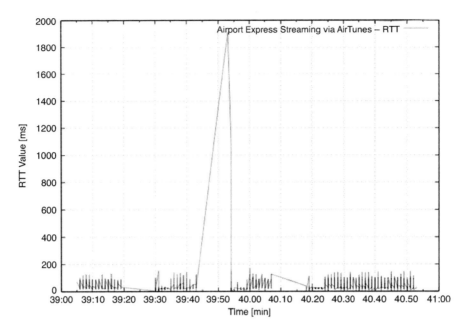

Figure 17.6   RTT trend of a streaming transmission of music with iTunes and Airport Express. The PRIMUS system has been turned-off from 39:42 to 39:53, simulating a malfunctioning. From 39:53 to 39:55 the system reconfigured itself and restarted to operate properly.

In order to evaluate the effectiveness of the proposed solution, an extensive set of trials on a prototypal system, coded mainly in Java, has been performed. The simulation campaign has been conducted by using the network simulator NCTUns [29]. It has been necessary to modify part of the original NCTUns code to integrate our framework, as NCTUns is coded in C++. To simulate the dynamic mapping between the MAC layer priorities and overlay ones, we have to adopt an Inter-Process Communication (IPC) mechanism, implementing the Message Queues technique. When a node increases its application layer priority, the application process leaves a message in a shared queue. This information is then received by the simulator process, parsed and exploited to assign the proper MAC layer priority to the same node.

NCTUns has been chosen because it is capable of simulating both wired and wireless environments. It adopts a particular simulation methodology called kernel reentering. Basically, the running simulation executes operations related to the network layer by injecting packets directly into the kernel, thus using the native Linux TCP/IP stack. The distributed communication exploits the TUN/TAP pseudo interface mechanism, allowing NCTUns to send/receive data from real running applications. When these are on the same machine, only the data link (including the MAC protocol mechanism) and physical layers need to be simulated. NCTUns supports also the emulation method. In this example, it is possible to plug in a real node to the computer where the simulation is going on and to use it like a properly emulated node. This feature has been exploited to emulate a 'fictitious' Access Point (AP) to reach an 'external' node. The function of the latter is only to connect the simulated environment to a machine acting as network monitor.

Our working scenario has been set up by using two machines connected to each other through an Ethernet interface. In the first one the whole wireless ad-hoc network has been simulated, with all nodes sending their traffic to the second machine, where data have been collected by using Wireshark, a popular packet sniffer. It is worth noting that the behavior of the overlay network is barely independent of the configuration of the underlying ad-hoc physical network. In particular, we consider a distance of one hop among nodes in the overlay; this may correspond to a single-hop or to multiple hops in the physical network. In the second case, an appropriate routing protocol, which is already defined in the standard (for instance by IEEE 802.11s), must be active. We point out that its operation is substantially transparent to the overlay.

The proposed overlay scheme designed to provide QoS guarantees over service-oriented wireless networks is then evaluated. We discuss some of the collected numerical results in the following.

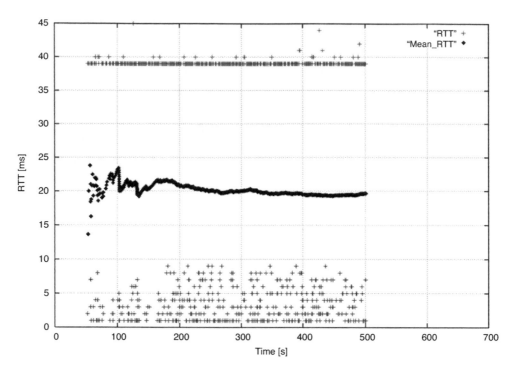

Figure 17.7   RTT of a node with priority 7, while sending traffic in the Best-Effort Zone.

The simulation scenario that we consider will show the behavior of the proposed framework in the presence of Mix-Mode hosts. Moreover, it will highlight the case of a massive bandwidth request made by a QoS node. The scenario is composed of 14 best-effort hosts and five QoS hosts. The latter try to communicate when all the bandwidth is used by the best effort hosts. On average, the available bandwidth is set equal to 700 kbyte/s. However, 30 kbyte/s are reserved for the overlay traffic. Thus, the maximum available bandwidth for every host is set to 670 kbyte/s.

Nodes from 2 to 11 have the maximum priority in the Best-Effort zone and are saturating the channel using 67 kbyte/s each. Thus, no best effort node can obtain bandwidth from them. Node 12 is the Mix-Mode one. It starts with priority 5 and requests 67 kbyte/s at instant 300. In this case the algorithm will start a priority 'climbing' for node 12. We underline that when the Mix-Mode host starts transmitting, it has an application layer priority of 8. This means that the corresponding MAC layer priority has been mapped to IEEE 802.11e UP 4, that is, the lowest priority in the QoS zone. Thus, as expected, the average RTT of the Mix-Mode node is lower than a simple Best-Effort node: in fact, a host transmitting with a priority belonging to the QoS zone range has a privileged channel access, as defined in the IEEE 802.11e standard. Figures 17.7 and 17.8 depict the RTTs of the TCP traffic transmitted by a Best-Effort node and a Mix-Mode node operating in the QoS Zone, respectively.

17.5   Conclusions and Future Developments

In this chapter we discussed the usage of overlays to provide QoS over IEEE 802.11b/g/e networks, which are a basic setting designed to pursue the pervasive and mobile paradigm. In detail, we discussed two kinds of overlay-based mechanisms. The first one, called PRIMUS, provides 'loose-QoS', whereas the second one is able to map application requests over MAC-supported QoS priorities. In this case, we define such differentiated services as 'strict-QoS'. Both frameworks are P2P-based and are independent of lower layers.

Future work will aim at refining the prototypes in order to port the mechanism to consumer devices equipped with IEEE 802.11 connectivity, which is the de-facto standard in the consumer market.

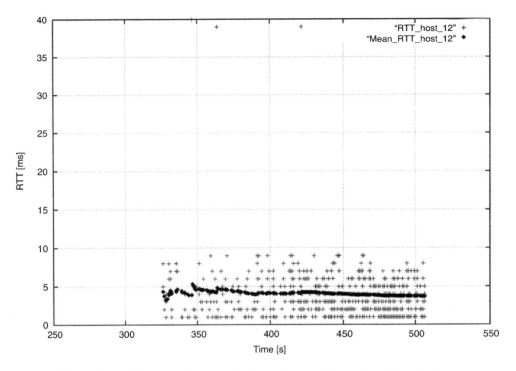

Figure 17.8 RTT of a node with priority 8, while sending traffic in Mixed Mode.

Appendix I. The Distributed Algorithm for Bandwidth Management

The distributed bandwidth management algorithm has basically to address the following three events:

1. A bandwidth request/release is performed by processes running inside a node.
2. A bandwidth request arrives from a remote node.
3. An answer is sent and received by a remote requester.

Thus, we introduce the following variables:

B_i, the bandwidth available to the i-th node.

U_i, the bandwidth required by the i-th node.

$R_i = B_i - U_i$, the residual bandwidth of the i-th node.

In order to cope with the first event, let us denote with RQ_i the amount of bandwidth requested/released by internal application processes. We have to consider the following two cases:

$$RQ_i \begin{cases} < 0 & \textit{bandwidth release} \\ > 0 & \textit{bandwidth request} \end{cases} \qquad (A1)$$

When $RQ_i < 0$ a process within the i-th node performs a bandwidth release: the procedure is handled locally, without any external signaling flows. When $RQ_i > 0$ a process running in the i-th node is requesting more bandwidth. If there is not sufficient 'internal' bandwidth, a request is sent to the network. Such 'pending' bandwidth request is waiting for either a network response or another internal update.

The other state variables are updated as follows:

$$\begin{aligned}
B_i^{new} &\leftarrow B_i^{old} \\
U_i^{new} &\leftarrow U_i^{old} + RQ_i \\
R_i^{new} &\leftarrow B_i^{new} - U_i^{new} = B_i^{old} - U_i^{old} - RQ_i = R_i^{old} - RQ_i
\end{aligned} \tag{A2}$$

The pending status is also recognizable by analyzing the sign of the residual bandwidth: if $R_i^{new} < 0$ there is a pending request and, hence, RQ_i is not fully satisfied. The amount of bandwidth still needed is then

$$\overline{RQ_i} \leftarrow |R_i^{new}| \tag{A3}$$

Such request is sent to all peers of the community.

Let us now see what happens when such a request arrives by considering the second event: an external request arriving by the generic peer j.

Each peer with unused bandwidth ($R_i > 0$) updates its own state as follows:

$$\begin{aligned}
B_i^{new} &\leftarrow B_i^{old} - R_i^{old} \\
U_i^{new} &\leftarrow U_i^{old} \\
R_i^{new} &\leftarrow 0
\end{aligned} \tag{A4}$$

This means that each peer adopts a sort of 'hot potato' approach and sends *all* its unused bandwidth to the requester immediately (independently of the actual required amount of bandwidth requested): this moves available resources quickly to portions of the network where they are greatly needed.

If we come back to the peer that issued the bandwidth request previously we have:

$$RP_i = \sum_{j \neq i} R_j^{rcvd} \tag{A5}$$

where RP_i holds the total amount of bandwidth received. The available bandwidth is then updated according to:

$$B_i^{new} \leftarrow B_i^{old} + RP_i \tag{A6}$$

and the used bandwidth becomes

$$\begin{aligned}
U_i^{new} &\leftarrow U_i^{old} & \text{if } R_i^{old} \geq 0 \\
U_i^{new} &\leftarrow U_i^{old} - R_i^{old} & \text{if } R_i^{old} < 0 \wedge |R_i^{old}| \leq RP_i \\
U_i^{new} &\leftarrow B_i^{new} & \text{if } R_i^{old} < 0 \wedge |R_i^{old}| > RP_i
\end{aligned} \tag{A7}$$

Finally, the residual bandwidth is updated in the usual way:

$$R_i^{new} \leftarrow B_i^{new} - U_i^{new} \tag{A8}$$

If $((R_i^{old} \geq 0) \vee (R_i^{old} < 0 \wedge |R_i^{old}| < RP_i))$, no further request is issued and the algorithm stops; otherwise, the data transfer is performed with the available bandwidth, and the query is reiterated after a given amount of time: the amount of bandwidth to be requested by node i is still given by (A3).

References

[1] A. Greenfield (2006) *Everyware: The Dawning Age of Ubiquitous Computing*, New Riders Publishing, Indianapolis, IN, USA, 2006.
[2] I. Stoica, R. Morris, D. Karger, M. Kaashoek and H. Balakrishnan (2001) 'Chord: a scalable peer-to-peer lookup service for internet,' In *Proceedings of ACM SIGCOMM 2001*, San Diego, CA, pp. 149–60.
[3] Kademlia: A Peer-to-Peer Information System Based on the XOR Metric, http://kademlia.scs.cs.nyu.edu/.
[4] T. Karagiannis, A. Broido, N. Brownlee, K.C. Claffy and M. Faloutsos (2004) 'Is P2P dying or just hiding?,' In *Proceedings of IEEE Globecom 2004 – Global Internet and Next Generation Networks*, Dallas TX, 3: 1532–8.

[5] L. Caviglione, F. Davoli (2005) 'Introducing P2EP: the fusion between peer-to-peer architectures and performance enhancing proxies,' In *Proceedings of the 2005 Internat. Symp. on Performance Evaluation of Computer and Telecommun. Syst. (SPECTS'05)*, Cherry Hill, NJ, pp. 42–7.

[6] L. Caviglione, F. Davoli (2005) 'Peer-to-peer middleware for bandwidth allocation in sensor networks,' *IEEE Commun. Letters* **9**(3): 285–7.

[7] L. Caviglione, F. Davoli (2004) 'A peer-to-peer bandwidth allocation scheme for sensor networks,' In *Proceedings of the Internat. Symp. on Performance Evaluation of Computer and Telecommun. Syst. (SPECTS'04)*, San Jose, CA, pp. 197–204.

[8] *IEEE Network* (2004) 'Middleware technologies for future communication networks', vol. 18, Jan.-Feb.

[9] P. Garg, R. Doshi, R. Greene, M. Baker, M. Malek and X. Cheng (2004) 'Using IEEE 802.11e MAC for QoS over wireless,' In *Proceedings of IEEE IPCCC 2004*, Phoenix, AZ, pp. 10–18.

[10] M. Malli, Q. Ni, T. Turletti and C. Barakat (2004) 'Adaptive fair channel allocation for QoS enhancement in IEEE 802.11 Wireless LANs,' In *Proceedings of the IEEE Internat. Conf. Commun. (ICC 2004)*, Paris, France 6: 3470–5.

[11] I. Aad, Q. Ni, C. Barakat and T. Turletti (2004) 'Enhancing IEEE 802.11 MAC in congested environments,' In *Proceedings of the IEEE Workshop on Applications and Services in Wireless Networks (ASWN 2004)*, Boston MA, pp. 82–91.

[12] H. Zhu, M. Li, I. Chlamtac and B. Prabhakaran (2004) 'A survey of QoS in IEEE 802.11 networks,' *IEEE Wireless Commun. Mag.* **11**(4): 6–14.

[13] Y.-L. Hsu, Y.-K. Huang and A.-C. Pang (2008) 'QoS enhancement for co-existence of IEEE 802.11e and Legacy IEEE 802.11,' In *Proceedings of IEEE Globecom 2008*, New Orleans, LA, pp. 1–5.

[14] D. Gao, J. Cai and C.W. Chen (2008) 'Admission control based on rate-variance envelope for VBR traffic over IEEE 802.11e HCCA WLANs,' *IEEE Trans. Vehic. Tecnol.* **57**(3): 1778–88.

[15] C.-L. Huang; W. Liao (2007) 'Throughput and delay performance of IEEE 802.11e enhanced distributed channel access (EDCA) under saturation condition,' *IEEE Trans. Wireless Commun.* **6**(1): 136–45.

[16] L. Caviglione, F. Davoli and M. Polizzi (2007) 'An overlay scheme to exploit IEEE 802.11e-based QoS,' In *Proceedings of IEEE Wireless Rural and Emergency Commun. Conf. (WRECOM 2007)*, Rome, Italy.

[17] L. Caviglione, F. Davoli, M. Polizzi and S. Bellisario (2007) 'Handling local user mobility and QoS in a controlled ad-hoc environment,' In *Proceedings of the Australasian Telecommun. Networks and Appl. Conf. (ATNAC 2007)*, Christchurch, New Zealand, pp. 117–22.

[18] L. Caviglione, L. Veltri (2006) 'A P2P framework for distributed and cooperative laboratories', in F. Davoli, S. Palazzo, S. Zappatore, eds., *Distributed Cooperative Laboratories – Networking, Instrumentation, and Measurements*, Springer, New York, NY, pp. 309–19.

[19] L. Caviglione, L. Veltri (2007) 'Using SIP as P2P technology', *African J. Inform. & Commun. Technol. (AJICT)* **1**(1): 38–43.

[20] L. Zhang (1990) 'VirtualClock: a new traffic control algorithm for packet switching networks,' *ACM SIGCOMM Computer Commun. Rev.* **20**(4): 19–29.

[21] C. Cramer, K. Kutzner and T. Fuhrmann, 'Bootstrapping locality-aware P2P networks,' In *Proceedings of the IEEE Internat. Conf. on Networks (ICON 2004)*, Singapore 1: 357–61.

[22] M. Heusse, F. Rousseau, G. Berger-Sabbatel and A. Duda (2003) 'Performance anomaly of 802.11b,' In *Proceedings of IEEE Infocom 2003*, San Francisco, CA, pp. 836–43.

[23] Official eMule WebSite: http://www.emule-project.net.

[24] D. Eastlake, P. Jones (2001) 'US Secure Hash Algorithm 1 (SHA1),' IETF, RFC 3174.

[25] The Java Programming Language, http://www.java.sun.com.

[26] iTunes by Apple Computers Inc., http://www.apple.com/itunes.

[27] AirPort Express Base Station by Apple Computers Inc., http://www.apple.com.

[28] L. Caviglione, F. Davoli (2009) 'Using P2P overlays to provide QoS in service-oriented wireless networks,' *IEEE Wireless Commun. Mag.* **16**(4): 32–8.

[29] The NCTUns Network Simulator and Emulator. Available on-line at: http://www.nsl.csie.nctu.edu.tw/nctuns.html.

18

Performance Evaluation of Pervasive Networks Based on WiMAX Networks

Elmabruk Laias[1] and Irfan Awan[2,3]

[1]*Informatics Research Institute, University of Bradford, Richmond Road, Bradford, West Yorkshire, BD7 1DP, Bradford, UK.*
[2]*Informatics Research Institute, University of Bradford, UK.*
[3]*Computer Science, KSU, Saudi Arabia.*

18.1 Introduction

In this chapter, we provide a brief overview of broadband wireless. The objective is to present the the background and context necessary for understanding WiMAX (worldwide interoperability for microwave access). We review the history of broadband wireless, specify its applications and discuss the business drivers and challenges.

Broadband wireless as it relates to WiMAX can be traced back to the desire to find a competitive alternative to traditional wireline-access technologies. Spurred by the deregulation of the telecom industry and the rapid growth of the Internet, several competitive carriers were motivated to find a wireless solution to bypass incumbent service providers. During the past decade or so, a number of wireless access systems have been developed, mostly by start-up companies motivated by the disruptive potential of wireless.

These systems varied widely in their performance capabilities, protocols, frequency spectrum used, applications supported and a host of other parameters. Some systems were deployed commercially only to be decommissioned later. Successful deployments have so far been limited to a few niche applications and markets. Clearly, broadband wireless has until now had a checkered record, in part because of the fragmentation of the industry due to the lack of a common standard. The emergence of WiMAX as an industry standard is expected to change this situation.

Given the wide variety of solutions developed and deployed for broadband wireless in the past, a full historical survey is beyond the scope of this section. Instead, we provide a brief review of some of the broader patterns in this development. WiMAX technology has evolved through four stages, albeit not fully distinct or clearly sequential: (1) narrowband wireless local-loop systems; (2) first-generation line-of-sight (LOS) broadband systems; (3) second-generation non-line-of-sight (NLOS) broadband systems; and (4) standards-based broadband wireless systems.

Naturally, the first application for which a wireless alternative was developed and deployed was voice telephony. These systems, called *wireless local-loop* (WLL), were quite successful in developing countries such as China, India, Indonesia, Brazil and Russia, whose high demand for basic telephone services could not be served using existing infrastructure. In fact, WLL systems based on the digital-enhanced cordless telephony (DECT) and code division multiple access (CDMA) standards continue to be deployed in these markets.

Pervasive Computing and Networking, First Edition.   Edited by Mohammad S. Obaidat, Mieso Denko, and Isaac Woungang.
© 2011 John Wiley & Sons, Ltd. Published 2011 by John Wiley & Sons, Ltd.

In markets in which a robust local-loop infrastructure already existed for voice telephony, WLL systems had to offer additional value to be competitive. Following the commercialization of the Internet in 1993, the demand for Internet-access services began to surge, and many saw providing high-speed Internet-access as a way for wireless systems to distinguish themselves.

For example, in February 1997, AT&T announced that it had developed a wireless access system for the 1900 MHz PCS (personal communications services) band that could deliver two voice lines and a 128 kbps data connection to subscribers. This system, developed under the code name 'Project Angel,' also had the distinction of being one of the first commercial wireless systems to use adaptive antenna technology. After field trials for a few years and a brief commercial offering, AT&T discontinued the service in December 2001, citing cost run-ups and poor take-rate as reasons.

During the same time, several small start-up companies focused solely on providing Internet- access services using wireless. These wireless Internet service provider (WISP) companies typically deployed systems in the license-exempt 900 MHz and 2.4 GHz bands. Most of these systems required antennae to be installed at the customer premises, either on rooftops or under the eaves of their buildings. Deployments were limited mostly to select neighborhoods and small towns. These early systems typically offered speeds up to a few hundred kilobits per second.

As DSL and cable modems began to be deployed, wireless systems had to evolve to support much higher speeds to be competitive. Systems began to be developed for higher frequencies, such as the 2.5 GHz and 3.5 GHz bands. Very high speed systems, called *local multipoint distribution systems* (LMDS), supporting up to several hundreds of megabits per second, were also developed in millimeter wave frequency bands, such as the 24 GHz and 39 GHz bands. LMDSbased services were targeted at business users and in the late 1990s enjoyed rapid but short-lived success. Problems obtaining access to rooftops for installing antennae, coupled with its shorter range capabilities, squashed its growth.

In the late 1990s, one of the more important deployments of wireless broadband happened in the so-called *multichannel multipoint distribution services* (MMDS) band at 2.5 GHz. The MMDS band was used historically to provide wireless cable broadcast video services, especially in rural areas where cable TV services were not available.

The advent of satellite TV ruined the wireless cable business, and operators were looking for alternative ways to use this spectrum. A few operators began to offer a one-way wireless Internet-access service, using the telephone line as the return path. In September 1998, the Federal Communications Commission (FCC) relaxed the rules of the MMDS band in the United States to allow two-way communication services, sparking greater industry interest in the MMDS band. MCI WorldCom and Sprint each paid approximately $1 billion to purchase licenses to use the MMDS spectrum, and several companies started developing high-speed fixed wireless solutions for this band.

The first generation of these fixed broadband wireless solutions were deployed using the same towers that served wireless cable subscribers. These towers were typically several hundred feet tall and enabled LOS coverage to distances up to 35 miles, using high-power transmitters.

First-generation MMDS systems required that subscribers install at their premises outdoor antennae high enough and pointed toward the tower for a clear LOS transmission path. Sprint and MCI launched two-way wireless broadband services using first-generation MMDS systems in a few markets in early 2000. The outdoor antenna and LOS requirements proved to be significant impediments. Besides, since a fairly large area was being served by a single tower, the capacity of these systems was fairly limited. Similar to the first-generation LOS systems were deployed internationally in the 3.5GHz band.

Second-generation broadband wireless systems were able to overcome the LOS issue and to provide more capacity. This was done through the use of a cellular architecture and implementation of advanced-signal processing techniques to improve the link and system performance under multipath conditions. Several start-up companies developed advanced proprietary solutions that provided significant performance gains over first-generation systems. Most of these new systems could perform well under non-line-of-sight conditions, with customer-premise antennae typically mounted under the eaves or lower. Many solved the NLOS problem by using such techniques as *orthogonal frequency division multiplexing* (OFDM), *code division multiple access* (CDMA), and multi antenna processing. Some systems, such as those developed by SOMA Networks and Navini Networks, demonstrated satisfactory link performance over a few miles to desktop subscriber terminals without the need for an antenna mounted outside. A few megabits per second throughput over cell ranges of a few miles had become possible with second generation fixed wireless broadband systems.

18.2 IEEE 802.16 Architecture and QoS Requirements

The PHY and MAC layer contribute in achieving QoS provisioning in a WiMAX network. Time division duplexing (TDD), frequency division duplexing (FDD) and orthogonal frequency division multiplexing (OFDM) are the standardized technologies for QOS provisioning at the PHY layer. The TDD scheme is involved in QoS provisioning by separating the downlink and uplink bandwidth with a period of time though the probability of congestion is high compared to other schemes. Different channels are assigned to downlink and uplink bandwidth mostly separated by 50 to 100 MHZ in FDD, which results in no collisions. However, OFDM is the preferred scheme as it is based on CDMA which uses multiple access techniques and provides distinct QoS for each subscriber.

The OFDM scheme addresses three subschemes for QoS provisioning called Fast Fourier Transform (FFT) which reduces the lengthy mathematics after conversion of analog to digital signal, or Forward Error Correction (FEC) that corrects error or lost bits by repeating the information bits, a use of multiple subcarriers is specified in multiple subcarrier subschemes to carry identical bits in various subcarriers in case loss occurs [1].

The 802.16 MAC differs from the rest of 802 MAC in the sense that 802.16 MAC is connection-oriented whereas the others are random access based. Connection oriented simplifies the work by mapping the specific service flow to the corresponding QoS. This transport and transformation is made at the convergence and common part sublayer. The convergence sublayer is responsible for the classification and transport of the traffic whereas the common part sublayer does the fragmentation, scheduling and QoS.

The IEEE 802.16 standard defines four types of QoS for different types of application. These types of QoS are defined as follows [1]:

1. *Unsolicited Grant Service (UGS):* is meant to support Real time data streams with constant bit rate services with fixed data packets that are generated on a periodic basis such as Voice over IP (VoIP) without silence. There are no contention and piggyback request allowed in this service flow, and the BS does not provide unicast request opportunities.
2. *Real-time Polling service (rtPS):* unlike UGS, rtPS supports real time data streams with variable size data packets on a periodic basis such as Moving Picture Experts Group (MPEG) video and VOIP with silence suppression. Similar to UGS, contention and piggyback requests are prohibited for transmission as it consists of nominal polling interval, tolerated poll jitter and minimum reserved traffic rate.
3. *Non real-time Polling Service (nrtPS):* is defined for non real time applications and generates variable size data grant on a regular basis with a minimum data rate. The file transfer protocol (FTP) is one examples supported by nrtPS piggyback request. Nominal polling interval, minimum reserved traffic rate and traffic priority are the key parameters for this service flow.
4. *Best Effort (BE):* there are no specific requirements and quality constraints defined in this service flow.

The standard addresses a fifth type specified in the mobile version which is under revision known as extended real time polling service (ertPS). It provides QoS to applications that are in need of handover.

There are different solutions for ensuring quality of service but where and when to use such mechanisms and methods rests with service providers.

Despite the fact that 802.16 takes account of QoS support, CAC and scheduling are still open issues. Figure 18.1 shows an 802.16 QoS architecture where the blocks with dashed lines represent the undefined parts.

In this chapter, we have proposed a new QoS framework for PMP 802.16 systems operating in TDD mode over Wireless MAN-OFDM physical layer. The proposed framework includes a CAC module and an uplink scheduling structure. The proposed CAC module changes the grants boundaries of the connections' QoS requirements to ensure efficient and fair use of the bandwidth resources. The uplink scheduling depends on frame-by-frame allocations to the current needs of the connections and the QoS requirements for the real time applications.

18.3 Related Work

There have been many attempts to provide a solution for QoS provisioning in WiMAX networks. Most of them focus on scheduling algorithms while a few have addressed the Connection Admission Control problem. Among the existing scheduling algorithms presented, the majority of the work assumes unrealistically the presence of a simple CAC. Based on the fact that CAC and scheduling handle the QoS level jointly, a suitable CAC algorithm

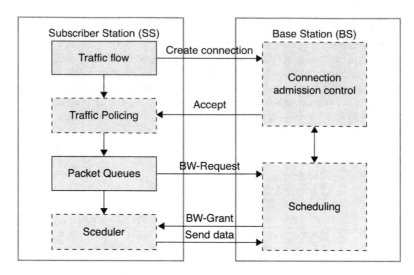

Figure 18.1   802.16 QoS architecture.

is needed in order to guarantee the required QoS. A failure of QoS provision occurs when CAC and scheduling algorithm operate under unbalanced and different criteria. This often leads to interference. Therefore, striving for a better scheduling algorithm together with a dependent CAC algorithm is expected to enhance the QoS management process.

Thus, many scheduling schemes are proposed [2–5]. However, none consider traffic data rate in real time or the queue size in the BS Uplink scheduler.

In [3], a two-layer scheduling structure for bandwidth allocation was proposed in order to support all the existing types of service flow. In the first layer, the Deficit Fair Priority Queue (DFPQ) was used to distribute total bandwidth among various flow services in different queues. Although the DFPQ suits scheduling for multi services, it presents the probability of being closed to null for servicing due to the rtPS deadline of packets and an increase in delay by an amount of at least the duration of one frame. Several packet scheduling algorithms for broadband wireless networks were published [6,7]. In [7] the authors introduced a scheduling scheme that considers the burst nature of real-time and non real-time traffic. A hierarchal scheduling scheme is also proposed with two levels, inter-class and intra-class. The first level provides service differentiation of rtPS and nrtPS-time classes. The second determines the service priority of messages within the same class. This scheduler considers delay for rtPS traffic and prioritizes it over nrtPS traffic in the uplink direction only. UGS and BE traffic classes are not treated. This work may be regarded as the special case of a scheduler that treats some classes while not taking into account the entire status of a WiMAX network.

In general they introduce complex scheduling schemes, composed of hierarchies of known schedulers, such as Earliest Deadline First (EDF) and Weighted Fair Queuing (WFQ). Simpler solutions are desired, since the scheduler executes at every frame, which in OFDM-based systems can occur at a frequency of 400 frames per second [1].

In [18], there is proposed a novel adaptive scheduling algorithm for WiMax wherein a SS sends a request for extra bandwidth to BS by speculating on the rtPS traffic patterns. As the arrival patterns and service time are random, it is complex to predict the future need of bandwidth at SS.

On the other hand, [19] and [20] have considered either only real-time traffic or only best-effort TCP traffic. In [2], there is proposed a hierarchical structure for bandwidth allocation to decide whether QoS for a particular connection can be satisfied at the BS. It is based on a simple connection admission control mechanism as described in the following equation:

$$BR \leq \beta - (B_{UGS} + B_{rtPS} + B_{nrtPS}) \tag{18.1}$$

where B_{UGS}, B_{rtPS} and B_{nrtPS} are the rates reserved for connections already admitted of type UGS, rtPS and nrtPS respectively and BR is the rate requested by the new flow and β is the total link bandwidth.

This solution enables providing bandwidth guarantee to admit flows, but cannot satisfy delay constraints for real time application.

[8] presented an admission control schema and analytical evaluations have been introduced for existing WiMAX networks. What is not provided are any constraints on degrading the low priority service flow, whereby the proposed schema doesn't consider the scheduler which may lead to scheduler overload.

The mechanism proposed here supports all four service types defined by the standard. To our knowledge, no other fully standard-compliant mechanism has so far been proposed.

With respect to CAC optimization, previous studies focus on the optimal revenue policy only in consideration of the profits for service providers. To improve on this, in this chapter we also take into account the requirements of WiMAX subscribers and develop a policy with satisfying tradeoffs between service providers and subscribers.

18.4   Proposed QoS Framework

When designing the 802.16 QoS framework, the major considerations and components are presented in this section.

18.4.1   Customized Deficit Round Robin Uplink Scheduler (CDRR)

An enhancement of MDRR with a Strict Priority algorithm [13], which is based on the DRR mechanism, is proposed in this section to be used as an uplink scheduler. Contrary to MDRR, the proposed scheme (CDRR) has the capability to minimize delay and improves the management of handling different classes laid on top of the MDRR mechanism. The proposed scheme is based on a single queue for both UGS and unicast polling and one queue for a BE. Moreover, a list of queues for both rtPS and nrtPS is provided. We should note that grouping multiple rtPSs connections into a single queue under EDF algorithm fails to guarantee the minimum reserved traffic rate of individual rtPS connections; this might lead to an unbalanced sharing of the available bandwidth since one rtPS connection with tight delay budget and extra amount of traffic may consume the entire bandwidth and starve all the other rtPS connections in the same queue. In the proposed scheme each queue in the list represents a single connection, as seen in Figure 18.2. This list is updated for every frame by adding new queues and removing empty queues from the list.

Bandwidth requirements can be measured by the maximum sustained traffic rate (r_{max}) and the minimum reserved traffic rate (r_{min}), depending on the service flow scheduling type. Each queue in the list is attached with a deficit counter variable to determine the number of requests to be served in the round and this is incremented in

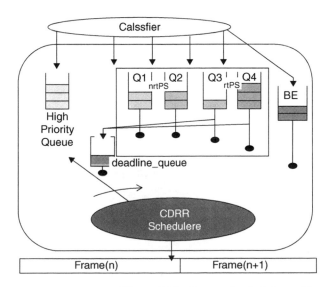

Figure 18.2 Proposed scheduler architecture.

every round by a fixed value (*quantum*) and the quantum is computed as follows [13]:

$$Quantum = MTU + (w - 1) \times \Delta \tag{18.2}$$

where MTU is a Maximum Transmission Unite on the link and Δ is a fixed increment (512 bytes) and the queue weight: $w_i = \frac{r_{\min[i]}}{\sum_{i=0}^{i=N} r_{\min[i]}}$, where $r_{\min[i]}$ is the minimum rate for the connection i and N is the number of all connections.

Thus, a queue with the lowest weight (equal to 1) is allowed to send an MTU on each round.

In this scheme, an extra queue has been introduced to store a set of requests whose deadline is due to expire in the next frame. Every time the scheduler starts the scheduling cycle, this queue will be filled by all rtPS requests which are expected to miss their deadline in the next frame. An rtPS request is said to be subject to deadline when the request connection is engaged but will probably fail to meet the expected deadline. In other words, we define a request packet as due to deadline in a certain frame if its deadline expires before the next frame is totally served and at the same time, the remaining capacity of the current frame will not be enough to service this packet after the service flow currently being serviced or about to be serviced uses up its Quantum.

More formally, an rtPS data grant G is said to be it is transmit time due to deadline if equations (18.3) and (18.4) are satisfied:

$$deadlineG < (frame_n + 1) \times frame\_duration \tag{18.3}$$

$$sizeG > frame_{aval} - DC_i \tag{18.4}$$

where $frame_n$ is the current frame number, $frame_{aval}$ is the available capacity in the current frame, $sizeG$ is the size of the current packet G and DC_i is the deficit counter of the i_{th} service flow belonging to the rtPS queue which is about to be serviced.

In the proposed scheme, it is assumed that the deadline of a request should be equal to the sum of the arrival time of the last request sent by the connection and its maximum delay requirement.

In the next scheduling cycle, the scheduler will first check if any request has been added to this extra queue. If the extra queue is not empty, the scheduler will serve this queue after the UGS and polling queue. Once the extra queue becomes empty and there are available BW in the UL_MAP, the scheduler will continuing serving the polling services (PS) list, using DRR with quantum as expressed in equation (18.2). Note that, it is well known that lower priority PS queues can actually be starved when extra queue has packets to be served. Therefore, PS queues cannot be guaranteed a minimum service rate or a maximum latency, unless constraints on the extra queue traffic are enforced. In this chapter, it is assumed that when the connection gets a polling opportunity, more requests are expected to be sent in the next uplink frame so the scheduler will share the polling interval time and the sum of deficient left as a result of moving a due to deadline requests from rtPS connections, in adding extra value to the deficient counter for the such connections to increase the number of requests to be served in this round under the following conditions:

(a) The connection has been just polled in this frame.
(b) The queue size of this connection is exceeding its maximum threshold:

$$queue\_ \max \_threshold = r_{\min} \times (frame\_duration) \tag{18.5}$$

For BE queues, the scheduler will assign the remaining bandwidth using the FIFO method since there is no QoS boundaries.

Next, we show the Pseudo code for the uplink scheduling algorithm which will act for the support of QoS requirements and also allocate bandwidth to the SSs. The proposed scheme is fully standard-compliant and can be implemented easily at the BS.

Initialize: (Invoked when the scheduler is initialized)
1. Active_list = 0;
2. Extra_queue = 0; exceed_deficent = 0;No_nrtPS//number of nrtPS connections in the active list.
For (i=0; i<0; i++)

```
        {
    Deficent_counter(i)=0;
        }
    Enqueue: (Invoked when a packet arrives)
3. Let J be Queue In Which Packet Arrives;
4. If (ExistsInList(J) = = FALSE) then
        {
AddQueueTo Active _List [CID]
        }
5. Insert the UGS grants and unicast polling into their corresponding queue that must be sched-
uled in this frame;
6. Insert the BW requests in rtPS,nrtPS,BE queues;
7. For each request i in rtPS queues
do
        {
    if ((deadline[i] − current_time)+ frame_duration >0) && (Number of Request [CID] >
max_threshold[CID]) then
        {
// which rtPS requests must be sent in the next frame in order to satisfy the maximum latency;
    extra_queue = extra_queue + dequeue.rtPS[i];
    exceed_deficent = exceed_deficent+size_of (req[i]);
        }
end do;
8. Schedule the requests in the UGS queue using First In First Out (FIFO);
9. if Avalible _slots>0 && extra_queue_is_empty = Flase
    then
        {
schedule the requests in the extra_queue using Early Deadline First (EDF);
        }
10. For each connection CID of type nrtPS do
        {
if Poll[CID].time==current_time && rmin[CID] ≤ granted_BW[CID]&& available_Slots>0;
    then
        {
Deficient _counter (CID) = = Deficient _counter (CID)+ (exceed_deficient / No_nrtp);
exceed_deficent = exceed_ deficent − (exceed_ deficent / No_nrtps);
        }
        }
end do
11. if available_slots >0
        then
        {
Schedule the requests in the rtPS, nrtPS queues starting from the head of the queue;
        }
12. if available_slots >0 && rtPS,nrtPS queues are empty then
        {
Schedule the requests in the BE queue starting from the head of the queue;
        }
```

In Line 7, examining deadline requests, each individual step requires $O(1)$ time and they are all executed for all the requests in the rtPS queues for r requests thus, the procedure requires $O(r)$ time. The add extra Deficient procedure (line 10) takes $O(n)$ time where n, represent the connections which just polled in the last frame, since in the worst case the number of rtPS and nrtPS connections is equal to n. As DRR complexity equal $O(1)$, so the time complexity of the proposed algorithm is $O(1+ n+ r)$.

18.4.2   Proposed Call Admission Control

Recalling equation (18.1), the maximum traffic rate is the necessary bandwidth which must be satisfied for both UGS and rtPS connections in order to meet QoS requirements. On the other hand, owing to the properties of nrtPS flow, the required bandwidth of an nrtPS flow may vary within the range of $[b_{nrtp}^{min}, b_{nrtp}^{max}]$ where b_{nrtp}^{min}, and b_{nrtp}^{max} are the minimum and maximum bandwidth required for the nrtPS flow, respectively. If sufficient bandwidth is available (i.e. fewer connections), each nrtPS flow can be transmitted at a higher rate.

Since the number of connections increases, the existing nrtPS flows can give up some bandwidth to new connections in order to have more connections in the system.

An nrtPS connection said to be non active if it has a connection queue size less than the fixed threshold calculated using equation (18.5). We call this model an interactive model, in which α is the greatest possible amount of bandwidth that can be taken from all ongoing non active nrtPS connections (*Nrtps*). Thus, the current reserved bandwidth for each non active nrtPS connection is b_{nrtps}^{max}.

In what follows, we describe our proposed call admission control algorithm:

```
Accept = false;
DSA = new connection request;
b_nrtps^min = Service Flow ->min_rate
b_nrtps^max = Service Flow ->max_rate
if DSA.Type = BE
Accept = true;
if DSA.Type = rtPS AND BUGS < B
Accept = true;
//calculate the number of all ongoing non_active nrtPS connections.
i from 0 to number of nrtps connection in the active list
if CID[i].Length < queue_max_threshold
Nrtps++,
```

$$\alpha = \frac{\sum_{i=0}^{i=N_{nrtps}} [b_{nrtps}^{max} - b_{nrtps}^{min}]}{N_{nrtps}} \quad // \text{ calculate possible degrade}$$

```
Endfor;
if DSA.Type = UGS AND BUGS < B
  do Accept = true;
else
  do check_scheduler_status (frame_number) // Function call
  if DSA.Type = UGS AND B_UGS < β +α AND Queues_delay = false
    do Accept = true;
else
  do Accept = false;
if DSA.Type = rtPS AND B_rtPS < β
  do Accept = true;
else
  do check_scheduler_status (frame_number) // Function call
  if DSA.Type = rtPS AND B_rtPS < β +α AND Queues_delay = false
  do Accept = true;
else
  do Accept = false;
function check_scheduler_status (frame_number)
{
while (frame_number MOD 1000) = 0
  do if extra_queue <threshold
    Queues_delay = false;
else
```

```
        Queues_delay = true;
        Quite;
    return Queues_delay;
    }
```

When the request for a BE connection arrives at the BS, the request is always admitted, but BS will not set aside any bandwidth for such a connection. In the 802.16 MAC layer, the BS connections get transmission opportunities only when other service connections do not transmit. Generally, BS connections have long idle periods and the data in each transmission are relatively small, especially in the uplink direction. Therefore, the QoS of BE can be easily satisfied.

18.4.3 Proposed Frame Allocation Scheme

To integrate all features in WiMAX PHY and QoS service classes, a well-designed algorithm is required to satisfy the following metrics. First, service classes must be satisfied for the requirements of QoS parameters such as minimum reserved rate, priority and maximum latency. The maximum latency guarantee is most important for the real time application in rtPS. Second, for fairness, the allocation algorithm should serve the service classes fairly to avoid starvation of low priority service classes. The problem leads to a dynamic downlink and uplink bandwidth allocation in a WiMAX BS.

Based on an uplink scheduler in Section 18.4.1 a basic mechanism is designed in order to allow the uplink subframe to make use of resources from the DL subframe, if the number of due-to-deadline requests in the uplink scheduler is high.

The initial size of both subframes is set to 70% and 30% for DL and UL subframes respectively and it is used as a reference to reset the system when the DL subframe 'needs' to have its resources back, giving the DL subframe bigger size than the UL subframe owing to the fact that BS wants to serve as much as possible of the DL traffic so as to avoid congestion at the BS scheduler. Also, the load in downlink is higher as compared to uplink, thus the downlink subframe is expected to have a longer duration.

A threshold based on number of deadline requests is used to reset the partitions to their initial settings.

1. Sum up the number of deadline requests so as to reserve bandwidth for those that must be served in this frame.
2. When a new frame starts, the required data/request size is first translated into number of slots as

$$Number\ of\ slot = \frac{Request\_Size}{Capacity\_of\_Slot} \tag{18.6}$$

Since a slot contains N number of data sub-carriers and the modulation decides the number of bits carried in a sub-carrier, we can have:

$$Capacity\_of\_slot = \frac{(N * X * codeing_{rate})}{8} \tag{18.7}$$

Where X represent the number of bits carried in a sub-carrier.

By summing up the number of reserved slots calculated from deadline requests and dividing them by the number of UL sub-channels in slot duration one obtains the amount of symbols to be reserved for the deadline requests.

3. If the amount of the symbols needed by the number of the deadline requests is greater than 30% of the frame size then the initial size of both subframes is set to 50% for DL and UL subframes; otherwise the amount of remaining symbols is calculated by subtracting the number of reserved symbols from the total number of symbols in a frame by allocating the remaining symbols for the DL and UL according to the amount of bandwidth requested by data/requests. Letting Req_DL and Req_UL represent the requested bandwidth for DL and UL, respectively, the proportion can be derived as:

$$\frac{Req\_DL}{Req\_UL} = \frac{Sym_{rem} - (Sym_{DL} \times Z)}{Sym_{UL} \times Z} \tag{18.8}$$

Where Sym_{rem} indicates the number of remaining symbols and Sym_{DL} and Sym_{UL} means the number of symbols in a DL and UL slot duration, respectively. Z shows the average number of slot durations for both DL and UL.

In short, the proposed scheme reserves symbols for the requests which must be served in this frame, then allocates the remaining symbols by the data/requests to determine the DL/UL sub-frame size.

18.5   Simulation Experiments and Numerical Results

We test the operation of the proposed QoS framework by a set of simulations have been carried out using the OPNET Modeler 14 simulator [12].

Several modifications have been added to the WiMAX model in order to implement the proposed uplink scheduler such as the deadline checker function and the functions responsible for redistributing the amount of the deficient counter which resulted from moving deadline requests from rtPS queues to the introduced *extra_queue*.

In order to realize an interactive_CAC it was necessary to make possible:

- the bandwidth release of an ending flow;
- the Subscriber Station activation in every instant and not only at the start of simulation;
- a real time invocation for the admission control function.

Simulations have been realized with the following configurations at MAC as well as the PHY layer; they make use of the Wireless MAN-OFDMA physical interface between BS and SSs, with the TDD duplexing technique and the NLOS configured for lower frequency at 5 GHz and channel bandwidth of 20 MHz and 2048 sub-carriers. The frame duration is 5 *ms* and the OFDMA symbol duration is 102.86 *μsec*. The modulation chosen for downlink and uplink direction is QPSK.

Different simulation scenarios have been conducted throughout the following section, and an analysis of the obtained results is provided. An assumption considered here for every scenario is that each SS can only transmit a single application to establish a connection with the BS which will not impact the simulation results in any way.

In the following simulation experiments, the number of SSs increased from 15 to 150 in steps of 10 units (one for each type of service). Each simulation ran 10 times with different seeds. The mean and the 95% confidence interval are shown in graphs.

Here, a network consisting of one BS serving a number of SSs, as depicted in Figure 18.3, has been simulated with different traffic flows, as shown in Table 18.1. For the incoming and outgoing video streams, the details are given in [11].

The interval between the unicast request opportunities of the rtPS service is 20 ms and the interval of the nrtPS service is 1 sec. For rtPS service, the delay requirement is 100 ms and each connection has its own maximum bandwidth requirement 64 Kbps.

The nrtPS service minimum bandwidth requirement is set to 64 Kbps, and the BE service would not be required for any QoS guarantee. The simulation compares the delays incurred in the proposed scheme with those incurred in the Modified Deficit Round Robin (MDRR) implemented in OPNET 14 WiMAX module [12]. The delay of the UGS class is negligible because UGS has the lowest and constant delay due to its higher priority and stringent QoS requirements. The video delay of rtPS is observed compared to the MDRR scheduler and this is shown in Figure 18.4.

From the graph, it can be seen that video packets encounter a higher delay when using MDRR. On the contrary, with the proposed scheduling schema, the delays are lower. This is due to the fact that real-time requests closer to the deadline are given high priority. We also consider increasing the opportunity for the service for both rtPS requests and nrtPS requests whenever their connections get a polling chance. Since every queue in PS List represents one connection, the queuing delay has been investigated and the results are depicted in Figure 18.5, storing due to deadline requests in a separate queue can be seen as a solution leading to a reduction in the requests' drop probability which in turn will improve rtPS queue delay.

Figures 18.6 and 18.7 show the throughput for rtPS connection for both MDRR and CDRR respectively; the throughput is lower than the load in time when we use MDRR since the rtPS connection configured as 64 Kbps, but it receives 96 Kbps in load. The difference is made up by overhead between the application layer and the MAC layer.

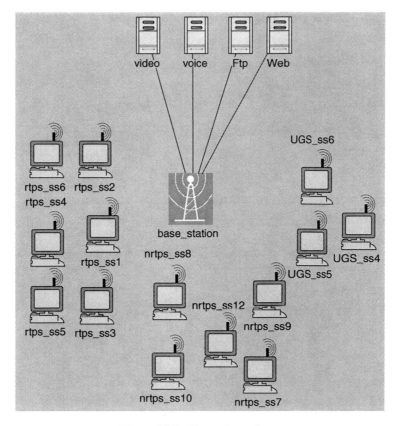

Figure 18.3   Network topology.

The throughput of rtPS for both MDRR and CDRR algorithms is given by the following formula [2]:

$$T = \sum_{j=1}^{j} Map(i, j) \tag{18.9}$$

where j is the total number of frames for a service flow i, and $Map(i,j)$ is the size of the data serviced for the service i in frame j.

The graph in Figure 18.8 shows four curves that represent the throughput of the rtPS and BE services in both the MDRR algorithm and the CDRR algorithm. We chose to compare rtPS and BE services to show that the higher priority service (rtPS) will enhance its throughput using the CDRR algorithm, on the other hand, the BE services will experience a relatively decreased throughput using the CDRR than when the MDRR algorithm is applied. However, the decrease in the throughput BE service is relatively low, since it will remain bounded, and it is guaranteed that the lower priority queues will not experience starvation, and they will be serviced eventually.

Table 18.1   Traffic flows

Traffic Type	Scheduling Type	Traffic Model	Distribution	Traffic Rate (λ)
Voice	*UGS*	*On/off*	*(On/:1.2s) (Off:1.8s)*	*66 bytes every 10 ms*
Video	*rtPS*	*Video Coference*	*Exponential Distribution*	*10 frames/s*
Web	*nrtPS*	*Light browsing*	*Hybrid longnormal/perto*	*Head (7247 bytes)*
				Tail (10558 bytes)
FTP	*BE*	*Medium Load*	*Exponential Distribution*	*512 Kbytes*

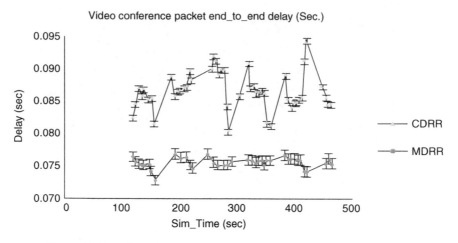

Figure 18.4   Delay of video packets in both MDRR and proposed scheme.

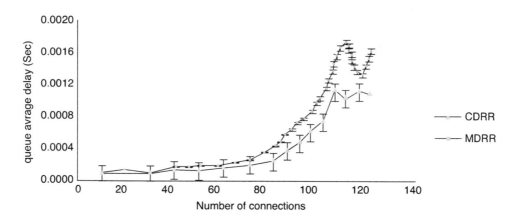

Figure 18.5   Average delay for the video conference queue.

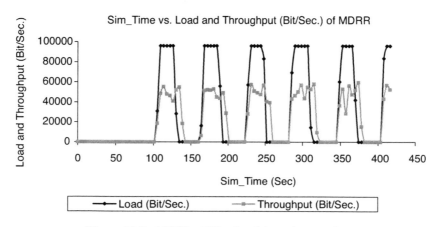

Figure 18.6   MDRR rtPS load and throughput vs time.

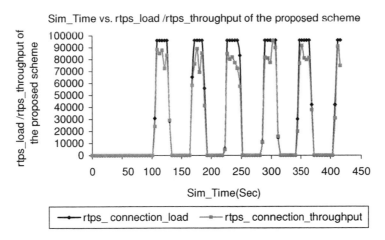

Figure 18.7   Proposed scheme (CDRR) rtPS load and throughput vs time.

We prove the fairness of the performance of our scheme using the definition of fairness provided in [9].

$$Fairness = \left| \frac{T_{rtps}}{S_{rtps}} - \frac{T_{nrtps}}{S_{nrtps}} - \frac{T_{be}}{S_{be}} \right| \qquad (18.10)$$

where T_{rtPS}, T_{nrtPS} and T_{BE} parameters represent the throughputs which were allocated to rtPS, nrtPS and BE flow services respectively, and S_{rtPS}, S_{rtPS} and S_{BE} parameters denote the total traffic of rtPS, nrtPS and BE services respectively.

In Figure 18.9 we study the long-term fairness where the simulation results are collected at load greater than 100 Kbit/fram; the results demonstrate that the CDRR scheme is fairer for the lower load than MDRR, which means that MDRR gives more service to the higher priority flows when the load is low until the overall load is around 112 Kbit/frame. When the overall load exceeds 112 Kbit/frame, both schedulers are likely to have the same effect on fairness This means throughput allocated to one flow will not exceed throughput allocated to the other flows of the same type.

Figure 18.10 shows the change in flow acceptance ratio as the average connection arrival rate changes. The flow acceptance ratio is almost 100% until the connection arrival rate is around 38. Thus, if one 802.16 network is to be deployed for voice only (UGS) traffic, then the network administrator should make sure that new connections arrive at a rate less than 38 connections per second.

According to standard IEEE 802.16, UGS connections have a fixed resource reservation. The CAC module and the scheduler module act according to the standard and guarantee such resources for every VoIP connection. This simulation demonstrates that every UGS connection has a constant delay on the WiMAX link. Moreover,

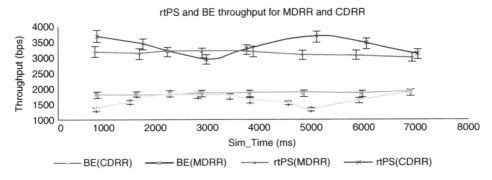

Figure 18.8   rtPS and BE throughput for MDRR and CDRR.

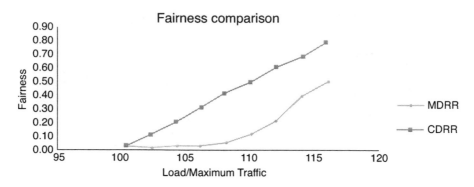

Figure 18.9   Fairness of the two schemes versus the increased traffic load.

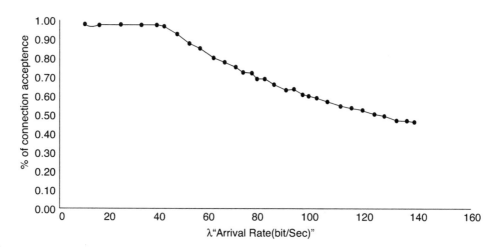

Figure 18.10   Acceptance ratio for UGS class using interactive_CAC.

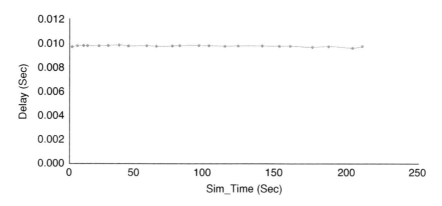

Figure 18.11   UGS delay (sec.).

Figure 18.11 shows that the delay is constant during the entire simulation time, thus demonstrating that the acceptance of a new UGS flow does not impact on the delay of the other already admitted UGS flows.

In the second simulation scenario, only rtPS connections are considered with maximum latency of 30 ms.

As depicted in Figure 18.12, BW_CAC is not able to provide delay guarantees, because it accepts new flows until there is enough free capacity in the link. On the contrary, the proposed CAC enables guaranteeing delay

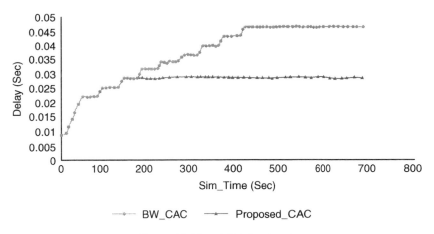

Figure 18.12   rtPS delay (sec.).

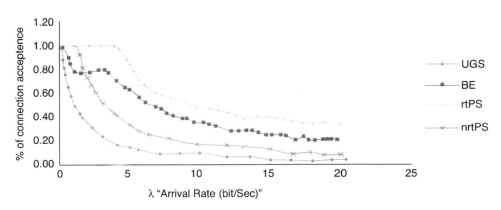

Figure 18.13   Acceptance ratio for all classes of traffic proposed_CAC.

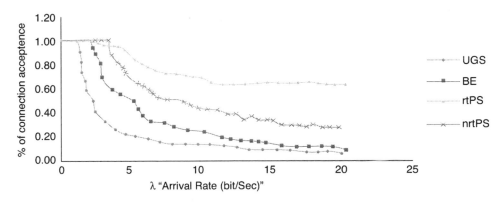

Figure 18.14   Acceptance ratio for all classes of traffic using using BW_CAC.

constraints by considering the number of the due deadline requests in the scheduler using a delay check process after the bandwidth check. In the same plot, it is possible to see how the proposed algorithm enables one to respect the limit of 30 ms required by the application over the WiMAX link.

Figures 18.13 and 18.14 show the Flow-Acceptance ratio of each class of traffic as the arrival rate increases, when each rtPS and nrtPS connection is allocated its maximum rate.

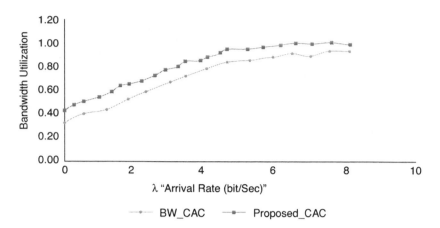

Figure 18.15   Bandwidth utilization comparison.

Figure 18.14 shows the Flow-Acceptance ratio of each class of traffic when the proposed CAC is used. When this is compared with the case when allocation is done based on BW-CAC (see Figure 18.13) it may be noted that the Flow-Acceptance ratio improved for all classes of traffic except nrtPS. Thus, nrtPS is similar to a rtPS connection except that the parameters have larger values (e.g. the min_rate and max_rate are larger than rtPS connection). Thus, slots (bandwidth) which were taken out using the degradation of nrtPS could not be used to admit more nrtPS connections because of their large bandwidth requirement for the other types of connection, and because of their small parameter values, were able to take up the saved slots and improve their acceptance ratio.

Although BW-CAC admits connections based solely on availability of bandwidth (slots), it will typically incur west of bandwidth. The proposed CAC, on the other hand, will have better utilization, and will have less bandwidth west since it admits connections only when delay of the connection can be met. Figure 18.15 compares the bandwidth utilization using both modes, the proposed CAC mode and the constant rate mode. When working in the constant rate mode, the nrtPS flows can only transmit at the maximum rate. Thus, we define a Utilization performance as defined below:

$$Utilization = \frac{[[(No\_of\_Slots[i] \quad total\_Slots) \times (No\_of\_admited\_conn)] - No\_of\_rejected\_conn)]}{(total\_No\_req\_conn)}$$

No_of_slots and total_slots represent the number of slots assigned to the connection i and total number of slots available respectively.

Other simulation experiments, not presented here owing to limitations of space, indicate that the rtPS connections acceptance ratio can be increased by tuning the value of the time interval used to check scheduler status.

In order to examine the frame allocation scheme a network is composed of one WiMAX cell with four SS nodes with traffic configuration such that all SS nodes have an uplink application load of 250 Kbps for a total of 1 Mbps and at specific times, the server generates 600 Kbps of application traffic directed to SS_0 and SS_1; this creates a total downlink application load of 1.2 Mbps.

Here an OFDMA frame with 512 subcarriers and 5 milliseconds duration is used. An uplink subframe is set to 12 symbol times and a downlink subframe is assigned 34 symbol times as an initial setting,for QPSK $\frac{1}{2}$, the capacity expected is Uplink: ~ 600 Kbps and Downlink: ~ 2.5 Mbps.

As shown in Figures 18.16(a) and 18.16(b), comparing the downlink throughput using the proposed scheme, the downlink traffic performs the same as in the constant subframe cases, up to 1.2 Mbps MAC throughput (e.g. no change in downlink throughput).also the application delay ~ 8 milliseconds (see Figure 18.17 which explains that the downlink traffic remain unaffected by the new scheme).

In the other hand, for the uplink traffic in Figure 1.18(a), (b), (c) the behaviour is different (compared to the DL traffic); note here that 'the period of time (0 – ~ 230) is the time needed to exchange initial massages between BS and SS's'. For uplink traffic it can be seen in the proposed subframe an allocation scheme for time periods

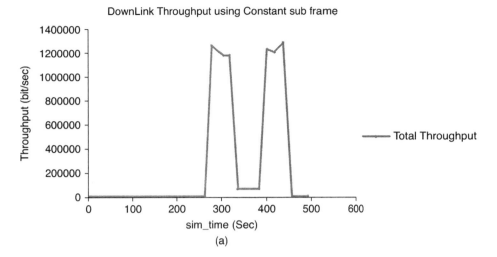

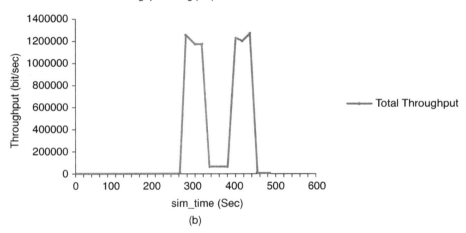

Figure 18.16 (a) Down link throughput using constant subframe (b) Down link throughput using proposed allocation schema.

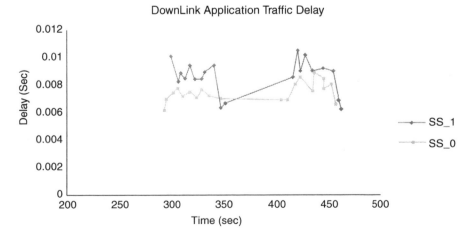

Figure 18.17 DL application delay using the proposed scheme.

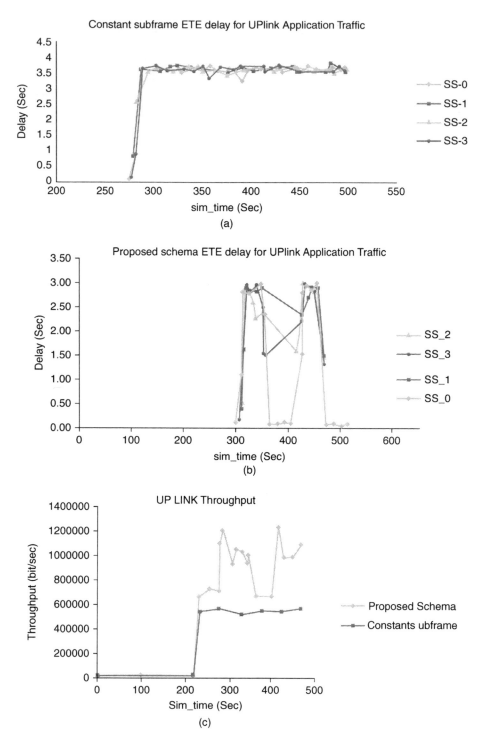

Figure 18.18 (a) Constant subframe ETE delay for UP link application traffic (b) Proposed scheme ETE delay for UP link application traffic (c) UL Throughput for both proposed and constant scheme.

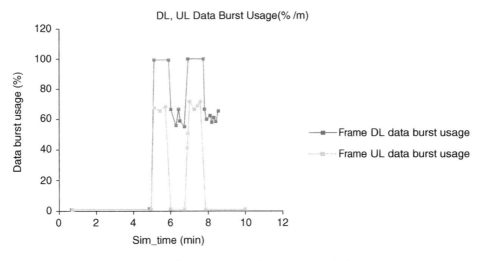

Figure 18.19   DL, UL data burst usage (% /min).

(230–320 seconds) and (400–440 seconds), the MAC throughput is ~700 Kbps and delays are ~3 second. On the other hand, for time periods (320–400 seconds) and (440–500 seconds), the MAC throughput is ~1 Mbps and delays are ~50 milliseconds, which in turn is compared with the constant subframe case, where throughput was ~560 Kbps and delay was ~3.7 seconds (between 280–500 seconds) So, there is a big improvement in the uplink performance owing to the reallocation of subframe resources.

Figure 18.19 shows the consumption of both the downlink (Figure 18.19(a)) and uplink (Figure 18.19(b)) sub-frames, as well as their available symbols during the simulation.

From Figure 18.20(a) and Figure 18.20(b), it can be seen clearly how the mechanism alters the subframe sizes (e.g. Subframe Usable Size). UL burst usage is high (100%) in the time pewriods (5–6 mints) and (7–8 mints) when it cannot increase its size to much because the deadline requests' size is less than 30% (average). However, there is a small increase in the UL subframe, most likely owing to DL sub-frames that are idle for very short periods of time.

For time periods (5–6 mints) and (7–8 mints), UL subframe size grows significantly and the corresponding utilization value increases. This is because the size needed by the deadline requests is over 30% during those periods.

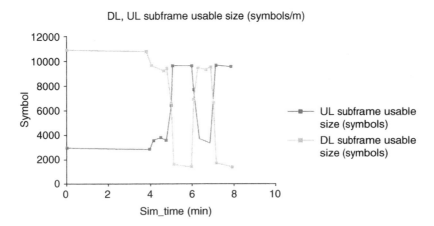

Figure 18.20   DL, UL subframe usable size (symbols/min).

18.6 Summary

In this chapter we have focused on the dependencies and the relationships between the Scheduler and CAC in 802.16 networks, and have outlined the BS model.

Simulation studies investigated the effect of various strategies involving the Scheduler and CAC algorithms, and their effect on network productivity in terms of real time applications of delay, frame throughput, provider return throughput, bandwidth use and other performance.

In this chapter, a new solution to the problem of Connection Admission Control algorithm and frame allocation schema in WiMAX network is presented.

With respect to frame allocation schema in the proposed mechanism the performance of downlink traffic remained unaffected; this was achieved by monitoring the number of due-to-deadline requests and returning the DL subframe to its original size if resources are required.

With the use of the OPNET simulator we demonstrated the effectiveness of the proposed algorithms, as it enables the provision of a probabilistic QoS guarantee to nrtPS flows admitted in the network, while a deterministic QoS guarantee is provided to UGS and rtPS flows.

The proposed mechanism is based on the IEEE 802.16 standard; the service level ertPS (extended real time Polling Service) specified in the IEEE 802.16e standard is not considered but the proposed framework can still be extended easily to support the ertPS service by admitting and allocating bandwidth to the ertPS connections in the same way that it allocates bandwidth to UGS connections.

Further research will be done in order to design an algorithm to fulfill all of the QoS constraints required by applications.

References

[1] 'Air Interface for Fixed Broadband Wireless Access System', *IEEE STD 802.16 – 2004*, October.
[2] J. Chen, W. Jiao, H. Wang (2005) 'A service flow management strategy for IEEE 802.16 broadband wireless access systems in TDD mode,' In *Proceedings of the IEEE International Conf. on Communications, (ICC 2005)*, pp. 3422– 6.
[3] K. Wongthavarawat, A. Ganz (2003) 'Packet scheduling for QoS support in IEEE 802.16 broadband wireless access systems' *International Journal of Communication System* 16(1): 81–96.
[4] H. Safa *et al.* (2007) 'New scheduling architecture for IEEE 802.16 wireless metropolitan area network', *AICCSA*, May.
[5] N.Ruangchaijatupon (2007) 'A traffic-based adaptive deficit scheduling for QoS support in IEEE 802.16e TDD mode,' *ICNS'07*.
[6] Y. Cao, Li Vok (2001) 'Scheduling algorithms in broad-band wireless networks,' In *Proceedings of IEEE 2001* 89(1): 76–86.
[7] M. Ma, B.C. Ng (2006) 'Supporting differentiated services in wireless access networks', *10th IEEE International Conference on Communication Systems*, p. 1, IEEE, Washington, DC.
[8] H. Wang, L. Wei, P. D.P. Agrawal (2005) 'Dynamic admission control and QoS for 802.16 wireless MAN,' In *Proceedings of IEEE 2005 Wireless Telecommunications Symposium (WTS 2005)*, April.
[9] Jianfeng Chen, Wenhua Jiao, Qian Guo (2005) 'An integrated QoS control architecture for IEEE 802.16 broadband wireless access systems,' Lucent Technologies, Bell Labs Research China.
[10] P. Brady (1969) 'A model for generating on-off speech patterns in two-way conversations' *Bell Systems Technical Journal* 48: 2445–72.
[11] P. Seeling, M. Reisslein, and B. Kulapala (2004) 'Network performance evaluation. using frame size and quality traces of single-layer and two-layer video: a tutorial,' *IEEE Communications Surveys and Tutorials* 6(2): 58–78.
[12] Opnet Simulator, http://www.opnet.com.
[13] 'Understanding and configuring MDRR/WRED on the Cisco 12000 Series internet router', doc. ID 18841, http://www.cisco.com/warp/public/63/ mdrr_wred_overview.html.
[14] C. Cicconetti, L. Lenzini, E. Mingozzi, and C. Eklund (2006) 'Quality of service support in IEEE 802.16 networks,' *IEEE Network* 20: 50–5.
[15] M. Shreedhar, G. Varghese (1996) 'Efficient fair queuing using deficit round robin,' *IEEE Transactions on Networking* 4(3): 375–685.
[16] M. Katevenis, S. Sidiropoulos, and C. Courcoubetis (1991) 'Weighted round-robin cell multiplexing in a general-purpose ATM switch chip,' *IEEE Journal on Selected Areas in Communications* 9(8): 1265–79.
[17] WiMAX Forum, 'Business case models for fixed broadband wireless access based on WiMAX technology and the 802.16 standard,' October 2004, http://www.wimaxforum.org/technology/downloads/WiMAX-The_Business_Case-Rev3.pdf.
[18] H. Lee, T. Kwon, and D. Cho (2005) 'An enhanced uplink scheduling algorithm based on voice activity for VoIP services in IEEE 802.16d/e system,' *IEEE Communication Letters* 9(8): 691–3.

[19] O. Yang, J. Lu (2006) 'A new scheduling and CAC scheme for real-time video application in fixed wireless networks,' In *Proceedings of 3rd IEEE CCNC*, pp. 303–7.

[20] S. Kim and I. Yeom, 'TCP-aware uplink scheduling for IEEE 802.16,' *IEEE Commun. Letters*, **11**(2): 146–8.

[21] Lin, Y. N., C. W. Wu, *et al.* (2008) 'A latency and modulation aware bandwidth allocation algorithm for WiMAX base stations,' *Wireless Communications and Networking Conference, 2008. WCNC*.

19

Implementation Frameworks for Mobile and Pervasive Networks

Bilhanan Silverajan and Jarmo Harju
Department of Communications Engineering, Tampere University of Technology, P. O. Box 553, Fin-33101, Tampere, Finland.

19.1   Introduction

Today, an extensive assortment of portable and powerful computing devices exists that is geared for mobility and pervasive connectivity over different types of networks. Such wireless networks include 802.11-based Wi-Fi LANs, WCDMA and EDGE-based cellular networks and short range wireless communication with NFC, Bluetooth or IrDA. Wide-area metropolitan networks, such as 802.16 WiMAX, higher speed cellular networks based on HSDPA/HSUPA/HSPA+ and low-power energy efficient radio networks such as ZigBee are all widely expected to play dominant roles in the medium to long-term future for wireless mobility. This wide variety of mobile environments presents a daunting challenge for the development of advanced pervasive services and protocol implementations, and has propelled research into architectural support for software development for mobile and pervasive networks at a relentless pace.

Such architectural research for the development of network components, services and protocols can be perceived to fall into one of two main categories. The first category focuses on languages, tools and formal methods which support the conception and validation of the system design under consideration. The second category focuses on tools, middleware and software frameworks that center on the implementation of these systems by providing reusable methodologies and components that could be assembled or extended as necessary to result in the desired system. Software development activities today continue to yield system artefacts that are custom designed as tightly coupled components with low reusability. Instead, using implementation frameworks is more effective at achieving quality software while avoiding the traps and pitfalls of writing and maintaining monolithic software [1].

A framework can be defined as a skeletal group of software modules that may be tailored for building domain-specific applications, typically resulting in increased productivity and faster time-to-market [2]. In [3], a framework is also described as a reusable, 'semi-complete' application that can be specialized to produce custom applications. In this chapter, we investigate the role of implementation frameworks for services in the domain of mobile and pervasive networks. We choose to focus our attention on implementation frameworks aimed at the creation of communication services, network components and protocols.

In devices, service interaction with transport-layer protocols is performed via system calls and APIs to the underlying operating system. However, service interaction with higher-level protocols occur using API calls to library implementations or a language-level module. Consequently, developers tie in the implementation of communications services very closely with the design and implementations of the utilized communication protocols themselves.

Pervasive Computing and Networking, First Edition.   Edited by Mohammad S. Obaidat, Mieso Denko, and Isaac Woungang.
© 2011 John Wiley & Sons, Ltd. Published 2011 by John Wiley & Sons, Ltd.

As it is important to understand exactly how these communication protocols are developed, Section 19.2 discusses activities related to how network services and protocols are specified, designed and finally implemented. Section 19.2 also takes a look at these activities in the context of pervasive networking and current practices for implementation. Section 19.3 expands on some of the biggest challenges that implementation frameworks have to address for mobile and pervasive networks. Section 19.4 looks at the current state of the art in using frameworks to implement communication services and protocols. Section 19.5 highlights our contribution towards creating an implementation framework for pervasive networking aiming to overcome many of the challenges discussed. Section 19.6 provides a look at how evaluation can be performed, both for implementations as well as frameworks. Section 19.7 summarizes the future outlook and concludes the chapter.

19.2   Correlating Design to Implementations

The methodology of applying software engineering principles in combination with formal methods for the design of communication protocols and services is called Protocol Engineering. Protocol Engineering was a discipline that was first recognized in the early nineties [4, 5].

Figure 19.1 shows that the main activities comprising the Protocol Engineering lifecycle can be summarized as:

- *Service specification phase,* which defines the functionalities the protocol can offer to a user.
- *Protocol specification phase*, which details the protocol behavior and the structure and content of the protocol data units.
- *Validation and verification phase* which uses formal methods aimed at non-ambiguous protocol analysis, performing reachability analysis, eliminating issues in systems interactions such as deadlocks, livelocks and non-deterministic state traversals as well as protocol simulation.
- *Protocol synthesis phase* which aims at using formal specification languages to design the protocol unambiguously.
- *Implementation phase* which starts with partially generated code from specification tools or custom frameworks and results in a full platform-specific running implementation.
- *Testing phase* which determines both the capabilities and the behavior of an implementation, ensuring it conforms and interworks with other independent implementations by using standard test suites.
- *Deployment phase* which brings the protocol into actual production usage.

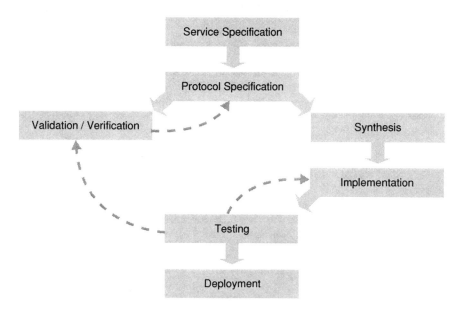

Figure 19.1   Protocol engineering complete lifecycle.

Protocol Engineering was heavily influenced by two Geneva-based standardization bodies: Comité Consultatif International Téléphonique et Télégraphique (CCITT) which was later renamed ITU Telecommunications Standardization Sector (ITU-T), and the International Organization for Standardization (ISO). The CCITT contributed effort towards the usage of the Specification and Description Language (SDL), and together with ISO, contributed the Tree and Tabular Combined Notation (TTCN) for conformance testing as well as Abstract Syntax Notation One (ASN.1) for the representation of data structures and encoding rules.

The ISO was responsible for providing two formal languages for specification and description: The Language of Temporal Ordering Specification (LOTOS) and the Extended State Transition Language (ESTELLE). However, the biggest contribution the ISO made towards protocol engineering was in the heavy influence of the OSI 7-layer reference model as well as its associated terminologies.

The OSI model was developed in 1977, but its design was based upon experience drawn from an earlier French network from 1972, CYCLADES [6]. CYCLADES was a packet-switched R&D network aimed at interconnecting research institutes, similar to the early ARPANET model. The OSI model was used as a basis for Public-Switched Telephone Networks (PSTNs), which adopted a technology-driven, circuit-switched approach. There were other implications on network design. Most notably, apart from being connection-oriented and end-to-end, such a design structure assumes hosts possess similar interface characteristics, having the ability to be constantly reachable via a predefined end-point address. This is in large part due to a limited mobility, arguably even assuming a static or wired topology. PSTNs have also had the luxury of dedicated networks, allowing controlled architectural deployment, vertical integration of service and call control layers, greater guarantees of QoS and limited heterogeneity of end-devices.

Such engineering activities as well as the OSI model have persevered for decades. They have served as both pedagogical as well as practical points of reference for all activities related to the design of communications software and protocols. Today, the OSI 7-layer model has been largely superceded by the more prevalent, simpler, Internet 5-layer model comprised of just the Application, Transport, Internet, Link and Physical layers. The deployment of the nodes may be random, or the nodes may be placed in specific points of interest, depending upon the application at hand and the ease of access to the terrain.

19.2.1 Evolving to meet Mobile and Pervasive Networks

Although drastic changes are not foreseen in the core packet-switched Internet, the various access networks that will interconnect with the global Internet are seen to be highly dissimilar in nature. Today, devices contain multiple disparate radio interfaces. Although simultaneous communication among endpoints using two or more radio interfaces is not yet easily achievable today, it is highly conceivable that multipath device communication would become a reality, forming and dissolving associations and ad-hoc connection paths as necessary, but belonging to the same context of usage. Advances in delay-tolerant networking call for inordinate periods of disruption, in which end-hosts may not even be reachable end-to-end at any period in time [7]. Wide-scale mobility also implies significant changes in the way physical, link and network layer characteristics need to be exposed to higher layer protocols.

Owing to the heavy influence from the CCITT and ISO, service and protocol specifications in classical protocol engineering have been performed with formal specification languages, especially SDL. This has also changed radically today, with the majority of applications and protocols now being designed for the Internet, the protocols of which are described informally as text-based standards documents by the Internet Engineering Task Force (IETF). With the circle of involvement for protocol engineering expanding greatly beyond the telecommunications domain, specifications are being issued in other formal or semi-formal ways, such as the use of XML by the World-Wide Web Consortium (W3C) or with IDL by the Object Management Consortium (OMG).

Perhaps the most significant development today is the need for the user to be formally taken into account when developing network applications and protocols. With the current OSI and Internet models, the highest abstraction possible is the application layer. This carries with it the implicit assumption that the needs of the user would be well reflected in the application logic and behavior with respect to the communication architecture.

While the 5-layer model is perfectly fine for modeling traditional IP-based client-server communication such as email, file transfers and web browsing, the majority of today's and tomorrow's breed of Internet applications can exhibit a protocol stack number greater than just five, with most of them being modeled as one application over another. A few examples follow:

- Using security mechanisms such as Secure Sockets Layer (SSL) or Transport Layer Security (TLS) [8] for secure web browsing or virtual private networking, the notion of a session becomes apparent. Similarly, emerging protocols for end-hosts, such as the Host Identity Protocol (HIP) [9] induce an additional protocol layer into the Internet model.
- Delay Tolerant Networking (DTN) [7] addresses the necessity for communication in climates which induce intermittent connectivity, low throughput, high error rates and highly disrupted communications such as under-water, deep space or extremely harsh terrestrial environments. The architecture introduces a store-and-forward Bundle Layer between Applications and the Transport Layer. In addition, the architecture takes into account different kinds of transport layers (not necessarily IP-based) and networks.
- Overlay networks, tunnel brokers, IPv6 transition mechanisms such as 6to4 [10] and Teredo [11] as well as P2P networks are designed to use the existing core Internet just as a link-level layer, onto which they add their own specific routing, network, transport and application layers.
- The rise of social networks and collaborative Web 2.0 content and document sharing, reveals a need for both sessions as well as a presentation layer.

19.2.2 Moving from Models to Implementations

Popular implementation solutions for communications services and protocol stacks in existence today contain a heavy influence from the field of software engineering, as opposed to classical protocol engineering. This results in activities such as domain-specific analysis of requirements and the need for prototyping. This has also resulted in modern proponents of communication protocol engineering to advocate a model-driven approach as an initial step to develop implementations, as opposed to a specification-driven approach. For example, [12] discusses requirement and analysis for protocol engineering using a Unified Modelling Language (UML)-driven approach, with modeling toolkits like the IBM Rational TAU [13] providing explicit support for the creation of communications software and protocols using both UML as well as the Systems Modelling Language (SysML).

Raising the abstraction level to a model-driven approach also allows developers to incorporate more modern techniques into protocol engineering. As shown in Figure 19.2, the engineering activity can now commence with developing a requirements model through a thorough domain analysis, resulting in a set of use case diagrams which

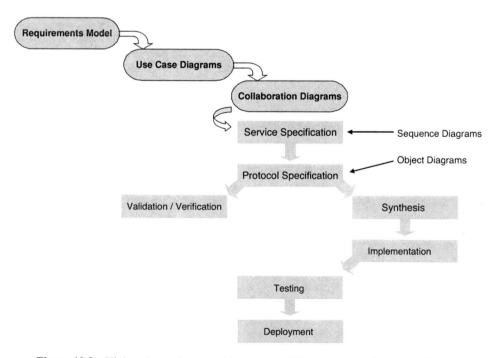

Figure 19.2 Higher abstractions provide better modeling of communications activities.

in turn facilitates the creation of collaboration diagrams outlining the communication amongst various components. This would also be advantageous in having a knock-on effect towards the later activities, as current paradigms such as object-oriented development, aspects, model driven architectures and UML can be brought into the fold easily, allowing specifications to be better correlated with object and sequence diagrams.

19.3   Challenges for Implementation Frameworks

A fundamental aspect of many of today's implementation frameworks and network middleware has been the provision of functional, well-defined building blocks such as objects, components as well as protocol layers to compose resulting protocols and services. Many of these frameworks strive to mask the underlying technology heterogeneity by providing consistent, OS-neutral interfaces for higher level network services and application protocols to utilize lower-level platform services. Frameworks such as Sockets++ [14], ACE [15] and OVOPS [16] which provide wrapper classes to encapsulate system-specific calls, such as communication using sockets, are a good example. In so doing, developers are relieved of the need to specifically design and code the resulting system to cater for differences in hardware platforms, operating systems, compilers as well as heterogenous networks. Instead, developers focus on familiarization with the facilities of the framework itself to achieve the result on their behalf.

Learning generally induces a significant learning curve, although the cost of learning the framework is usually offset by using it in the development of several successive as well as separate systems. This results in a significant reduction in time for prototyping and deployment. In addition, this approach provides significant opportunities for code re-use and portability. In [1], it is asserted that developers take six to nine months to become productive with frameworks, but can mitigate the effects of the learning curve, firstly by prototyping and incrementally focusing on framework features immediately relevant to the task at hand and secondly, by amortizing the framework over many projects to increase cost effectiveness. Consequently, the use of implementation frameworks for the creation of new protocols and network services is as much a business strategy as it is a useful technology. Over time, the cost and time invested in training developers to become familiar with domain specific frameworks is easily recuperated.

With their usage infringing into new application spaces and network domains, implementation frameworks have begun to inherit new challenges that urgently need to be addressed. This is particularly so for pervasive systems. To play a central role for supporting development in mobile and pervasive networks, implementation frameworks need to directly tackle several issues to deliver efficiently implemented protocols and services for a wide plethora of networks and technology domains. The demands of executing protocols and network applications in dynamically changing network conditions would place a premium for such frameworks to evolve and actively deliver environmental data and timely semantic information to interested components, and not just provide a 'middleware' layer that irons out heterogeneity.

These main challenges are categorized into the following subsections and discussed in detail.

19.3.1   Obtaining and Providing Device Characteristics

The diversity of consumer devices continues to expand at a relentless pace. This is particularly the case with mobile and wireless devices. Owing to a much reduced cost of ownership of highly powerful devices that range from laptops to smart phones, the amount of hardware functionality and packed into them easily rivals that commonly found in desktop computers and workstations. In addition, wireless devices today often possess more than just a single wireless interface allowing protocol and application communication over a variety of disparate networks technlogies such as WiFi, WiMAX, Bluetooth, as well as 3G/HSPA.

Implementation frameworks need to firstly interact with the underlying operating systems and hardware to obtain information about possible network interfaces. This would be of great benefit to developing protocols and network services that are capable of understanding current points of network attachment to seamlessly perform various activities such as multihoming, multipath communication or vertical migration to new networks if necessary.

In addition to the existence of physical network interfaces, many operating systems also supply logical network interfaces. Such interfaces tend to be highly OS-centric, but nevertheless, provide important information to implemented protocols and services about the communication capabilities of the device itself. Good examples of logical interfaces would be a virtual interface for carrying VPN traffic, loopback interfaces as well as overlay tunnels for ferrying IPv6 traffic in IPv4 packets.

Apart from such communication interfaces, an implementation framework should secondly be able to supply rich information about the current execution platform in a consistent manner to running protocols and services. Microchip technologies have advanced rapidly to the point where most of today's laptops already use multicore processors, with research efforts continuing to be expended for Network on Chip (NoC) design. An implementation framework allowing the development of protocols and services which are aware of the hardware architecture can prove highly beneficial to the development of high-performance systems. Likewise, protocols and network services that implement caching and data-centric operations, will benefit from obtaining information about the current methods of storage technologies available so that they can optimally adjust their strategies.

19.3.2 Supporting Context Awareness

Closely linked to the provision of device characteristics mentioned in the previous section is the provision of context awareness by implementation frameworks. Many definitions of 'context' abound, but a simple, yet effective definition denotes 'context' as a combination of elements of the environment that the user's computer knows about [17]. Hence we can describe context awareness for protocols and services as a means to obtain, understand and use information that represents the immediate operating environment of the device. Such context can include sensory information obtained, such as movement, temperature, or humidity. It could also result from access to obtaining location or positioning information via the functionalities of the device itself, such as GPS coordinates, Bluetooth beacons and WiFi access points. Detection of other peers can also be performed this way.

Frameworks having the facility of extracting this information and furnishing them to resulting implementations offer a broader horizon in terms of implementation functionalities. This is an especially salient challenge for the implementation of protocols and services for fixed and mobile ad-hoc networks, self-organizing topologies, peer-to-peer networks as well as sensor networks.

Support for context awareness is also deemed a precursor for providing fully-fledged dynamic adaptation by pervasive services and mobile applications to changing environmental conditions. Such demands would require implementation frameworks to provide the ability to design a flexible system infrastructure. This ability can also facilitate, at runtime, inspection of the current execution state of pervasive protocols and services, as well as a consistent and predictable transition to a new state triggered by the change in context.

19.3.3 Managing User and Quality of Service Policies

Context awareness and dynamic adaptation depend heavily on environmental information. However, any decision-making algorithm or inference engine which is utilized cannot behave autonomously to pre-empt the user as well as user preferences. Today's user is capable of high degrees of mobility, directly resulting in work being done in vertical handovers, and possesses multiple devices that may have overlapping functionalities. Since the user is a vital source of information, especially in relation to mobility, implementation frameworks must provide a means of user interaction to extract and understand user preferences and acceptable QoS levels for mobile and pervasive networks. Important user policies include:

- *Outlining hardware functionality.* Users may define usage profiles and policies controlling hardware functionality to conserve energy. Such preferences may be coarse-grained, such as turning a Bluetooth radio interface on or off, or fine grained, such as explicit control over transmit and receive power for Wi-Fi communication.
- *Issues related to pricing and quality.* In the presence of several available network technologies, users may want to state specific network usage with respect to price and expected quality and available bandwidth.
- *Device associations.* Personal and body area networks tomorrow may consist of a multitude of communicating applications working in tandem on behalf of the user. As users move into different technology domains, the roles of each device in the Personal Area Network (PAN) or Body Area Network (BAN) may change based on user policies, whilst preserving their overall association.

19.3.4 Discovering Resources and Services

With the increase in mobility, the way service discovery and lookups are performed would also become significant, as devices would need to search for essential services upon movement. Often, when devices roam into network spaces for the first time, they will have no previous knowledge of the types of services and other devices resident in the new network space. How broadcast and multicast capabilities can be utilized for active service discovery hinges on understanding both the link level and network level characteristics. Even though both a mesh wireless network and a Wi-Fi network may provide IP connectivity, broadcasting and multicasting capabilities in both networks would be vastly different. At times, such as with Mobile IPv6, active discovery into new networks may not be possible, as IP mobility is hidden from transport-layer protocols. Therefore, armed with sufficient link and network information, discovery mechanisms may seek to either utilize broadcast and multicast capabilities present, or choose to rely on point-to-point communication with a lookup service.

In terms of applications, new breeds that are location and context sensitive are rapidly being developed. On the micro level such as in sensors, the boundaries that define applications, protocols and operating systems are distinctly blurred. At the macro end of the scale large applications consisting of many components may need to be constantly aware of their movement at all times, bypassing the modularity proposed by the OSI model for low-level access to network and link level characteristics of the host they execute on. Often, such communication may not even occur asynchronously using packets and service access points. Instead, low-level libraries provide blocking function calls that applications can explicitly use to bypass the stack entirely.

19.3.5   Interworking and Interoperability

End developers often find a need to work with more than one technology or framework. Multi-threading, Inter-Process Communication (IPC) techniques, as well as remote procedure call (RPC)-style communication can be used for such interworking among separately executing threads, services or protocols.

The problem becomes more apparent if the facilities offered by two or more frameworks need to be interwoven into a single resulting executable, especially on platforms in which the usage of threads is either too expensive or impossible. For example, let's assume the existence of two implementation frameworks, Framework f_A and Framework f_B. A protocol developed for network adaptation using fA might need to be enhanced with a notion of geographical location offered by f_B. Points of commonality need to be found that allow a single executable to accomplish this. Should interworking be required among two or more event-driven frameworks or toolkits, the design of the framework itself should facilitate allowing the event handling among all participants. For example, a communication service derived from a message based Framework f_C may have to interact not only with the network, but also with human users using a Graphical User Interface (GUI) designed with an event-driven graphical toolkit.

Consequently, interworking and interoperability among implementation frameworks is poised to be a big challenge for protocol and communication service development. This is particularly so in the area of pervasive networking and mobile computing, which is still witnessing a rapid proliferation in the amount of frameworks, libraries, middleware platforms and toolkits that are being created, used and extended.

In essence, a fully interoperable implementation framework ideally allows a developer to seamlessly access features offered by a separate framework within the same executable. Such interworking can be classified into roughly two techniques: Horizontal interworking and vertical interworking.

Horizontal interworking can be performed by the framework with proper decoupling, offering code hooks, providing callbacks and communications interfaces that can reliably transfer method calls and data structures back and forth between itself and the other framework. The developer is consequently directly exposed to the functionalities and execution logic of the various frameworks to be used by the implementation. Extra time and effort is needed on the part of the service developer for familiarization with all the frameworks being used. Additionally, although a framework f_A may offer interoperability possibilities, should a need to interoperate with another framework f_B arise, it remains the service developer's responsibility to actually integrate and co-ordinate communication among the service logic, f_A and f_B in the resulting implementation. Horizontal interworking is often applied when requiring functionality from several independently developed implementation frameworks for a resulting implementation, where no distinction of one clear choice of framework or technology exists.

Vertical interworking on the other hand describes a scenario where an implementation derived from framework f_A interacts with another framework f_B implicitly by objects, method calls and data structures provided by f_A. The result would be that the implementation would communicate solely with framework f_A. One can therefore perceive the relationship between the two frameworks f_A and f_B in one of two ways. Either framework f_A would provide a bridging and translating facility for using framework f_B (a 'layered' approach where f_B would reside below f_A), or framework f_A needs to wholly or partially absorb framework f_B (an 'encapsulated' approach where f_B would reside within f_A). Vertical interworking is especially common if the designers of one framework take an explicit decision to use, extend or enhance the features of another. In so doing, the implementation developer is shielded from needing to understand or learn the details of accessing information from both frameworks. However, the responsibility of ensuring that interworking proceeds reliably now shifts to the developers of the framework. Care must be taken to ensure that dependencies are properly checked, and new versions of the various frameworks offering potentially different communication techniques are accounted for.

19.4   State of the Art in Implementation Frameworks

Today, a number of popular implementation frameworks exist which allow the creation of communication protocols and network services. Here, we examine some example frameworks that have been utilized. These frameworks employ different methods and strategies to allow implementers to conceptualize, prototype, implement and test the resulting network components. The differences in implementation approach can be classified into several categories: those relying on high-level specifications, those using a micro-protocol strategy, those employing software and communication design abstractions to realize the created service or and those protocol adopting a repository-like approach.

19.4.1   Adopting High-Level Specifications

The IBM Rational SDL suite (formerly known as the Telelogic SDL Suite) [18] comprises a set of tools as well as a framework aimed at the construction of protocols and network components beginning with a high-level specification of the entire system. A complete development environment is provided to specify the system to be developed, using SDL. Graphical editors are provided that allow a developer to visualize, conceptualize and specify the complete system at the system, block and process level. Message sequence chart (MSC) editors are provided to augment the SDL diagrams with graphical notations depicting use cases and interaction scenarios.

Because of the usage of a formal language, the specification of the completed design lends itself easily for formal analysis and verification. Other components are also provided to allow the verified model to undergo further testing and simulation to remove any design errors. Once the process of design and testing is completed, the developer uses the SDL suite to generate executable C or C++ code that can be directly compiled for the target execution platform.

19.4.2   Adopting a Micro-Protocol Approach

Microprotocol frameworks lend an innovative approach towards implementing protocols. Each protocol to be constructed is perceived as a set of its various properties. These frameworks then allow a developer to implement any protocol by modeling them from smaller building blocks called micro-protocols. Each micro-protocol implements a single functionality or property that is required in the final protocol. Examples of such functionality implemented as micro-protocols include error detection, reliability, ordering, caching, link monitoring and so on. By organizing and configuring a set of such micro-protocols, a developer can create protocols with different properties.

Appia [19], Cactus [20], CORDS [21], Coyote [22] and Ensemble [23] are such protocol implementation frameworks predicated on composing protocols from smaller, functional building blocks. All these frameworks stem from earlier efforts in two independent efforts. Appia, CORDS and Ensemble are derived from an earlier framework called the x-kernel [24], while Cactus and Coyote trace their ancestry to the Horus [25] framework. These frameworks largely support the implementation of group communication protocols for the purpose of distributed services, applications or middleware for the Internet. However, they can be easily adapted for use to design and implement protocols in mobile and pervasive networking too. For example, [26] describes how protocol decomposition can be performed for routing protocols implemented with micro-protocol frameworks in Ad-hoc networks, while [27] discusses their usage and evaluates resulting protocol implementations for both ad-hoc and wireless sensor networks. Dynamic as well as context-aware adaptation of micro-protocol frameworks for use in mobile environments is discussed in [28].

19.4.3   Adopting a Design Pattern Approach

Adaptive Communication Environment (ACE) [15] is a C++ object-oriented framework used widely in network programming. Its features include the ability to execute in resource constrained environments, mission critical and real time systems, high performance platforms, as well as support for concurrency, multiprocessing and multithreading.

ACE provides C++ wrapper classes and an OS adaptation layer, which encompass various network and operating system specific details. It also provides event demultiplexing components, service initialization components, service configuration components, stream components, and CORBA ORB adapter components. While service configuration components aid in allowing implementations to assemble services and protocols at install as well as run-time, stream components assist in the development of user-level protocol stacks.

The distinctive feature of ACE is that, instead of a specification language, it makes extensive use of software design patterns to provide a higher-level conceptual abstraction of how such protocols and network services can be constructed. The patterns are utilized to explain object interaction and the inner workings in ACE. They are also employed as a means of guiding programmers to become familiar with the various framework features to develop their protocols and services, by deriving them from the objects and classes found in the ACE framework.

The event-handling mechanism in ACE is documented in the Reactor [29] design pattern, whereby a single handler monitors resources and registered objects for arrival of events. Upon such arrivals, the handler subsequently demultiplexes incoming requests and delegates individual event handlers to process associated events. ACE also provides the Proactor [30] design pattern for leveraging operating systems which provide asynchronous I/O support. The proactor involves an event handler performing an asynchronous I/O operation and registering with a completion dispatcher to obtain completion events. When the I/O operation is completed, the operating system notifies the dispatcher which subsequently notifies the event handler.

ACE has also been extended to encompass other application domains. JAWS [31] is an adaptive web server which uses ACE, while TAO (The ACE ORB) was developed as an ACE extension to support real-time CORBA communication.

19.4.4   Adopting a Protocol Repository Approach

An alternative to developing protocols and communication software is to employ frameworks with which entire, ready-made protocols could simply be added to result in an implementation. The Twisted [32] networking framework is one such framework. Twisted is written entirely in Python. Hence, Twisted is able to execute atop any platform which supports Python. It is also asynchronous and event-driven. Callback functions are implemented in Twisted using objects called Deferreds. Just like ACE, each Twisted-based implementation uses a Reactor which is responsible for dispatching events to their appropriate handlers.

The framework has an active developer community and has subsequently amassed an extensive range of protocols that are ready to be imported as modules for immediate use in Python applications. This renders it easy for a developer to quickly create prototype systems based on existing protocols implementation. These protocols are largely IP-centric, have well-defined interface, and can simply be imported into an end-user python implementation as modules. However, classes are also provided that developers can use to derive and create new protocols if necessary.

Twisted is an appealing choice of technology to work with, given the rapidly scriptable, interpreted nature of Python. Quick modifications and fixes can be made to the code being interpreted, and code modules can be reloaded with minimal effort. This ability to modify parts of a Python application while it is running, together with the ease with which it can be debugged, renders Twisted highly popular, both when designing custom protocols from scratch or when extending an existing protocol which already exists in some package. As with the previous categories, the framework renders itself easy to extend into new application domains as well. Such activities include Twisted being used for network communication by the Lycaon [33] evaluation framework for protocols in ad-hoc networks, as well as in the Simple Sensor Syndication [34] project for wireless sensor networks.

19.5   Current Frameworks Research for Network Protocols and Applications

In this section, we present our own research activity in addressing and alleviating many of the concerns discussed earlier in Section 19.3. We discuss building a highly flexible object-oriented implementation framework called DOORS [35]. Apart from protocol implementations, the framework can also be used for developing network services, ranging from simple socket-based systems, to distributed applications, object-based gateways as well as general event-based client-server applications. This highly generic framework is founded on the event-driven programming paradigm, in which callback functions are written by the developer and invoked by event handlers. DOORS is single-threaded and capable of being compiled and used across virtually all UNIX and Linux variants.

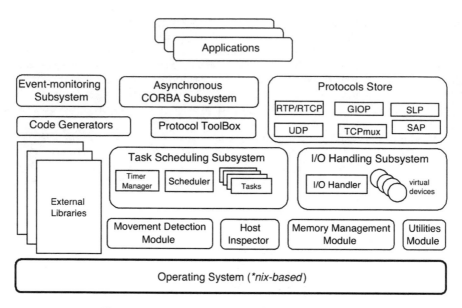

Figure 19.3   Architecture of the DOORS framework.

It is also lightweight enough to encompass and execute natively on embedded systems, particularly on the Maemo platform used by the Nokia N800 Internet Tablet and OpenWRT, a Linux-based firmware distribution for wireless access points.

19.5.1   DOORS Framework Architecture

DOORS is a white-box framework that traces its lineage to the OVOPS framework, mentioned in Section 19.3. In order to fully exploit its potential, a developer would need to become familiar with the object classes provided by the framework as well as their methods of interaction with other parts of the framework before subsequently extending them towards the implementation of the resulting system. Conceptually, the framework exists at development time as a set of C++ classes and a set of binary tools. At runtime, it exists purely as one or more shared libraries linked against the developed implementation. DOORS can be perceived as residing above the host operating system whilst supporting the derived service or protocol. Figure 19.3 is an architectural depiction of the modules, subsystems and tools of DOORS framework. The architectural model also reflects the relative placement of the framework's various parts with respect to a device's operating system and external libraries.

One of the major driving principles behind the organization of the various subsystems, is that they are highly decoupled from each other. This is to ensure as minimal a dependency as possible exists among the various subsystems within the framework both at compile-time and at run-time.

At compile-time, the framework is built as a set of binaries as well as several small shared libraries. This provides resource and computation-constrained devices the possibility to use as minimal a number of subsystems running as possible in order to have a functioning implementation. The framework uses a modified GNU Autoconf build system. This modified subsystem allows fine grained options to tune the build process for specific needs, enabling and disabling the compilation of certain subsystems. This modification was also done so that compilation for multiple target architectures could be processed concurrently on the same source tree, something that is impossible with a standard autoconf configuration. This provides a significant time savings advantage, allowing a single source-code tree to be mounted to multiple hosts, and the existence of a separate build directory that is unique to each target architecture.

The event-driven nature of communication extends beyond the interaction between the framework and the resulting application code. Virtually all the run-time interaction among the set of objects within every subsystem is message-based and asynchronous. The many subsystems provide convenience classes and tools for developing any communications service or protocol. However, at runtime, the only mandatory dependencies a minimal

DOORS-derived implementation would have are to the Task Scheduling Subsystem and to the I/O Handling subsystem. Even then, because the source code tree logically reflects the structure of the subsystems and components, an adept developer should be able to further minimize the size of the resulting shared library simply by quickly creating a custom build and omitting any unneeded objects at compile-time.

19.5.2 Description of Selected Framework Components

Event-based behavior is represented in DOORS as base objects called Event Tasks, while protocols are implemented based on a further derivative of the Event Task, called the Protocol Task. Protocol Task objects reside in the Protocol Toolbox which also contains other specialized objects for aiding protocol development such as finite state machine implementations, Service Access Points, and encoding/decoding of Protocol Data Units, bi-directional Ports used by Protocol Tasks for message passing and multiplexers that allow M:N communications between Protocol Tasks.

The Task Scheduling Subsystem consists of a scheduler handling requests from Tasks, and providing execution turns to these Tasks using a priority-based callback mechanism. Each task primarily contains the behavioral logic representing either a protocol layer or sub-layer. The Subsystem also contains timers that can be used for providing soft real-time guarantees together with a timer manager.

The I/O Handling Subsystem supplies event-driven tasks and applications a consistent interface to communicate with external entities (devices, applications, as well as other DOORS UNIX processes) by introducing the concept of a virtual device. Virtual devices model aspects of communications activities such as network endpoints or operating system inter-process communication (IPC) APIs as objects within DOORS. Such virtual devices currently include socket-based facilities to perform both unicast and multicast communication over IPv4 and IPv6, stream I/O facilities such as files and pipes as well as OS services such as terminals.

The Protocols Store contains several existing protocol implementations that a developer can directly use. Implementations currently available include the Real-Time Transport Protocol (RTP) and RTP Control Protocol (RTCP) [36] for audio streaming, an IPv6-enabled Service Location Protocol (SLP) [37] for service discovery, Session Announcement Protocol (SAP) [38] for session announcements as well as UDP and TCP helper classes.

The Asynchronous CORBA Subsystem comprises a lightweight, library-based CORBA solution for asynchronous communications which consists of an Object Request Broker (ORB), a Portable Object Adapter (POA) and dynamic interfaces for client-side and server side in the forms of the Dynamic Invocation Interface (DII) and the Dynamic Skeletal Interface (DSI), which run atop IIOP.

The Event-monitoring Subsystem consists of facilities used at runtime to monitor various parts of a running system such as the state of tasks, internal variables, the number of messages processed, their types, timers as well as the scheduler load. Using a hierarchically organized symbol handling meta-interface, a local event monitor has the capability of inspecting intra-process events in a locally running application. A prototype distributed event monitor is also provided which uses SLP and CORBA communication, to search for DOORS-based applications in a network and inspecting events, tools for logging events into files as well as a through which the event monitors communicate with the application.

The Utilities Module provides basic support for implementation offering rudimentary datatypes, structures and class templates. The Memory Management Module provides support for runtime memory management that is often faster than using standard memory allocations and deallocations offered by C++. At present, it features basic, statistical and block memory allocations.

19.5.3 Supporting Service Discovery for Mobile Devices

Mobile devices need rapid service discovery mechanisms that are neither predicated on the presence of lookup mechanisms nor on specific network topologies; the ability for mobile applications to detect movement, dynamically discover services and potentially even offer services in network spaces ranging from ad-hoc to infrastructure oriented environments presents an interesting research challenge.

The SLP implementation in DOORS comprises part of the Protocols Store and is able to provide support for service discovery to other interested components and applications. SLP is a very versatile protocol, and it supports a high flexibility in configuration. Services can be discovered and advertised via multicast dynamically using agents, or agents' behaviors can be configured statically, to avoid using multicast, and so on. However, SLP does not support the notion of automatic service rediscovery, nor changes in network topologies, and relies on external methods for providing these facilities. Consequently, its use has been confined to fairly fixed network topologies.

The DOORS implementation of SLP extends the standard SLP towards usage in Mobile IPv6 in two significant ways [39]. Firstly, it understands and handles movement of the mobile node by having an SLP agent residing in the mobile node listen or actively probe for changes in specific service advertisements that emanate from deployed, existing SLP agents in visited networks.

When information pertaining to movement and a change in the network topology is received, the protocol is able to take corresponding steps to automatically discover services in the new network. Secondly, the protocol now supports the notion of persistent and transient services. This allows pervasive network applications offering services, the freedom in selecting whether the service will be always reachable regardless of the mobile node's current location, or whether the service will only be offered and reachable at the current point of network attachment.

The SLP implementation in DOORS has also been extended to garner movement information messages from the Movement Detection Module. Obtaining movement information this way reduces the latency of the protocol in having to wait for changes in advertisements in new networks and improves the time taken to discover and announce services in new network spaces. The Movement Detection Module is discussed in the next subsection.

19.5.4   Support for Basic Movement Detection

The Movement Detection Module obtains movement information by interacting asynchronously with low-level libraries and mobility protocol APIs that monitor the device's network address and default routes. Information is exported in a consistent manner for use by the other subsystems in the framework. The module works in IPv6 and Mobile IPv6 networks, by monitoring changes in the host IPv6 address on behalf of all the various protocols and services using or comprising the DOORS framework. Address changes are propagated to all other interested components by supplying the information in DOORS-specific messages.

Currently no POSIX-compliant APIs or methods yet exist in an operating system that allows any application to directly retrieve this information. In fact, kernel-space hooks are currently necessary to obtain the information directly from the TCP/IP protocol stack implementation. In the case of the Movement Detection Module, such low-level information is obtained by interfacing with a low-level library called libmobinfo [40], a small C-based library, whose sole purpose is to provide registered applications with changes occurring to the IP address as soon as the mobile node's IP layer itself becomes aware of it. Libmobinfo exports an API that can be used both synchronously and asynchronously.

Tasks send messages to the Mobility Module to register or unregister their interest in receiving movement notification. Upon movement, the Mobility Module converts the movement information received from libmobinfo into a message for processing by other DOORS Tasks.

Using the Mobility Module this way, it becomes easy to provide movement detection information to applications from the framework level: The applications themselves receive the information in a consistent manner of communication, and need not be concerned if changes occur to mobinfo. The Mobility Module on the other hand can easily be modified if such changes occur, without affecting the application logic of the other Tasks. Extra messages can also be defined in the future containing richer information such as GPS data, sensor-based contextual information, and router prefixes.

19.5.5   Addressing the Need to Furnish Platform Characteristics

The Host Inspector is a single object instance that can be queried by all running DOORS-based protocols and services for information pertaining to the device itself and its associated interfaces. Such information includes processor type and endian architecture, host operating system and version, list of all fixed, virtual and radio interfaces present as well as their currently assigned addresses.

The Host Inspector gathers this information in two main ways. It firstly parses the platform configuration file autogenerated by the Autoconf build system that DOORS employs, to obtain platform specific information which reflects what the target architecture is, the endianness, whether cross compilation was involved, if IPv6 is supported, the existence of a Bluetooth stack and so on. It then uses the API of the libpcap [41] packet capture library at start-up to obtain more information about existing fixed, wireless and virtual interfaces, their currently assigned addresses. In the case of the non-virtual interfaces, the Host Inspector also extracts their hardware properties.

19.5.6   Analysis of the DOORS Framework

Explicit design decisions undertaken during the development and evolution of the DOORS framework ensures that it remains easy to learn for new end-users wishing to use the framework for implementing protocols and communications software. It was also aimed to be of maximum benefit to developers producing or extending other frameworks, who may wish to incorporate certain features that DOORS provides, or allow their frameworks to interwork with DOORS.

Many core aspects of the DOORS framework can be easily modified. For example, the default priority-based round robin scheduler can be easily substituted by a framework user providing his own scheduling algorithm. Such an algorithm can range from a simple subroutine, to a complex hierarchy of schedulers, each implementing its own algorithm with a top-level scheduler responsible for overall execution turns.

19.5.6.1   Extending the I/O Handling Subsystem

The I/O handling subsystem can be easily extended to encompass new kinds of wireless networks. The DOORS I/O Handler incorporates the Reactor design pattern, allowing it to interwork with other event-driven, Reactor-based frameworks such as ACE and Twisted. Currently, the I/O Handler waits for events from the network by blocking on select(). However, work is being done to allow this to be substituted by the highly flexible event dispatching library, libevent [42]. Libevent is a low level library which is highly portable across Solaris, Mac OS X, the various BSD operating systems, Linux and Windows. It selects the most scalable event notification present. Currently libevent supports /dev/poll, kqueue(), event ports, select(), poll() and epoll().

Additionally, each I/O event handler is represented by a virtual device object which contains callback functions for read and write operations as well as for error conditions. This mechanism enabled the rapid prototyping and implementation of virtual devices capable of Bluetooth communication with Radio Frequency Communication (RFCOMM) [43] and Logical Link Control and Adaptation Protocol (L2CAP) [44]. The prototype was developed in Linux-based devices (both laptops as well as the Nokia N800), that use the BlueZ official Linux Bluetooth stack. Bluetooth APIs are not standardized across different operating systems. However, the framework's Bluetooth virtual devices can be easily extended towards other Bluetooth stacks and operating systems that DOORS can run atop of. Because the virtual devices export a consistent interface to applications and tasks in DOORS, it provides a very portable means to users of the framework to develop Bluetooth-based communications.

19.5.6.2   Usage of High-Level Specifications

Just like ACE, many components and interactions in DOORS can be expressed using design patterns. In contrast with ACE and other implementation frameworks though, DOORS offers developers a possibility to specify protocols and communications software with a high-level specification language.

Information flow and service models in DOORS are consistently expressed using terminology from the OSI layered model, allowing many developers to quickly grasp core communication concepts when using DOORS. For example, messages are usually indicated as Service Data Units (SDUs) and are represented entering and leaving Service Layers and objects via Service Access Points (SAPs). Peer abstraction as well as invoking encoding and decoding rules upon Protocol Data Units (PDUs) are instilled in a separate Peer class.

The initial specification of implemented messages and task behaviour in DOORS that eventually are deployed as protocols, applications or network services, are written in XML by the developer. This is in contrast with the IBM Rational SDL suite using SDL. The XML specifications are then parsed by code generators in DOORS at compile time to produce corresponding framework specific C++ classes such as Messages, Service Access Points, Service Primitives as well as extended finite state machines. These can be directly used as part of an implementation.

Using XML for message and protocol specification serves as a technology neutral platform for expressing many of the ITU-T or IETF standardized protocols and communications mechanisms. It is a natural springboard for developers when designing communication architectures to extend its usage from simply being an 'easy to read and write' specification document towards more powerful XML paradigms to encompass document templates, transformations and scripting with other XML technologies for the future.

This allows for a portrayal of the resulting software architecture which renders itself easy for future extensions as well as interoperability with other potential external services and applications. Any relevant kind of network or even simulation framework capable of producing events or traces as an XML output can be transformed for usage into DOORS. This also provides a natural way of expressing users, user policies, preferences and events to be fed as messages into a running DOORS system and vice versa.

19.6   Evaluating Frameworks and Implementations

For new protocols and communications services, evaluation often begins at the simulation stage. Simulation tools and packages are often employed to simulate protocols or network services being designed to understand their characteristics and performance under various parametric changes in networks. However, practical evaluation of communications protocols and software are also performed with testbeds while software metrics can be used to evaluate implementation frameworks. Subsections 19.6.1 and 19.6.2 explore how evaluations can be performed on implementations, while subsection 19.6.3 discusses how implementation frameworks themselves can be evaluated.

19.6.1   Simulation-Based Evaluation

Since developing implementations aimed at heterogenous pervasive networking is an inherently complex exercise, using simulation tools and software is of particular benefit, as the simulated model could be quickly exposed to various mobility models efficiently and results can be garnered in a verifiable and repeatable manner. However, comprehensive studies have indicated that simulations in certain domains suffer from a lack of credibility. For telecommunication networks, studies [45] and [46] state that the large majority of published simulation results cannot be trusted as they contain systematic flaws. In addition, [47] and [48] report that owing to imprecise or simplistic assumptions, the credibility of simulation work is heavily compromised for Mobile Ad-Hoc Networks (MANETs).

Among the reasons cited for this are a failure to account for protocol stack interactions and the lack of usage of real data to be used as input to simulators. It was also shown in [49] that a simple flooding protocol that was implemented in three commonly used network simulation tools, NS-2 [50], OPNET Modeler [51] and GloMoSim [52]), exhibited large differences. This suggests that while logically only one of the simulations should be the correct one, the divergent results indicated that all three could be wrong. The authors of [48] surmise that accuracy of simulations can only be achieved with a proper validation of the simulation model against the intended or real-world protocol implementations. However, they also acknowledge the difficulty in validating the simulation of a newly developed protocol for which no actual implementation or testbed exists to serve as a baseline.

Simulation-based evaluation therefore, can be mitigated if real data can be used as input where possible, and implementations of the system can be evaluated in a testbed either as an extension of the simulation code, or in conjunction with it. Such is the case of the Opportunistic Networking Environment (ONE) [53] simulator designed for routing and protocol evaluation in Delay-Tolerant Networking (DTN). The simulator allows users to create scenarios based upon different synthetic movement models and real-world traces and offers a framework for implementing routing and application protocols. Interactive visualization and post-processing tools support evaluating experiments and an emulation mode allows the ONE simulator to become part of a real-world DTN testbed.

19.6.2   Testbed-Based Evaluation

Testbed-based evaluations of protocols and communications software offer an alternative approach to a simulation-based approach discussed in the previous subsection. Three key requirements of scientific experimentation with MANET testbeds are identified by [54] are:

- *Reproducibility.* Independent research groups must be able to reproduce the results of an experiment. For mobile ad-hoc networks, reproducibility is a significant challenge due to the complex impact of radio propagation and node mobility on the results of an experiment.
- *Comprehension.* A scientist conducting an experiment must be able to access all relevant information to comprehend and explain the results of the experiment. There is a need for tools that collect information on different layers and combine this information to allow a detailed analysis.
- *Correctness.* Any experiment may suffer from broken tools, errors with the setup and problems when conducting the experiment. While reproducibility and comprehension will most likely reveal these problems, it is vital to

the efficiency of a researcher to be able to verify whether any given experiment has produced valid results. This can be supported by an established methodology and a selection of appropriate tools.

Naturally, the biggest drawback of a testbed-based approach is the time, effort and cost expended in procuring or setting up a testbed. However, this approach has the ability to overcome some shortcomings seen earlier. For instance, testbeds provide better accuracy in portraying the characteristics of lower protocol layers. At times, deploying an implementation into a real environment for testing may not be feasible for various reasons. It is frequently impossible to deploy wireless sensor networks into real networks for testing and evaluation, owing to the difficulty involved in selecting suitable terrains or reprogramming the sensor nodes. In this sense, testbeds can be used to construct realistic environmental conditions and scenarios against which protocols and communications software can be evaluated rigorously. Additionally, it is increasingly becoming possible to combine a testbed-based evaluation with that of a simulation-based evaluation. The ONE simulator discussed earlier is a good example.

Apart from the ONE simulator, other network emulation tools exist. The W-NINE [55] emulation platform aims at mobile and wireless systems, attempting to deal with accuracy with the addition of a simulation stage before emulation so as to use accurate models to produce the emulation parameters. MoteLab [56] is a wireless sensor testbed platform which provides the ability to investigate resource limitations, communication loss and energy constraints at scale in a sensor network under test. At the same time, it exports a web interface allowing users to create and schedule experiments as well as interact with individual nodes while also automating testbed reprogramming. Consequently, it simplifies and accelerates the deployment as well as evaluation of wireless sensor network applications and protocols.

MoteLab is an indoor, laboratory-based testbed. Other wireless and sensor testbeds that fall into this category are APE [57], Kansei [58], Mirage [59], Mint [60] and ORBIT [61]. Open research testbeds such as PlanetLab [62] and Emulab [63] also exist which facilitate large scale tests of novel protocols, communications services and architectures.

19.6.3 Metrics-Based Evaluation

Undertaking the evaluation of any implementation framework is a highly daunting challenge, considering the variance and availability of programming technologies as well as applicability to specific hardware platforms. In addition, relative comparison is often difficult considering that the effectiveness of many frameworks is scoped for a narrow, specific range of domain usage. Consequently, evaluation of frameworks is often performed based on their own merits and limitations. However, we can identify certain key metrics for evaluating frameworks:

- *Suitability for a specific domain of usage.* For example, a small, Java-based communications service desiring a rough notion of location-awareness might simply need to use the Java Micro Edition (ME) Location API [64], whereas a platform such as CARMEN [65] introduces a larger storage and performance overhead, but provides rich location information to wireless mobile devices whose connectivity stands a better than average chance of disruption and migration to new locations. Similarly, a framework might be limited by the choice of the offered programming platform to support concepts such as reflection, aspect orientation and dynamic adaptation.
- *Capacity of the framework to interwork and interoperate with other frameworks, toolkits and applications.* A framework that offers a tightly coupled, highly integrated development environment may not be as effective as a framework offering a cleanly decoupled model with clear callback interfaces and communication infrastructure.
- *Costs of learning and maintenance.* How steep the learning curve of the framework is, particularly in relation to the anticipated lifetime of the resulting implementation is important. Should the need to rapidly prototype, code and evaluate an implementation be necessary, using an interpreted framework such as Twisted might be far more effective than using a compiled framework such as ACE. Both the learning cost as well as the cost of maintenance are heavily influenced by the amount of documentation present for the framework. The cost of maintenance is also influenced by the ability of the framework to provide a reliable, consistent and bug-free platform.
- *Framework evolution and portability.* This pertains to the capability of the framework to evolve and adapt to new pervasive networks, incorporate support for new hardware platforms and functionality and possessing an active, healthy and sustainable development effort.

19.7 Conclusion

The convenience with which connectivity can be achieved with wireless networks partially accounts for its rapid adoption into the mass market. With the average user possessing more affordable powerful wireless portable equipment than even half a decade ago, the uptake in wireless communication has been considerably staggering. As device vendors begin to offer programming support and open platforms for their hardware, implementation frameworks have begun to increasingly gain prominence for their roles in facilitating the development of communication protocols and services.

This chapter reviewed some of the major challenges for implementation frameworks that can be anticipated when used for pervasive networking. A survey of some approaches of using frameworks in implementing protocols and services was presented, which caters to different developer needs. Our own research activity was also outlined with respect to the challenges ahead as well as with other popular implementation frameworks. Finally, how evaluations of implementations as well as frameworks are performed was also explained.

As frameworks continue to evolve and address new kinds of networks and usage scenarios, new challenges would inevitably unfold. What is apparent already today is that it is unlikely that any one framework would single-handedly be suitable for all implementation efforts. How interoperability among frameworks will improve, grafting new functionality onto existing ones remains an important question.

References

[1] D.C. Schmidt, A. Gokhale, and B. Natarajan (2004) 'Leveraging application frameworks' *ACM Queue* **2**(5): 66–75.
[2] G. Fregonese, A. Zorer, and G. Cortese (1999) 'Architectural framework modeling in telecommunication domain,' In *Proceedings of the 1999 International Conference on Software Engineering*, Los Angeles California USA.
[3] M. E. Fayad, D. C. Schmidt (eds) (1997) 'Object-oriented application frameworks,' *Communications of the ACM* **40**(10).
[4] M.T. Liu (1989) 'Protocol engineering' *Advances in Computers* **27**: 79–195.
[5] M.T. Liu (1991) 'Introduction to special issue in protocol engineering', *IEEE Trans on Computers* **40**(4): 373–5.
[6] L. Pouzin (1975) 'The CYCLADES network-present state and development trends,' In *Proceedings of IEEE Computer Society Symposium on Computer Networks*, Maryland.
[7] V. Cerf, S. Burleigh (2007) 'Delay-tolerant networking architecture,' *IETF RFC 4838*, April.
[8] T. Dierks, E. Rescorla (2008) 'The Transport Layer Security (TLS) Protocol Version 1.2,' *IETF RFC 5246*, August.
[9] R. Moskowitz, P. Nikander (2006) 'Host Identity Protocol (HIP) Architecture,' *IETF RFC 4423*, May.
[10] B. Carpenter, K. Moore (2001) 'Connection of IPv6 domains via IPv4 clouds,' *IETF RFC 3056*, February.
[11] C. Huitema (2006) 'Teredo: tunneling IPv6 over UDP through Network Address Translations (NATs),' *IETF RFC 4380*, February.
[12] M. Popovic (2006) *Communication Protocol Engineering*, Taylor & Francis Publishing.
[13] IBM Rational TAU, http://www.ibm.com/developerworks/rational/products/tau/ accessed on 31 August 2009.
[14] S. Boecking (2000) *Object-Oriented Network Protocols*, Addison-Wesley Publishing.
[15] D.C. Schmidt (1993) 'The ADAPTIVE Communication environment: an object-oriented network programming toolkit for developing communication software,' In *Proceedings of 11th Sun User Group Conference*, Brookline Massachusetts USA.
[16] J. Harju, B. Silverajan and I. Toivanen (1997) 'OVOPS – experiences in telecommunications protocols with an OO based implementation framework,' In *Proceedings of ECOOP '97 Workshop on Object Oriented Technology for Telecommunications Services Engineering*, Jyvaskyla Finland.
[17] P. Brown (1995) 'The stick-e document: a framework for creating context-aware applications' *Electronic Publishing* **8**(2/3): 259–72.
[18] IBM Rational SDL Suite, http://www.ibm.com/developerworks/rational/products/sdlsuite/ accessed on 31 August 2009.
[19] H. Miranda, A. Pinto, and L. Rodrigues (2001) 'Appia, a flexible protocol kernel supporting multiple coordinated channels,' In *Proceedings of the 21st Int'l Conf. on Distributed Computing Systems (ICDCS-21)*, Phoenix Arizona USA.
[20] M. Hiltunen, R. Schlichting, X. Han, M. Cardozo, and R. Das (1999) 'Real-time dependable channels: Customizing QoS attributes for distributed systems,' *IEEE Transactions on Parallel and Distributed Systems* **10**(6): 600–12.
[21] F. Travostino, E. Menze, and F. Reynolds (1996) 'Paths: Programming with system resources in support of real-time distributed applications,' In *Proceedings of the 2nd IEEE Workshop on Object-Oriented Real-Time Dependable Systems*, Laguna Beach California USA.
[22] N. Bhatti, M. Hiltunen, R. Schlichting, and W. Chiu (1998) 'Coyote: A system for constructing fine-grain configurable communication services,' *ACM Transactions on Computer Systems* **16**(4): 321–66.
[23] M. Hayden (1998) 'The Ensemble System', PhD thesis, Cornell University.
[24] N.C. Hutchinson and L. Peterson (1991) 'The x-kernel: An architecture for implementing network protocols,' *IEEE Transactions on Software Engineering* **17**(1): 64–76.

[25] R.V. Renesse, K. P. Birman, and S. Maffeis (1996) 'Horus: a flexible group communication system,' *Communications of the ACM* **39**(4): 76–83.

[26] F. Bai, N. Sadagopan, and A. Helmy (2004) 'BRICS: a building-block approach for analyzing RoutIng protoCols in ad hoc networks – a case study of reactive routing protocols,' *IEEE International Conference on Communications (ICC)*.

[27] F. Bai, G. Bhaskara and A. Helmy (2004) 'Building the blocks of protocol design and analysis – challenges and lessons learned from case studies on mobile ad hoc routing and micro-mobility protocols' *ACM Computer Communication Review* **34**(3): 57–70.

[28] P. Grace (2009) ' Dynamic adaptation', Chapter 13 in *Middleware for Network Eccentric and Mobile Applications*, B. Garbinato, H. Miranda, and L. Rodrigues (eds), pp. 285–304, Springer Publishing.

[29] D.C. Schmidt (1995) 'Using design patterns to develop reusable object-oriented communication software,' (1995) *Communications of the ACM* **38**(10): 65–74.

[30] J.C. Hu, I. Pyarali, and D.C. Schmidt (1998) 'Applying the proactor pattern to high-performance web servers,' In *Proceedings of the 10th International Conference on Parallel and Distributed Computing and Systems*, October.

[31] D.C. Schmidt, J. Hu (1998) 'Developing flexible and high-performance web servers with frameworks and patterns,' *ACM Computing Surveys* **30**.

[32] A. Fettig (2006) *Twisted Network Programming Essentials* O' Reilly Media Inc.

[33] P. Marti (2006) 'A framework for the evaluation of protocols and services in ad-hoc networks', M. Sc Thesis, Trinity College Dublin.

[34] M. Colagrosso, W. Simmons, and M. Graham (2006) 'Demo abstract: simple sensor syndication,' In *Proceedings of the 4th International conference on Embedded networked Sensor systems (SenSys '06)*, Boulder Colorado USA.

[35] B. Silverajan, J. Harju (2009) 'Developing network software and communications protocols towards the internet of things,' Iin *Proceedings of the 4th International Conference on COMmunication System softWAre and middleware (COMSWARE 2009)*, Dublin Ireland.

[36] H. Schulzrinne, S. Casner, R. Frederick, and V. Jacobsen (2003) 'RTP: a transport protocol for real-time applications,' *IETF RFC 3550*, July.

[37] E. Guttman (2001) 'Service location protocol modifications for IPv6', *IETF RFC 3111*, May.

[38] M. Handley, C. Perkins, and R. Whelan (2000) 'Session announcement protocol,' *IETF RFC 2974*, October.

[39] B. Silverajan, J. Harju (2007) 'Factoring IPv6 device mobility and ad-hoc interactions into the service location protocol,' In *Proceedings of IEEE Conference on Local Computer Networks (LCN 2007)*, Dublin, Ireland.

[40] J. Kalliomäki, B. Silverajan, and J. Harju (2007) 'Providing movement information to applications in wireless and mobile terminals,' In *Proceedings of 13th Open European Summer School and IFIP TC6.6 Workshop (EUNICE 2007)* Enschede, Netherlands.

[41] Libpcap packet capture library, http://www.tcpdump.org.

[42] N. Provos, 'libevent – an event notification library,' http://www.monkey.org/provos/libevent.

[43] Bluetooth Special Interest Group (SIG) (2003) RFCOMM with TS 07.10. Bluetooth Specification Version 1.1. June.

[44] Bluetooth Special Interest Group (SIG) (2007) Core Specification v2.1 + EDR. July.

[45] K. Pawlikowski, H.-D.J Jeong, and J.-S.R Lee (2002) 'On credibility of simulation studies of telecommunication networks,' *IEEE Communications* **40**(1): 132–9.

[46] K. Pawlikowski (2003) 'Do not trust all simulation studies of telecommunications networks,' In *Proceedings of International Conference on Information Networking (ICOIN 03)*, Jeju Island South Korea.

[47] D. Kotz, C. Newport, R.S. Gray, J. Liu, Y. Yuan, and C. Elliott (2004) 'Experimental evaluation of wireless simulation assumptions,' In *Proceedings of the 7th ACM International Symposium on Modeling, Analysis and Simulation of Wireless and Mobile Systems (MSWiM '04)*, New York NY USA.

[48] T. Andel, A. Yasinsac (2006) 'On the credibility of MANET simulations' *IEEE Computer* **39**(7): 48–54.

[49] D. Cavin, Y. Sasson and A. Schiper (2002) 'On the accuracy of MANET simulators,' In *Proceedings of the 2nd ACM International Workshop on Principles of Mobile Computing*, Toulouse France.

[50] The network simulator – NS-2 http://www.nsnam.isi.edu/nsnam/.

[51] OPNET Modeler http://www.opnet.com/products/modeler/home.html.

[52] Global Mobile Information Systems Simulation Library – GloMoSim http://pcl.cs.ucla.edu/projects/glomosim/.

[53] A. Keränen, J. Ott, and T. Kärkkäinen (2009) 'The ONE simulator for DTN protocol evaluation,' In *Proceedings of the 2nd International Conference on Simulation Tools and Techniques (SIMUTools '09)*, New York NY USA.

[54] W. Kiess, S. Zalewski, A. Tarp, and M. Mauve (2005) 'Thoughts on mobile ad-hoc network testbeds,' In *Proceedings of IEEE ICPS Workshop on Multi-hop Ad hoc Networks: From Theory to Reality*, July.

[55] T. Perennou, E. Couchon, L. Dairaine, and M. Diaz (2004) 'Two-stage wireless network emulation,' In *Proceedings of the Workshop on Challenges of Mobility held in conjunction with 18th IFIP World Computer Congress*, Toulouse France.

[56] G. Werner-Allen, P. Swieskowski, and M. Welsh (2005) 'Motelab: a wireless sensor network testbed,' In *Proceedings of the 4th International Conference on Information Processing in Sensor Networks (IPSN'05), Special Track on Platform Tools and Design Methods for Network Embedded Sensors (SPOTS)*, April.

[57] C. Tschudin, P. Gunningberg, H. Lundgren, and E. Nordstrom (2005) 'Lessons from experimental manet research' *Elsevier Ad Hoc Networks Journal* **3**(2): 221–33.

[58] Kansei: Sensor testbed for at-scale experiments. http://ceti.cse.ohio-state.edu/kansei/.

[59] C. Ng, P. Buonadonna, B.N. Chun, A.C. Snoeren, and A. Vahdat (2005) 'Addressing strategic behavior in a deployed microeconomic resource allocator,' In *Proceedings of the 3rd Workshop on Economics of Peer-to-Peer Systems*, Philadelphia PA USA.

[60] Mint, an autonomic reconfigurable miniaturized mobile wireless experimentation testbed. http://www.ecsl.cs.sunysb.edu/mint.

[61] D. Raychaudhuri *et al.* (2005) 'Overview of the ORBIT radio grid testbed for evaluation of next-generation wireless network protocols,' In *Proceedings of IEEE Wireless Communications and Networking Conference (WCNC 2005)*, New Orleans LA USA.

[62] PlanetLab Consortium. Planetlab: An open platform for developing, deploying, and accessing planetary-scale services. http://www.planet-lab.org/.

[63] B. White, J. Lepreau, L. Stoller, R. Ricci, S. Guruprasad, M. Newbold, M. Hibler, C. Barb, and A. Joglekar (2002) 'An integrated experimental environment for distributed systems and networks,' In *Proceedings of the 5th Symposium on Operating Systems Design and Implementation*, Boston MA USA.

[64] Sun Developer Network (SDN). Java ME at a Glance. http://java.sun.com/javame.

[65] P. Bellavista, A. Corradi, R. Montanari, and C. Stefanelli (2003) 'Context-aware middleware for resource management in the wireless internet,' *IEEE Transactions on Software Engineering* **29**(12): 1086–99.

Index

6LoWPAN, 143

access control, 161–2, 167–8, 170
accounting, 167
active directory, 162
active sensors, 250–1, 253, 255, 257–9
actuators, 205, 208
ad hoc cognitive radio networks, 109
adaptation table, 237–46
adaptive architecture, 237
adaptive learning, 39, 43
ad-hoc system, 145
administration, 224, 227, 229, 231–2, 234
agent security, 11–12
agents, 207, 213–14
ambient intelligence, 207, 211
AODV, 182–3
application layer, 226–7
architectural, 172
architecture, 222–3, 225–6, 232, 310, 312–13, 315
artificial intelligence, 203–4, 207, 213–14
authentication, 162, 165–7
authorization, 167, 171
autonomic, 226–8, 230, 233
autonomic wireless mesh network (AWMN), 221–2, 224–34

backbone network, 137, 145
background infrastructure, 204–5, 208, 210, 212, 214, 216
bandwidth, 281, 283–4, 286–8, 293–4, 298
barcode, 205, 211
base station (BS), 282, 289
best effort (BE), 281–2
bluetooth, 114, 117, 119, 125–31, 139–40, 142–4, 148, 152, 154–5, 163–4, 166, 168
bluetooth SDP, 31, 33
boards, 205, 215
body area network (BAN), 114–16, 118–19, 124–5

BRAN HIPERACCESS, 141
broadband, 279, 280, 282

caching, 224–7, 233
capacity, 280, 284, 287, 292, 294
CARMEN, 169
cellular, 222–3
centralized authority, 175
certification authority, 175
challenges, 302, 305, 316
channel, 281, 287–8
choke points, 175, 179
code division multiple access (CDMA), 279–81
coexistence, 138, 140–5
collaborative attack, 175
collaborative defense, 189, 191–2, 194–5, 199, 201
communications protocols, 314
computing, 3–6
computing paradigms, 42
congestion, 281, 287
connected k-coverage, 252
connection admission control, 281–2, 298
context-aware, 227
context awareness, 27, 40, 45, 135–6, 138, 156, 214–15, 227, 233
continuous, 113, 116–18, 121, 124, 127–8
cooperative caching, 227, 233
coverage configuration protocol, 250, 253
cross-layering, 226–8
cryptographic, 177, 179–80

data encapsulation, 179
data link layer, 226–7
data ownership, 161, 170
decentralized, 183
deficit round robin (DRR), 283, 288
delay adaptation, 237–9, 247
Delay Tolerant Networks (DTN), 152–4
Denial of Service (DoS) attack, 175, 177

---

deterministic sensing model, 249, 255, 261
digital-enhanced cordless telecommunications
    (DECT), 139–41
directional antennas, 182
Discretionary Access Control, 167
diseases, 115–17, 119–22, 124, 129
distance education, 44
distributed tuples, 146
DOORS framework, 310, 312–13
DownLink (DL), 281, 287–8, 294–5, 297–8
dynamic reconfiguration, 9, 10, 13–14, 19, 23
dynamic spectrum access, 101

earliest deadline first (EDF), 282
e-health systems, 113, 115–16, 125, 128
e-learning, 42, 43, 45
encryption, 177
energy, 249, 251, 253, 260
energy consumption, 211
European Telecommunication Standards Institute
    (ETSI), 141–2
Extended Data Rates for GSM Evolution (EDGE), 141
extended real-time polling service (ertPS), 281, 298
Extra-BAN communications, 115, 119, 125

file transfer protocol (FTP), 281, 289
flooding attack, 175
forbidden list, 184
fragmentation, 279, 281
framework, 281, 283, 288, 298, 301–2, 305–16
frequency, 279–82, 288

Gateways, 137, 141, 145, 147, 149–53
General Packet Radio System (GPRS2), 141
Geographic Information System (GIS), 40
Global Cell Systems, 138, 140
GPRS, 119, 130
GPS, 177, 183, 185
Graph-based solutions, 182
Grid topology, 177
GSM, 139–41, 145, 162
Guard node, 178, 182

hashed messages, 178
HAVi, 146–8
healthcare, 143, 151, 154–5
HELLO PACKETS, 177, 184
Helly's theorem, 255
heterogeneity, 145–6, 149
hidden mode, 179
high-speed, 280
High Speed Circuit Switched Data (HSCSD3), 141
high speed wired links, 180
HIPERLAN, 139, 142
hop count, 176

hops, 175–7, 179–80, 183–5
hybrid, 225

IEEE 802.3, 169
IEEE 802.11, 142, 166, 169, 263–70, 272, 274
IEEE 802.15.4, 119, 125, 155
IEEE 802.16, 281, 291, 298
impersonation attack, 175
in band wormhole, 179, 186
INSTEON, 139, 144
intelligence, 38, 45
intelligent environment, 203–4, 207–16
interactivity, 43, 45
interconnectivity, 145, 151
internet of things, 144
internet worms, 194–5
interoperability, 9, 10, 13, 15–19, 21, 23–4, 234
Intra-BAN communications, 115, 119, 124–5
IPv6, 143–4, 152
ISA SP100, 144
ISP, 161, 170

Jini, 27, 29, 33, 148–51

K-coverage, 249–61
Keys, 178–9

LAN, 222–3, 230
latency, 284–5, 287, 292
LDAP, 162
learning, 37–40
leashes, 178, 181, 185–6
legitimate node, 179, 181, 186
lens, 257–9
load balancing, 233–4
location aware guard nodes (LAGN), 178
location awareness, 45, 208, 216
location Dependency, 170
long range directional wireless links, 180
loosely synchronized clocks, 185
LORAN-C, 185

macro-cell/cellular, 138–40
maintenance, 221–2, 224, 227, 231–2, 234
malicious node, 177, 179–81, 183
massive multiplayer online role playing games
    (MMORPG), 164
media access control (MAC), 281, 287–8, 294, 297
mediacup, 206–7, 212
Medium Access Control (MAC), 102–3, 105–11
mesh clients, 233
mesh routers, 231, 233
metropolitan area network (MAN), 281, 288
microelectromechanical systems (MEMS), 203, 205
middleware, 136–7, 145–6, 148–53

MMAC, 139, 142
mobile, 221–3, 225, 230–1, 233–4
mobile agents, 10–11, 13–14, 16
MobileIP, 162
mobile networks, 263, 265
mobility, 135, 136–7, 139, 141–2, 145
Modified Deficit Round Robin (MDRR), 283, 288–92
monitoring, 113–21, 124, 126, 128–30
Mote technology, 118
movement detection, 310, 312
multichannel multipoint distribution service (MMDS), 280
multi-hop cognitive radio networks, 102, 103
multimodal interfaces, 211, 214–15
multiparametric, 124

nanoNET, 139, 144
neighbor monitoring, 176–9, 181–4, 186
nervous system, 224
network connectivity, 250–53
networking, 3–6
network layer, 223, 225–7
network programming with DOORS, 308
neural networks, 161, 167
neuro-computing, 166
NLOS, 279–80, 288
nomadic system, 145, 151
non real-time polling service (nrtPS), 281–9, 291, 293–4, 298

Object Request Broker, 146
on demand routing protocol, 176, 186
ontology, 27–8, 33–5
Open Services Gateway Specification (OSGi), 149–50
operation, 221–2, 224, 231–2
OPNET, 288, 298
opportunistic pervasive networking, 101, 111
opportunistic spectrum access, 103
orthogonal frequency-division multiplexing (OFDM), 280–2, 288, 294
out band wormhole, 179, 186
outdoor antenna, 280
overhead, 184, 186
overlay, 263–70, 273

P2P, 164, 168
pads, 205, 215
participation mode, 179, 181, 184
path planning, 72, 82–3, 85–6
PawS, 171
PDA, 163, 170, 172
peer-to-peer (P2P), 263–4, 266–9, 274
PerSE, 168
personal access communications system (PACS), 141
personal area network, 139, 141–3

personal health systems, 114–15
personal networks, 172
pervasive, 3–6, 42–8, 161–5, 168–72
pervasive healthcare, 113, 115–17, 119, 121, 123–4, 128, 131
pervasive networking, 221–2, 224, 227, 234
pervasive wireless sensor networks, 249–52
PHY, 281, 287–8
physical layer, 226–7, 229
physiological signals, 115, 117–20, 122–3, 126, 129–31
Pico-cell network, 139
point-to-multipoint network (PMP), 281
privacy, 161–2, 164, 166, 169–72
pro-active countermeasures, 181
pro-active routing protocol, 177
probabilistic sensing model, 255
programmable matter, 210
protocol, 279, 281–3
protocol engineering, 302–4
provisioning, 228, 231–2
PSIUM, 170–1

QPSK, 288, 294
quality of information, 71, 75, 94
quality of privacy (QoP), 171–2
quality of service (QoS), 263–9, 272–4, 281–4, 286–8, 298

radio frequency identification (RFI), 40–4, 46
random deployment, 250, 261
reactive countermeasures, 181–2
real-time polling service (rtPS), 281–6, 288–94, 298
recognition, 114, 117, 119–24, 130
remote, 113, 119, 128–9, 131
remote procedure call, 146
reuleaux triangle, 255–6
RFID, 161, 165–6, 205, 207, 210–11, 214, 216
Role Based Access Control (RBAC), 162, 167
ROUTE REQUEST, 177, 183
routing, 225–6, 230–1, 234

scalability, 163, 166, 172
scenario, 288, 292
scheduler, 282–4, 286–8, 291, 293–4, 298
scheduling, 226–7, 231–4, 281–4, 288–9
security, 221–3, 228–30
security by contract, 164
self-CHOP, 222, 228, 233–4
self-configuration, 221, 230–1, 233
self-heal, 222, 228, 230, 233
self-management, 222, 228, 231
self-optimization, 222
self-organization, 222, 228, 230–1
self-protection, 222, 228–30, 232

semantic, 27–8, 33–5
sensing coverage, 250–1
sensor board, 118–20, 126, 130–1
sensor networks, 10, 13–18, 24, 230, 249–52, 254
sensor scheduling, 249–50
sensors, 203, 205, 207–8, 212–16
service advertisement, 53, 60–1
service component, 237
service composition, 53, 57, 63–7
service control, 146, 148
service description, 52–3, 59, 64–5, 67, 146
service discovery, 27–35, 54, 56–63, 66–7, 148–9,
     306–7, 311
service flow, 281–4, 286, 289
service implementation, 52
service management, 51–7, 67
service manager, 52–3
service oriented architecture (SOA), 210, 214
service provider, 53, 55–7, 60–1, 64–7, 279–81, 283
service requestor, 52–3, 55, 57, 60, 65–7
service requirement, 53
simulation, 288, 291–2, 294, 297
slice, 255–9
SLP, 27, 29–30, 32, 35
small-world, 191–3, 196, 201
smart appliances, 207
smart car, 204, 214, 216
smart clay, 205
smart devices, 203–8, 210–11, 215–16
smart dust, 205
smart environment, 207–8, 211, 215
smart everyday object, 206–7, 212, 216
smart home, 211–12, 216
smart items, 207
smart laboratory, 215
smart library, 215
smart mote, 210
smart phones, 163
smart room, 213–14
smart skin, 205
smart telephone, 205
Smart-Its, 212
SOA, 27–9, 34–5
spectrum, 279–80
spectrum management, 140
state of the art, 302, 308
statistical methods, 182
stochastic k-coverage protocol, 250, 254, 260
stochastic sensing range, 258
subcarrier, 281, 294
supply chain management, 214

tabs, 205
tag, 204–5, 207, 211, 213–14, 216
threshold probability, 254, 258
tightly synchronized clocks, 178, 185–6

time division duplex (TDD), 281, 288
tower, 280
tracking, 204, 211, 214–15
traffic, 281–4, 286–9, 291–6, 298
transceivers, 177, 179
transition diagram, 260
transport layer, 226
tunable sensitivity parameter, 184

UbiCOSM, 168–9
ubiquitous, 3, 5, 41–4, 45, 221–5, 228, 232–4
u-learning, 39, 44
Ultra Wide Band (UWB), 139, 142–3
UMTS, 119, 130
unicast, 281, 283, 285, 288
Universal Mobile Telecommunications System
     (UMTS), 139–42, 145
Universal Plug and Play (UPnP), 148–51
Universal Remote Console Control Protocol (URCC),
     149
Unsolicited Grant Service (UGS), 281–6, 288–9,
     291–3
un-trusted code, 164
uplink (UL), 281–4, 287–8, 294, 296, 298
usability, 222, 229–30
user interface, 211, 213, 215

virtual link, 179
virtual overlay networks, 151
vital signal monitoring, 113
VoIP, 281, 291
volcano monitoring, 18–20, 24

WAN, 221, 223
wearable sensors, 113–17, 120–1, 128–31
Web 2.0, 45–6
well-being learning, 46–7
Wibree, 139, 144
WiFi, 127, 163
WiMAX, 141, 152, 164, 222–4, 232, 279, 281–3,
     287–8, 291, 293–4
wireless access system, 279–80
wireless broadband, 280
wireless communication, 114–15, 117, 128–31
wireless local area network (WLAN), 263–5, 268–9,
     272
wireless mesh networks, 223, 226, 228, 230, 233
wireless sensor network, 71, 79, 85, 95, 139, 143–4,
     154–5, 204
wormhole, 175–86
WWVB, 185

XML, 169, 303, 313

ZigBee, 114, 125–8, 130, 139, 143, 163, 166
Z-wave, 139, 144

CPSIA information can be obtained at www.ICGtesting.com
Printed in the USA
270233BV00005B/3-42/P